KT-134-919

# Growth Regulation by Nuclear Hormone Receptors

# CANCER SURVEYS

## Advances and Prospects in Clinical, Epidemiological and Laboratory Oncology



Published for the

# Growth Regulation by Nuclear Hormone Receptors

Guest Editor
M G Parker

COLD SPRING HARBOR LABORATORY PRESS 1992

**CANCER SURVEYS**
*Growth Regulation by Nuclear Hormone Receptors*
Volume 14


Published by Cold Spring Harbor Laboratory Press
Printed in the United States of America
ISBN 0-87969-371-1
ISSN 0261-2429

*Cover and book design by Leon Bolognese & Associates, Inc.*



All Cold Spring Harbor Laboratory Press publications may be ordered directly from Cold Spring Harbor Laboratory Press, 10 Skyline Drive, Plainview, New York 11803-9729. Phone: Continental US & Canada 1-800-843-4388; all other locations (516) 349-1930. FAX: (516) 349-1946.

# Contents

# Growth Regulation by Nuclear Hormone Receptors

# Introduction

M G PARKER

Most reviews of hormones and cancer have, until recently, dealt almost exclusively with oestrogens and breast cancer or with androgens and prostate cancer. This is hardly surprising because oestrogens have been implicated in breast cancer for almost a century following the use of oophorectomy by Sir George Beatson for the treatment of metastatic breast cancer. Similarly, orchidectomy was found to be effective in the treatment of prostate cancer by Huggins and his co-workers in the 1940s. In addition, epidemiological studies have emphasized the importance of oestrogens as a major risk factor for breast cancer and for uterine cancer and of testosterone for prostate cancer. A major breakthrough in our understanding of steroid hormone action came about with the discovery of receptors for oestrogens in the 1960s by Gorski and Jensen and their colleagues: their work led to models for the mechanism by which steroid hormones regulate gene expression in target tissues. More recently, with the isolation of recombinant DNA clones for the individual steroid receptors, it has become clear that they represent members of a much larger family of proteins containing many other related receptors. These include the receptors for thyroid hormone and retinoic acid and many novel receptors whose hormonal ligand and target genes have not yet been identified. These so called "orphan receptors", which currently number in excess of 30, are likely to be involved in signalling pathways completely unsuspected so far. When the last issue of Cancer Surveys devoted to hormones and cancer was published in 1986, the term "nuclear receptors" did not exist, and in view of the expansion of this field, it now seems a good time to produce an issue dedicated to growth regulation by nuclear receptors.

This issue continues to emphasize the importance of oestrogen and androgen receptors but also includes articles on thyroid hormone and retinoic acid receptors and other related receptor proteins. It is fair to say that progress in analysing the structure and function of the receptors as transcription factors has been far more rapid than the identification of target genes involved in the hormonal control of cell growth, and this is reflected in the contributions. In view of the structural similarities between the receptors, it has been thought that the mechanisms by which they regulate gene expression will also be conserved. Although this is true to some extent, Fig. 1 shows that the receptors do exhibit a number of interesting differences that contribute to variations in their overall transcriptional activity.

The most obvious difference in the action of nuclear receptors is that the

*Cancer Surveys* Volume 14: *Growth Regulation by Nuclear Hormone Receptors*
 0-87969-371-1/92. $3.00 + .00

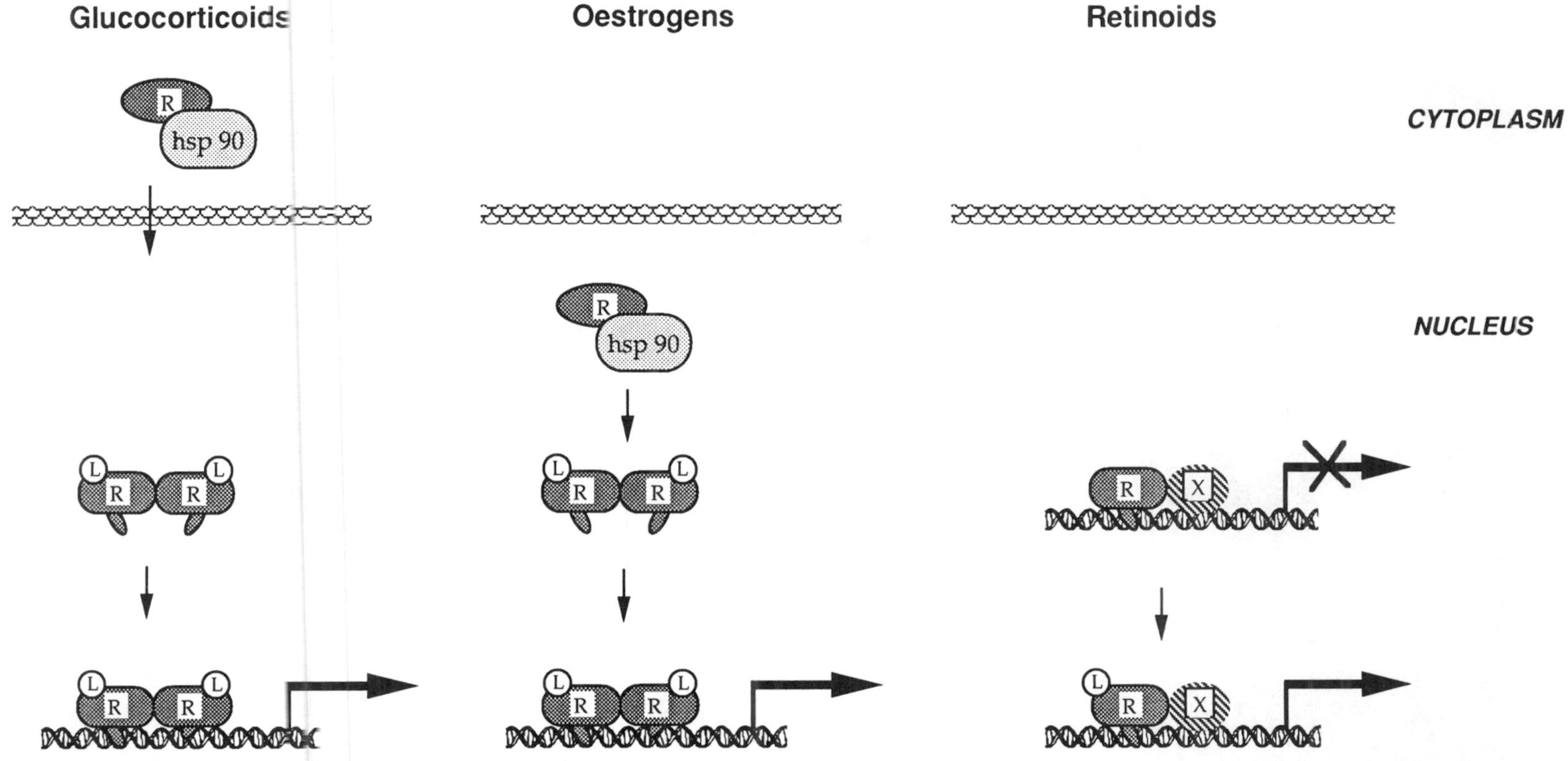


**Fig. 1.** Scheme to describe the mechanisms by which nuclear receptors activate gene transcription in target cells. The rate of transcription (arrow) depends on the binding of a hormonal ligand (L) with a soluble receptor (R). Steroid receptors bind to DNA as homodimers, whereas retinoid acid receptors (and thyroid hormone receptors) appear to bind to DNA preferentially as heterodimers in combination with another receptor, the retinoid X receptor

binding of a hormonal ligand is required to promote the DNA binding of all steroid hormone receptors but not for the thyroid hormone and retinoic acid receptors. This difference is probably a consequence of an interaction between steroid hormone receptors and heat shock proteins, which appears to block receptor dimerization and thereby inhibit DNA binding activity. In the case of the glucocorticoid receptor, the heat shock proteins may also mask nuclear localization signals that are presumably exposed or activated only upon hormone binding. The subcellular localization of receptors and the role of nucleocytoplasmic shuttling are discussed by Milgrom and colleagues. Steroid hormone binding is required not only to promote DNA binding but also to induce full transcriptional activity of the receptor. This appears to be the primary role of thyroid hormone and retinoids, since their receptors are capable of binding to responsive genes but are not transcriptionally active until stimulated by hormone. Another major difference is that steroid receptors bind to DNA as homodimers, whereas it now seems likely that a number of other nuclear receptors including those for thyroid hormone and retinoic acid are more likely to bind to DNA as heterodimers with another nuclear receptor termed the retinoid X receptor. Since many of these receptors exist in more than one form even in a single cell type, it is not yet clear which combinations actually exist in target cells and how this affects the regulation of hormone responsive genes.

Obviously, ligands that compete with the hormone for binding to the receptor and thereby interfere with DNA binding or inhibit transcriptional activity will function as hormone antagonists. Perhaps the best characterized antagonist is the non-steroidal anti-oestrogen, tamoxifen, used as adjuvant therapy for postmenopausal women with breast cancer. However, its been known for some time that despite the efficacy of tamoxifen in the treatment of breast cancer, especially when the primary tumour is oestrogen receptor positive, therapeutic failure often results from the development of drug resistance. Potential mechanisms for the development of breast cancer cells that are insensitive to hormone are discussed by King and Horwitz, and tamoxifen resistance is discussed by Jiang and Jordan. Alternative anti-oestrogens, reported to be devoid of any agonist activity, are described by Wakeling. Mutations in the oestrogen receptor are predicted to be one means of generating either tamoxifen resistance or hormone insensitivity; examples of these together with their potential consequences are reviewed by Horwitz and by McGuire and his colleagues. The occurrence of mutations in the androgen receptors found in prostate tumours has yet to be reported, but a growing number of inherited mutations have been identified that cause defective development and growth of sex accessory tissues, as discussed by Brinkmann and Trapman. The importance and potential role of androgens in prostate cancer are reviewed by Wilding.

There are now two examples of receptors that, when mutated, are associated with leukaemia. One involves the viral protein v-erbA, a mutant version of the thyroid hormone receptor α, which in combination with v-erbB, a truncated version of the epidermal growth factor/transforming growth factor α receptor, induces erythroleukaemias and sarcomas in chickens. The possibility that v-erbA is acting as a dominant negative version of the normal ligand dependent thyroid hormone receptor is discussed by Ghysdael and Beug. The second example is associated with a specific t(15;17) reciprocal translocation, which results in the formation of a fusion protein between the retinoic acid receptor α and a myeloid protein called PML (for promyelocytes). De Thé and Dejean discuss the possibility that it too is acting as a dominant negative oncogene. Both thyroid hormone and retinoic acid have extremely diverse actions in probably every tissue in the body: their role in normal growth and development is reviewed by Chatterjee and Tata and by Morriss-Kay, respectively. Roberts and Sporn discuss the intriguing possibility that transforming growth factor α might act as key regulators of retinoids and steroid hormones in the control of cell growth. Finally, a number of chemicals that act as carcinogens in the liver stimulate the proliferation of organelles called peroxisomes. It has now been shown that members of the nuclear receptor family mediate the action of peroxisome proliferators, and Green discusses a model for receptor mediated carcinogenesis.

The reviews in this issue clearly demonstrate the progress that has been made in identifying and characterizing many members of the nuclear receptor family. Current work is focused on elucidating the molecular mechanism of action of both individual and combinations of receptors and analysing how their transcriptional activity is modulated by other signalling pathways such as those mediated by the proteins c-jun or c-fos. Much of this work has been facilitated by the use of artificial model systems, and this type of work has certainly told us a great deal about what receptors are capable of doing. More now needs to be done to extend this work to more natural genes in target cells, particularly those encoding rate limiting factors involved in cell proliferation or cell differentiation, and to investigate whether the expression or these or of other genes can account for the development of drug resistance in tumours.

# Immunolocalization of Steroid Hormone Receptors in Normal and Tumour Cells: Mechanisms of Their Cellular Traffic

**MARTINE PERROT-APPLANAT • ANNE GUIOCHON-MANTEL**
**EDWIN MILGROM**
*Hormones et Reproduction (Inserm U135), Faculté de Médecine Paris Sud, 94275 Le Kremlin-Bicêetre*



## INTRODUCTION

Steroid hormone receptors are intracellular proteins that may be found in both the cytosolic and nuclear fractions of a tissue homogenate. Until 1984–1986,

*Cancer Surveys* Volume 14: *Growth Regulation by Nuclear Hormone Receptors*
 0-87969-371-1/92. $3.00 + .00

the classic model of action of steroid hormones was that cytoplasmic receptor proteins undergo nuclear translocation after interaction with their ligand. The ligand-receptor complex becomes associated with the chromatin, which results in stimulated or repressed transcription of discrete genes, leading to the specific biological effects exerted by steroid hormones. This early model was based on cell homogenization and fractionation studies (Jensen *et al,* 1968; Gorski *et al,* 1968). A major drawback of all assays performed on tissue extracts lies in their inability to provide information about intercellular and intracellular distribution of the receptor. The development of monoclonal antibodies against oestrogen receptor (ER) (Greene *et al,* 1980), progesterone receptor (PR) (Logeat *et al,* 1983; Sullivan *et al,* 1986) and glucocorticoid receptor (GR) (Okret *et al,* 1984) allowed immunocytochemical studies to be undertaken. This technology has been extended to other receptors. It displays several advantages over the steroid binding method: it does not depend on the association of radiolabelled steroid hormone with the receptor, and receptors can be detected, even at low concentration, in tissues taken from animals or patients with high endogenous hormone concentrations. The development of these immunocytochemical methods is valuable for the assessment of receptor content and distribution in different target tissues and for basic investigations on receptor localization and function as well as for clinical purposes.

This review discusses the detection of steroid receptors by light and electron microscopy in tissues and in normal and cancer cells, as well as some of the mechanisms of their intracellular traffic. We shall draw primarily from our experience with the progesterone receptor.

## STEROID HORMONE DETECTION BY LIGHT MICROSCOPY

### Introduction

Receptor immunocytochemistry with monoclonal antibodies started with the pioneering work of King and Greene (1984) who reported exclusive nuclear localization of the oestrogen receptor (ER) even in the absence of endogenous hormone. This work suggested that unoccupied oestradiol receptors are loosely bound to the nucleus and released only in the cytosolic fraction of cells by extraction in hypotonic buffer. This model also agreed with the biochemical studies of Welshons and Gorski (1984), which showed that cytochalasin induced enucleation of pituitary (GH3) cells led to partitioning of unoccupied ER almost exclusively in the nucleoplast fraction. When cells were exposed to hormone, the complexed receptor became tightly associated with the nucleus, and a high salt concentration was required for its extraction. Initial studies on the progesterone receptor (PR) and more recently on the androgen receptor (AR) also confirmed the exclusive nuclear localization of these receptors even in the absence of endogenous hormone (Perrot-Applanat *et al,* 1985; Tan *et al,* 1988). Thus, this group of steroid hormone receptors (ER, PR, AR) is transported, in the absence of ligand, from the cytoplasm to (or into) the nucleus, where the receptors remain in the inactive state. The hormone binding triggers an interaction of receptors with high affinity DNA binding sites, where the primary events involved in transcription activation occur. The glucocorticoid (GR) (Fuxe *et al,* 1985; Wikström *et al,* 1987) and probably mineralocorticoid (MR) (Lombes *et al,* 1990; Farman *et al,* 1991) receptors follow another pattern: (a) in the absence of hormone, they appear to be exclusively cytoplasmic or both nuclear and cytoplasmic and (b) in the presence of hormone, they appear to concentrate in the nucleus (Fuxe *et al,* 1985). As with ER and PR, the steroid bound GR becomes tightly associated with nuclear components and a high salt concentration is required for its extraction. How receptors are rapidly transferred from their site of synthesis in the cytoplasm to sites of action in the cell nucleus will be discussed below.

## Detection of Oestrogen and Progesterone Receptors

A detailed distribution of ER positive cells in the female reproductive tract and in the central nervous system can be found in Press and Green (1987). Specific immunoperoxidase staining for receptor in all oestrogen sensitive tissues and cells studied so far is exclusively nuclear.

Progesterone receptor localization with monoclonal antibodies produced in our laboratory (Logeat *et al,* 1983; Lorenzo *et al,* 1988; Vu Hai *et al,* 1989) and indirect immunoperoxidase techniques (the peroxidase-antiperoxidase technique or the streptavidin bridge method) have shown an exclusively nuclear distribution of this receptor in frozen sections of human breast tumours (Perrot-Applanat *et al,* 1987); uterus from human, primate, rabbit and guinea pig species (Perrot-Applanat *et al,* 1985; Garcia *et al,* 1988; Groyer-Picard *et al,* 1990); other mammalian reproductive tissues (Groyer-Picard *et al,* 1990); rabbit and guinea pig brain (Perrot-Applanat *et al,* 1985; Warembourg *et al,* 1986); in fixed T47D human breast carcinoma cultures (Perrot-Applanat *et al,* 1987) and in paraffin embedded sections of breast tumours, human endometrium and various rabbit reproductive tissues (Perrot-Applanat, 1985, 1989). Nuclear localization of PR has been confirmed by several other groups in the human and monkey uterus (Clarke *et al,* 1987; Press and Greene, 1988; Okulicz *et al,* 1989) and in several tissues of the chicken (Gasc *et al,* 1984; Ylikomi *et al,* 1985). Nuclear PR has also been described in the guinea pig hypothalamus (Blaustein *et al,* 1988, 1990) and in the ovary from human, monkey, rabbit and chicken (Isola *et al,* 1987a; Hild-Petito *et al,* 1988; Korte and Isola, 1988; Iwai *et al,* 1990). Both immunocytochemistry and autoradiography with [$^{3}$H]ORG 2058 (Gasc *et al,* 1986) have confirmed that the chick oviduct PR does not undergo a temperature dependent translocation from cytoplasm to nucleus on binding ligand. New target cells for ER and PR have also been identified with immunocytochemistry: these receptors are present at relatively high concentration in the muscular cells of uterine arterial walls (tunica media), a fact that may be related to physiological uterine vascular changes, such as those observed during the cycle or during pregnancy

(Perrot-Applanat *et al,* 1988). Modifications of uterine blood flow are hormonally regulated but remain poorly understood, and various indirect and complex mechanisms have been proposed (reviewed in Resnik, 1986). The presence of ER and PR in muscle cells of uterine arteries suggests that steroid hormone may exert a direct effect at this level.

The regulation of ER and PR content in different cell types in the target tissue has been examined in many immunocytochemical studies. Oestrogen (E2) has been shown to induce an increase in PR concentration in all cells (epithelial, stromal and myometrial smooth muscle cells, as well as arterial muscle cells) of uterine tissue of both rodents (Perrot-Applanat *et al,* 1985, 1988) and primates (Okulicz *et al,* 1989). Progesterone receptor induction by E2 was also observed during the proliferative phase of the menstrual cycle in the human. Early studies concerning PR regulation showed that it was decreased by progesterone or synthetic progestins in a variety of systems (Milgrom *et al,* 1973; Isomaa *et al,* 1979; Kreitmann *et al,* 1979; Horwitz *et al,* 1983). Immunohistochemical evidence now shows that there are cell and tissue specific differences in the ability of progesterone to reduce PR concentrations and that sustained progesterone exposure can be associated with maintenance or even increases in cellular PR levels (reviewed in Clarke, 1990). For example, in the human uterus, immunoreactive PR levels in endometrium, stroma and myometrium are increased during the proliferative phase of the cycle, with a maximum in the late proliferative and early secretory phase (Bergeron *et al,* 1988a; Garcia *et al,* 1988; Lessey *et al,* 1988; Press *et al,* 1988; Zaino *et al,* 1989). At ovulation, PR levels in epithelial cells drop precipitously and remain suppressed as the glandular elements regress at the end of the cycle. By contrast, there is moderate to strong PR staining in isolated glands in the basalis region of the endometrium during the mid secretory phase (Press *et al,* 1988), highlighting the heterogeneous response of PR in different populations of epithelial cells. In the myometrium and stroma, PR also persists throughout the secretory phase (Garcia *et al,* 1988; Lessey *et al,* 1988), consistent with the known role of progesterone in stromal decidualization and myometrial quieting. There is also persistence of PR in the cell types important in maintenance of pregnancy such as decidua (Perrot-Applanat *et al,* unpublished observations). The molecular mechanisms involved in these regulations have recently been studied (Savouret *et al,* 1991).

Immunofluorescence methods have also been developed to localize both ER and PR proteins in the same cells. They have been applied to the uterus (Zaino *et al,* 1989) and the rodent hypothalamus (Warembourg *et al,* 1989).

## Nuclear Detection of Androgen Receptor

The presence of AR is required to mediate transcriptional activation by androgen in the male reproductive tract and other tissues. Immunohistochemical localization of AR was performed with antibodies against synthetic peptides with sequences corresponding to different regions of the AR protein.

Androgen receptor was localized to specific cell nuclei in reproductive tissues (ventral prostate, seminal vesicle, testis, epididymis), submaxillary gland, anterior pituitary and brain (preoptic hypothalamic area) of the rat, as well as in human prostate (Tan *et al,* 1988; Sar *et al,* 1990). Regulation of AR protein and mRNA concentrations has also been investigated by immunohistochemistry and in situ hybridization (Takeda *et al,* 1991; Sar *et al,* 1990).

## Detection of Glucocorticoid and Mineralocorticoid Receptors

Several biochemical and immunocytochemical studies have supported the concept of a cytoplasmic localization of GR and its translocation to the nucleus in the presence of glucocorticoid hormones (Govindan, 1980; Papamichail *et al,* 1980; Antakly and Eisen, 1984). However, extensive evaluation of how the fixation, permeabilization and staining procedures by themselves affect the apparent localization and staining intensity of GR has only been reported by Gustafsson's group using a monoclonal antibody (Wikström *et al,* 1987). A cytoplasmic GR was found with all different fixatives used, both of the crosslinking type and of the precipitating type. This excludes the possibility of a diffusion of GR from the cell nucleus during fixation and incubation procedures. The cytoplasmic staining on ligand treatment was weakened but not abolished (Wikström *et al,* 1987), contrary to what has been reported by other investigators (Papamichail *et al,* 1980; Govindan, 1980).

By contrast with normal cells, mouse GR overexpressed in Chinese hamster ovary (CHO) cells was located in the nucleus (Sanchez *et al,* 1990) in the same manner as reported for ER and PR produced from normal cellular genes. In addition, confocal microscopy of this overexpressed GR shows a non-random distribution (mottled pattern) (Martins *et al,* 1991).

A monoclonal antibody generated by an autoanti-idiotypic procedure and directed at the aldosterone binding site of MR was used in immunohistochemical studies to detect MR in rabbit kidney (Lombes *et al,* 1989, 1990; Farman *et al,* 1991). At the cellular level, immunostaining occurred in the cytoplasm and, in the majority of cells, in the nucleus as well. The nucleocytoplasmic distribution of MR seemed to be unaffected by adrenalectomy or by administration of aldosterone.

## Immunocytochemistry of Steroid Receptors and Technical Considerations

Initial studies concerning the localization of ER, PR and GR have been the subject of conflicting results. For example, ER and PR were either found in the cytoplasm and translocated in the nucleus after hormone administration or always found in the nucleus in the absence or presence of hormone. Our experience, as well as the collective experience of the past decade of immunocytochemistry of steroid receptors, indicates that the evaluation of specificity of the staining remains the central difficulty. This includes the selection of well characterized antibodies, a critical approach to methodology and to the interpretation of results.

### *Antibodies*

Any immunological staining will be only as good as the antibody used.

*Specificity:* Because of the difficulty of steroid hormone receptor purification, receptor preparations used for immunization usually contain contaminating antigens that also provoke the formation of antibodies. The use of these polyclonal antisera in immunocytochemical or immunoblot experiments often yields a signal corresponding in part to receptor and in part to other antigens. Background has also been erroneously interpreted as showing cytoplasmic staining for ER. Monoclonal antibodies reduce this kind of confusion and provide well defined reagents.

*Affinity:* Antibodies that have a too moderate or low affinity for the antigen generally give a faint staining and represent another limitation for a valid immunocytochemical study. This may also happen with well characterized antibodies applied to species different from that which have been used for the preparation of the antigen. Analysis of the cross reactivity between species, including mammalian and birds, of 22 well characterized monoclonal antibodies prepared against human and rabbit PR in our laboratory has shown that the cross reactivity depends on the method used. For example, some antibodies cross react with receptor from other species using double immunoprecipitation, but not in immunocytochemical experiments. In our experience, there is a good correlation between results obtained from western blot analysis and those obtained with immunocytochemistry (Groyer-Picard *et al,* 1990). Moderate affinity towards the antigen may also occur using antibodies against synthetic peptides covering different domains of the receptor. Antibodies against synthetic peptides have been produced when complementary DNA (cDNA) sequences of ER, PR, AR and GR were established; some of these antibodies have been used for morphological studies. This approach is especially useful when purification of a sufficient amount of the receptor cannot be obtained (Sar *et al,* 1990). It is based on the selection of stretches of about 10–20 aminoacids that have a high antigenicity index, with the aid of a computor prediction program (for example, see Zegers *et al,* 1991). However, although most of these antibodies recognize the native receptor that is present in high concentrations in transfected cells, they may give lower (or no) staining in normal cells that express the receptor in lower concentrations.

Another practical consideration is that each antibody may react differently to a given fixation and tissue processing protocol. For example, some antibodies to PR have to be used on frozen sections, whereas others work on paraffin sections (Perrot-Applanat *et al,* 1989). A similar situation exists for monoclonal antibodies to ER (Shintaku and Said, 1987).

### *Selection of a Method for Tissue Preparation*

This is a compromise between conservation of morphological features of tissue and preservation of the protein antigenicity. For PR and ER, good results have been obtained with frozen and paraffin sections of tissues fixed in 4% formaldehyde or in a mixture of paraformaldehyde and picric acid (Zamboni's fixative). Glutaraldehyde can also be used for PR, but not for ER, in electron microscopy studies (Perrot-Applanat *et al,* 1986). Other fixatives, such as formol sublimate and Carnoy gave weaker staining. In cryostat sections, receptor preservation is optimal; however, morphological preservation is inferior to that seen in paraffin sections.

Immunocytochemistry studies on cultured cells differ in one important aspect from those mentioned above, since in this case, antibodies are applied on intact cells; no sectioning is therefore involved in the procedure. The localization of steroid receptors in this type of preparation does not raise any major new technical problems. For such studies, the cell membrane must be permeabilized to large molecules immunoglobulin by fixation and freezing or by using a detergent solution before the application of antibodies.

### *Routine Methods in ICC of Steroid Receptors*

The three most commonly used staining techniques are indirect immunofluorescence, the peroxidase-antiperoxidase method (Sternberger, 1970) and the method using the avidin (or streptavidin)-biotin complex (Hsu, 1981). In our opinion, the two latter methods are good candidates for staining steroid receptors present in tissue sections, although we have found immunofluorescence to be efficient on cell cultures. Both receptor staining and tissue morphology can be observed using the peroxidase method, while autofluorescence is often present in tissue sections.

## STEROID HORMONE DETECTION BY ELECTRON MICROSCOPY

Although data have accumulated on the binding of steroid-receptor complexes to regulatory regions of cloned genes (Beato, 1989), little is known about the topology of these regulations in the intact cell. Is receptor distribution random or non-random, with the existence of regions that do not contain receptors? Are unliganded receptors located (or not) near to the high affinity DNA acceptor sites or does the receptor undergo a spatial distribution after it binds hormone and changes from low affinity to high affinity DNA binding sites? Until now, only a few studies have examined the ultrastructural distribution of steroid receptors in tissue sections. This could probably be explained by the various methodological difficulties that had to be resolved to observe the in situ localization of steroid receptors. In each case, it is necessary to define the precise experimental conditions (fixation and embedding conditions, choice of the method of labelling) that give good preservation of ultrastructural details without impeding the immunological recognition of the receptor.

## Technical Considerations

Briefly, two types of methods can be used in immunoelectron microscopy: (a) pre-embedding techniques in which labelling of sections with an indirect immunoperoxidase method is performed before the embedding procedure (Sternberger, 1970) and (b) post-embedding techniques, which consist of incubating sections from embedded tissue (Epon, Araldite, LR White, Lowicryl) with monoclonal antibodies and gold-labelled reagents (protein A or secondary antibodies) (Roth, 1982). Through our experience with these techniques, we have developed a strong preference for immunogold labelling methods, which allow a fine identification of the labelled nuclear and cytoplasmic structures (Perrot-Applanat *et al,* 1986). Because of their high electron density, gold particles are easily detected in the electron microscope and, in contrast to the peroxidase-antiperoxidase technique, do not obscure the ultrastructural details of the labelled structures. This is especially important for the nucleus where physiological events are probably compartmentalized in the presence of a high structural complexity of interphase chromatin arrangement.

## Progesterone Receptors

The progesterone receptor has been localized in the rabbit uterus by immunocytochemistry at the electron microscopic level, using monoclonal antibodies and the protein A–gold technique (Perrot-Applanat *et al,* 1986). The PR in uterine stromal cells was mainly localized in the nucleus; however, a small fraction of antigen was present in the cytoplasm, where it was associated with the rough endoplasmic reticulum and with free ribosomes (Fig. 1). The plasma membrane was not labelled. In the nucleus, the receptor was always associated with condensed chromatin or areas surrounding condensed chromatin, whereas the nucleolus was not labelled. In the chromatin, receptor distribution varied according to the hormonal state: in the absence of progesterone, the receptor was randomly scattered over clumps of condensed chromatin (Fig. 1a), which is known to be for the most part transcriptionally inactive (Fakan and Puvion, 1980). Since receptors are easily extracted when the tissue is homogenized, they are probably loosely bound to some component of condensed chromatin. By contrast, after the administration of hormone, the intranuclear concentration and distribution of the receptor were changed in the uterine stromal cells (Fig. 1b). The labelling was decreased by hormone treatment, especially in condensed chromatin. Decondensation or dispersion of chromatin was also observed. Progesterone receptor immunoreactivity was detected mainly in the border regions between condensed chromatin and nucleoplasm and, to a lesser extent, over dispersed chromatin in the nucleoplasm. These localizations correspond to the regions known to be most active in extranucleolar gene transcription (Fakan and Nobis, 1978; Fakan and Puvion, 1980). The observed partial redistribution of the PR in the nucleus of stromal cells after administration of hormone may correspond to a translocation of the receptor from one site to another. Alternatively, it may be

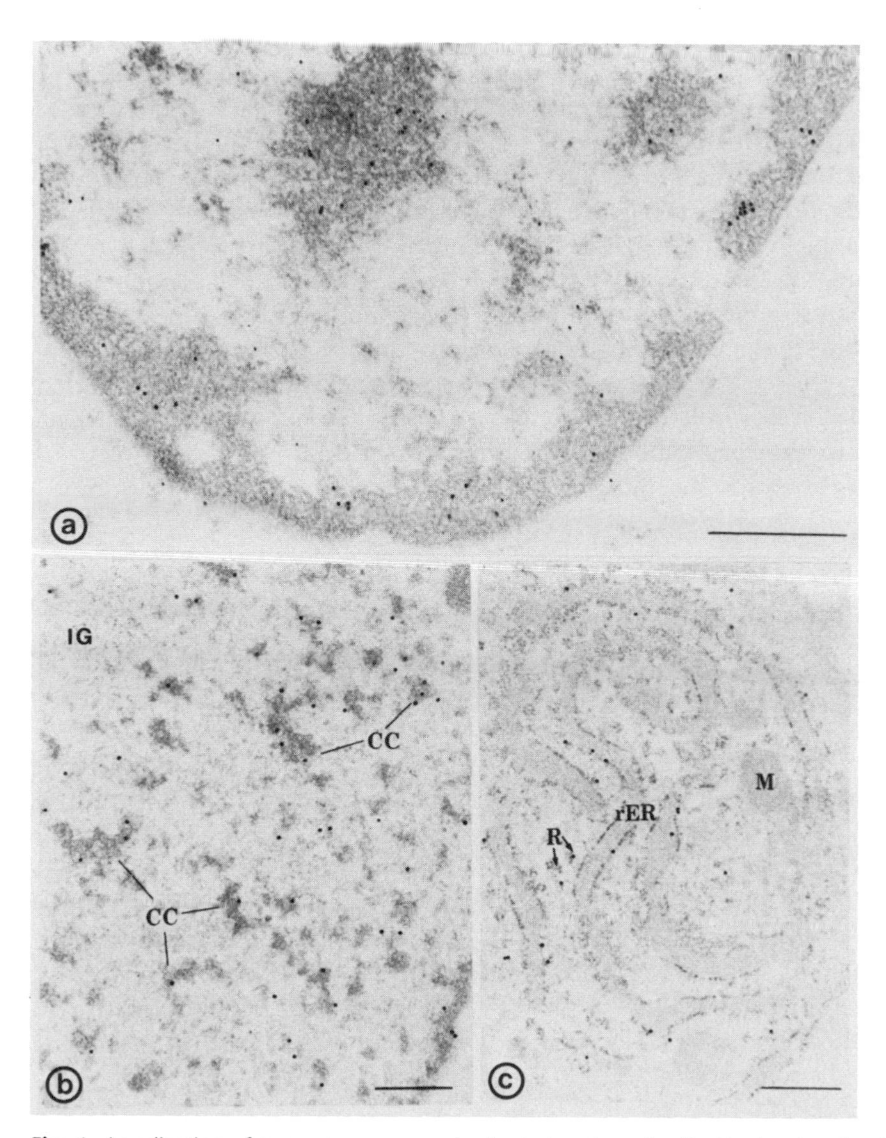


**Fig. 1.** Localization of progesterone receptor in uterine stromal cells. Nuclear localization of PR (a) in the absence of hormone and (b) after administration of the progestin R5020. (c) Presence of a small amount of PR in the cytoplasm, located along the rough endoplasmic reticulum (rER) membranes or associated with clusters of free ribosomes (R). CC, condensed chromatin; IG, interchromatinic granules; M, mitochondria. Bar = 0.5 μm

due to the fact that the receptor simply accompanies the change in localization of at least some specific genes.

Other electron microscopic analyses of PR immunoreactive cells have been completed in chick oviduct (Isola, 1987, 1987b), guinea pig hypothalamus and endometrium (Blaustein *et al,* 1988; Brown *et al,* 1990). These analyses always confirmed the predominantly intranuclear localization of PR and the absence of immunolabelling in the nucleolus. In agreement with our observations, some studies also indicate the presence of a small fraction of PR in the cytoplasm of oviduct cells and hypothalamic neurons (Isola *et al*, 1987b; Blaustein *et al*, 1988). However, differences in the intranuclear localization of occupied and unoccupied PRs have not been observed using peroxidase-antiperoxidase (PAP) treated material (Isola *et al*, 1987b; Brown *et al*, 1990), probably because of the properties of this method. As already discussed, the resulting intensely immunoreactive precipitate is rather diffuse and may cover the underlying chromatin structures in the nucleus; it also prevents the detection of changes in the intensity of PR labelling. Isola (1987) has compared the effect of progesterone on the intranuclear localization of PR in chick oviduct, using two different methodological conditions (either under pre-embedding conditions with the PAP method or with immunogold labelling on ultrathin sections). Different results were obtained according to the method used; a partial change in the localization of PR after hormone administration was described using the immunogold technique, a result that confirms our observations.

### Oestrogen Receptors

The subcellular identification of ER has been described in human and rat endometrium (Press *et al*, 1985; Vasquez-Nin *et al*, 1991) and in mice and human mammary carcinoma (Peralta Soler and Aoki, 1989; Walsh *et al*, 1990). As for PR, results clearly show a predominantly nuclear localization in different tissues. Most studies, except that of Press, described a low but significant labelling in the cytoplasm—namely in association with ribosome rich areas (Vasquez-Nin *et al*, 1991). However, different results were obtained for the intranuclear distribution of the receptor, a fact that can be explained by the various techniques, tissues or cell types and antibodies used. In mammary carcinoma, ER immunoreactivity was found over heterochromatin. By contrast, nuclear distribution of ER was observed in the euchromatin and did not seem to be modified by hormonal status or oestradiol injection in human uterine glandular cells (Press *et al*, 1985; Vasquez-Nin *et al*, 1991). However, differences in distribution seemed to take place according to the cell type (epithelial, muscle and fibroblast) (Vasquez-Nin *et al*, 1991). Interestingly, double labelling experiments (anti-ER and anti-nuclear ribonucleoproteins [hnRNP]) stained with the protein A–gold labelling procedure showed a close relation between the ER and RNP constituents, especially with perichromatin fibrils (Vasquez-Nin *et al*, 1991); these fibrils are known to contain heterogeneous nuclear (hn)RNA undergoing processing (Fakan *et al*, 1976, 1986). These immunocytochemical and other biochemical (Liao *et al*, 1973; Chong and Lippman, 1982) results provide support for the binding of oestradiol-ER complex to the premessenger RNA. A similar pattern of GR and small nuclear (sn)RNP distribution has also recently been observed by confocal microscopy in CHO cells that overexpressed the receptor (Martins *et al*, 1991). These results, however, remain to be confirmed and probably do not reflect the DNA binding activity of the receptor.

In conclusion, because of several methodological difficulties and the possible coexistence of different pools of a steroid receptor in the nucleus, other ultrastructural studies need to be completed, especially studies that use mutant forms of the receptor.

## CELLULAR TRAFFIC OF HORMONE RECEPTORS

### Mechanisms of the Nuclear Localization of Steroid Hormone Receptors

The nuclear localization of proteins has been shown to take place through two mechanisms (reviewed in Nigg *et al,* 1991). In the first, the protein diffuses through the nuclear membrane and is then trapped by binding to an intranuclear component. The second mechanism is an active process, is temperature dependent, requires ATP, and displays saturation kinetics. This active mechanism was initially shown for large molecules (>60 kDa). More recently, this type of mechanism has been implicated for small proteins such as histone H1 (Breeuwer and Goldfarb, 1990). Active transport into the nucleus requires that proteins contain suitable nuclear localization signals (NLS). They are mostly short basic sequence motifs, rich in arginines and lysines. The first to be described was the SV40 large T antigen signal (Kalderon *et al,* 1984; Lanford *et al,* 1986). Since then, numerous signals have been identified, and it has been shown, in the case of nucleoplasmin, that two interdependent basic domains may constitute a bipartite signal (Robbins *et al,* 1991). These NLS are recognized by NLS binding proteins (NBP), which are thought to function as adaptor molecules between the karyophilic protein to be transported and the transport machinery of the nuclear pore complex. Several candidate NBP have been identified by an affinity approach. Nuclear import of proteins takes place in at least two steps. The first step is the interaction between the protein and the nuclear pore complex through the NLS. The second step is the translocation to the nucleus. Only this second step requires ATP (Newmeyer and Forbes, 1988; Richardson *et al,* 1988). For steroid hormone receptors, both mechanisms can be implicated.

Picard and Yamamoto (1987) have used in vitro mutagenesis to study the mechanism of subcellular localization of the rat GR. They have shown that the main signal directing the receptor into the nucleus was localized in the steroid binding domain. A 28 aminoacid SV40 large T antigen like sequence, located in the hinge region (aminoacids 497 to 524), had a less important role and was also effective only after binding of hormone.

We have studied the signals responsible for the nuclear localization of the rabbit PR using a series of deletion mutants (Guiochon-Mantel *et al,* 1989). Its main NLS is a stretch of aminoacids located in the hinge region around position 638–642 and bearing similarities to the NLS present in the SV40 large T antigen. This putative signal is constitutive (acts in the absence of hormone), and if it is deleted, the ligand free receptor becomes cytoplasmic. The sequence of the SV40 large T antigen NLS and of the PR putative NLS can be aligned in two different ways, thus defining a sequence of 8 aminoacids in the PR (RKFKKFNK). This sequence is completely conserved between rabbit (Loosfelt *et al,* 1986) and human (Misrahi *et al,* 1987) PR and contains a single change (of a non-basic residue) compared with the chicken PR (Gronemeyer *et al,* 1987; Conneely *et al,* 1987). If other steroid hormone receptors, including GR, AR, and MR, are aligned through their DNA binding regions, homologous sequences are found at exactly the same position (10 aminoacids after the last conserved cysteine), except in the case of the ER, where a homologous sequence is found 11 aminoacids after the last conserved cysteine. For a given steroid receptor, this region is highly conserved in the different species studied.

A mutant deleted of the first NLS is cytoplasmic in the absence of hormone and is translocated to the nucleus in the presence of hormone. Thus, there is a second mechanism involved in the nuclear localization of the PR. It is located in the DNA binding domain activated either through the binding of the hormone or by deletion of the steroid binding domain (constitutive receptor). The receptor could passively cross the nuclear membrane, bind to the DNA, and accumulate in the nucleus by this mechanism. Alternatively, it was possible that a second karyophilic signal was intermingled with the DNA binding domain and could be unmasked by exactly the same mechanisms as those involved in the accessibility of the DNA binding site. Energy depletion experiments allowed us to distinguish between these two mechanisms (Fig. 2). Thus, we have shown that there is a second NLS, located in the second zinc finger (Guiochon-Mantel *et al,* 1991). This NLS is masked and unmasked by mechanisms identical to those regulating the activity of the DNA binding function. Examination of the sequence of this domain shows two stretches of basic aminoacids, which are candidates for being NLS: aminoacids 614–618 and aminoacids 624–627. Use of different energy inhibitors on different receptor mutants has allowed us to classify these two NLS: the first one (located in the hinge region) being more potent than the second one (located in the second zinc finger). These two NLS are additive. It has been shown in different models that proteins carrying two NLS are more efficiently transported into the nucleus than proteins carrying a single NLS (Lanford *et al,* 1986; Dworetzky *et al,* 1988).

The existence of multiple NLS is a frequent feature of nuclear proteins (Roberts, 1989; Silver and Goodson, 1989). A mutant deleted in the first NLS, cytoplasmic in the absence of hormone, retains a nearly complete biological activity when saturated with hormone, showing that initial nuclear localization is not a prerequisite of biological activity.

In the case of the human ER, a 48 aminoacid fragment located in the hinge region (aminoacids 256–303) mediates efficient nuclear localization of a β-galactosidase fusion protein (Picard *et al,* 1990). This portion of the receptor includes three basic stretches: aminoacids 256–260, 266–271 and 299–303. The first two basic stretches are almost perfectly conserved between the different species that have been sequenced. The third stretch is conserved only in

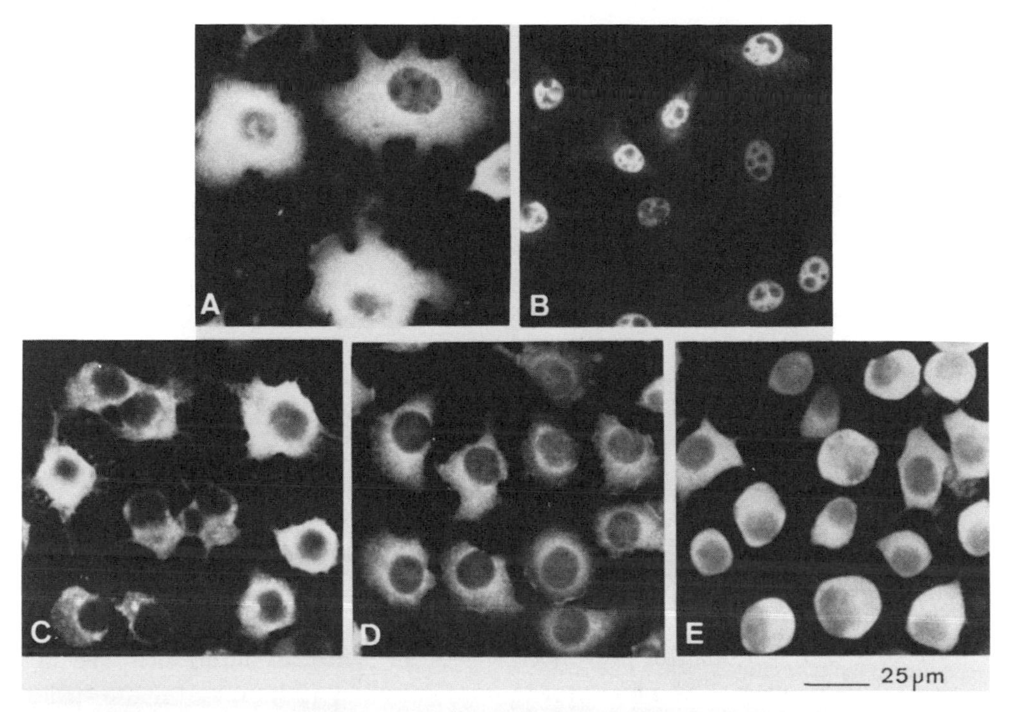


**Fig. 2.** Energy depletion inhibits hormone dependent nuclear transfer of Δ638–642 mutant progesterone receptor in L cells. (A) Cells incubated at 37°C; (B) cells incubated at 37°C with hormone for 4 hr; (C) cells incubated at 37°C with hormone and sodium azide and 2-deoxyglucose; (D) cells incubated at 37°C with hormone and atractyloside and 2-deoxyglucose; (E) cells incubated at 4°C with hormone

charge. None of these basic stretches is itself sufficient as a nuclear localization signal.

## Nucleocytoplasmic Shuttle of Steroid Hormone Receptors

### *Interactions between Receptor Monomers during Nuclear Transport*

A dimerization process has been observed in vitro by band shift electrophoretic analysis of receptor-DNA interactions (Tsai *et al,* 1988; Kumar and Chambon, 1988). By deleting epitopes recognized by monoclonal antibodies, it is possible to follow different receptor mutants within the same cells in vivo (Guiochon-Mantel *et al,* 1989). In the absence of ligand, the receptor is transferred into the nucleus as a monomer. After hormone administration in the presence of an inhibitor of protein synthesis, a "cytoplasmic" monomer is transferred into the nucleus through interaction with a "nuclear" monomer. This interaction takes place through the steroid binding domain of both receptors. Oligomer formation occurs independently of any DNA binding event.

### *Nucleocytoplasmic Shuttle of Steroid Hormone Receptor*

Oligomer formation indicates that both monomers have to contact each other; this could be explained by two mechanisms. Either the "cytoplasmic" form of receptor could passively diffuse into the nucleus but could not remain there, except if a "nuclear" form was present and formation of oligomers led to trapping of the mutant receptor in the nuclear compartment. Alternatively, contact between "nuclear" and "cytoplasmic" monomers of receptor might have been due to the fact that the "nuclear" receptor was continuously shuttling between the cytoplasm and the nucleus and was actively accumulated there. During this shuttle, it could contact the "cytoplasmic" monomer. By partial inhibition of energy formation, we could distinguish between both possibilities. We have shown that the mutated receptor, devoid of NLS, cannot cross the nuclear membrane, whereas nuclear receptor species are continually recycled through the cytoplasm (Guiochon-Mantel *et al,* 1991). Moreover, we have shown, with inhibitors of protein synthesis, that these transfers involve the same receptor molecules and not newly synthesized receptors.

To confirm the existence of a nucleocytoplasmic shuttle of receptor, we have studied the migration of progesterone receptor between nuclei in interspecies heterokaryons (Borer *et al,* 1989). We fused a mouse L cell line permanently expressing the progesterone receptor with human 293 cells devoid of receptor, in the presence of protein synthesis inhibitors. We observed the presence of the receptor in human nuclei 12 hours after the fusion (Fig. 3). Thus, the receptor had migrated from one nucleus to the other, indicating a shuttle through the cytoplasm.

The residency of the progesterone receptor in the nucleus seems to be a dynamic event resulting from the continuous active transport into the nucleus counterbalanced by some diffusion into the cytoplasm. It is not known if this diffusion is a totally passive phenomenon or if it necessitates the presence of a NLS in the protein (Mandell and Feldherr, 1990). Interaction with specific protein(s) in the pore could then take place during nuclear exit of the protein. Shuttle mechanisms have recently been proven for two nucleolar proteins (Borer *et al,* 1989) and discussed in other cases, including steroid receptors (Nigg, 1990). This mechanism of nuclear localization of the receptor explains some previous observations: as described earlier, ligand free progesterone or oestrogen receptors, which reside in the nucleus, are found in the cytosol after homogenization even when nuclear structures have been preserved. This is probably due to diffusion through nuclear membranes under conditions where active transport is blocked by dilution and low temperature. Incidentally, association of receptors with nuclei after homogenization of cells at 25°C has been described (Molinari *et al,* 1985). Moreover, localization of ligand free glucocorticoid receptor in cytoplasm or both cytoplasm and nuclei has been considered as completely different from that of oestrogen or progesterone receptors located in the nucleus. Many authors have been puzzled by the fact that proteins having such similar properties may exhibit such differences in their subcellular localization. However, if receptor continually shuttles between nucleus and cytoplasm, the case of the glucocorticoid receptor may be only quantitatively and not qualitatively different from that of sex steroid receptors. Less effective constitutive karyophilic signals would lead to an in-

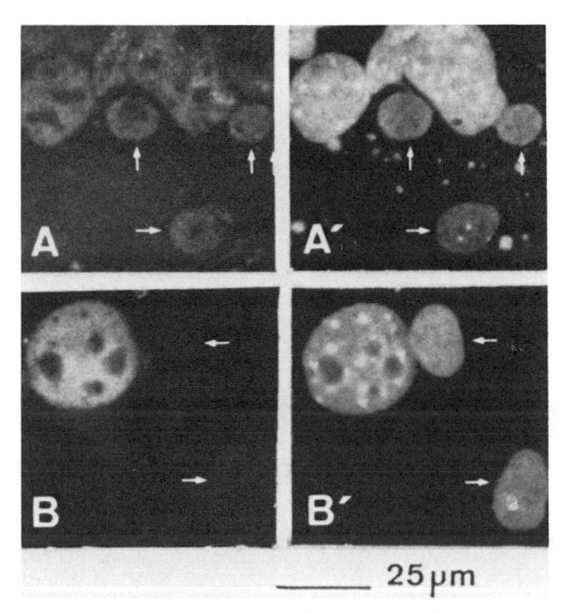


**Fig. 3.** Transfer of receptor from mouse to human nuclei in heterokaryons. (A, A′) Mouse L cells containing wild type progesterone receptor were fused with 293 human cells devoid of receptor; heterokaryons were observed. Cycloheximide was administered to prevent neosynthesis of progesterone receptor. Progesterone receptor was labelled by immunofluorescence 12 hr after the fusion (A). Human (arrows) and mouse nuclei could be distinguished by fixation of Hoechst 33258 (A′) and by their size. After the fusion, progesterone receptor appears in human 293 cell nuclei. (B, B′) Control experiment, in which cells have not been fused. Progesterone receptor was labelled with immunofluorescence (B). Coloration of Hoechst 33258 allows distinction between the two different cell types (B′). 293 cells are devoid of receptor (arrow)

creased duration of cytoplasmic localization and to an apparent distribution between the cytoplasmic and nuclear compartment. The results of studies of the SV40 large T antigen like NLS of the GR favour this hypothesis. This NLS is able to direct a cytoplasmic protein, like the β-galactosidase, to the nucleus (Picard and Yamamoto, 1987). It is also able to direct the GR itself if it is cloned in a region of the GR other than its natural region (Picard *et al,* 1988). Monoclonal antibodies directed against a synthetic NLS have been produced. These antibodies react with the activated form of the steroid-receptor complex but not with the unactivated form (Urda *et al,* 1989), showing that the NLS region is occluded in the unactivated complex but exposed as a result of activation.

The understanding of receptor function may also be modified by receptors shuttling between nucleus and cytoplasm. Receptors could thus interact with cytoplasmic components. For instance, many speculations have been published on the problem of receptor interaction with the 90 kDa heat shock protein (hsp90). This protein is found mainly in the cytoplasm (Lindquist and Craig, 1988) and has led, in some cases, to the conclusion that receptor binding to this protein was an artefact due to receptor extraction from the nucleus during cell homogenization. Other reports have tried to prove that a small fraction of the hsp90 is intranuclear (Gasc *et al,* 1990). Obviously, such discussions are meaningless if receptors cycle between cytoplasm and nucleus. Moreover, this mechanism is compatible with receptor exerting biological activities in the cellular cytoplasm. Such effects have indeed been described (Liao *et al,* 1980; Verdi and Campagnoni, 1990). The study of the cellular traffic of steroid receptors is of interest not only for the understanding of the mechanism of action of these hormones but also as a model to understand the mechanisms underlying the nuclear localization of the increasing number of proteins implicated in the regulation of gene transcription.

## RECEPTOR IMMUNOLOCALIZATION IN PATHOLOGICAL STUDIES

Oestrogen and progestin receptors can be observed in a proportion of the neoplasms arising in the female genital tract and in breast carcinomas. Inherent to the homogenization of the tissue specimens for the ligand binding assay is the problem of tissue heterogeneity and the subsequent contribution of various amounts of receptor-positive benign tissue elements, such as normal or hyperplastic glandular epithelium, stroma (in endometrial carcinoma) or myometrium to the total observed binding (receptor). Determination of these receptors in tissue sections is thus of great importance to analyse the tumour at the cellular level. Immunocytochemistry also used very small amounts of tissue that would be unsuitable for steroid binding assay for both receptors, including frozen tissue and paraffin embedded tissue, cytological imprint material and fine needle aspirates (Ricketts and Coombes, 1989, Charpin *et al,* 1989) from breast carcinoma tissue. Technical aspects of both ER and PR determination have been detailed in several reports (King *et al,* 1985; Perrot-Applanat *et al,* 1987, 1989). Best results were obtained with 3.7% formol (or 4% paraformaldehyde) and Zamboni's fixative (picric acid paraformaldehyde) and highly selected antibodies (H222 for anti-ER, Abbott Laboratories, USA; Let126, Li417 [Transbio, France], and KD68 [Abbott] for anti-PR). Immunohistochemical evaluation usually incorporated both intensity and distribution of staining. Because of the heterogeneity of staining and variations within individual microscope fields, a score system must be used in the reporting of results.

### Breast Carcinoma

Oestrogens have an important role in the genesis of experimental and human breast cancers. Once established, the tumours are frequently hormone dependent, ie they regress or stop growing if deprived of oestrogen. In humans, about one third of advanced breast cancers are thus susceptible to remission with endocrine therapy. Criteria for the selection of patients that will respond to hormonal treatment have been the subject of intense study during the past two decades. On the molecular level, this has led to the development of analytical tools for the study of steroid receptors. Initial studies correlated the presence (by ligand-binding assay) of ER and PR with (a) a high probability of response to endocrine therapy; 75–80% of the patients respond to hormone treatment when both receptors are present and (b) a more favourable prognosis, ie a greater disease free interval between PR positive and negative patients. Receptor status was the test most commonly used to predict response to therapy. After the first detection of immunocytochemical evidence of ER and PR using monoclonal antibodies (King *et al,* 1985; Perrot-Applanat *et al,* 1987), there was an increased number of reports that cannot be related in detail in this chapter.

Immunocytochemical techniques have shown an exclusively nuclear distribution of ER and PR in carcinoma breast cells and benign ductal or lobular epithelial cells. No specific staining is seen in the surrounding connective tissue. Specific ER and PR immunostaining was found to be very heterogeneous (Poulsen *et al,* 1981; King *et al,* 1985; Perrot-Applanat *et al,* 1987): PR positive tumours showed variations in intensity and distribution of staining among cells, as well as in different areas of the same sections or in different samples from the same primary lesion. Those variations were also noted among non-malignant cells of breast ducts and in T47D breast cancer cells. This heterogeneity may be attributed either to receptor content variations, which are cell cycle dependent, or to variations due to the presence of both progesterone responsive and non-responsive cells. Polyclonal distribution of hormone receptors may explain the clinical unresponsiveness seen in about 20% of patients with PR positive breast cancer. For future studies, it is likely that the most important feature of immunocytochemistry lies in its ability to provide information on the distribution of these receptors in cell populations.

Several studies showed a good correlation between steroid binding assay and immunoperoxidase staining for both ER and PR contents (King *et al,* 1985; Perrot-Applanat *et al,* 1987; Berger *et al,* 1989; reviewed in Ricketts and Coombes, 1989). Discordance rates among studies were influenced by the arbitrary cutoff levels that were used to define receptor positivity in the biochemical assay. Positive ER staining was shown by statistical analysis to correlate significantly with tumour size, tumour grade, tubule formation, number of mitosis, tumour necrosis and lymphocytic infiltration (Berger *et al,* 1987; Charpin *et al,* 1988; Walker *et al,* 1988). Positive PR staining was shown to correlate with tumour grade (Perrot-Applanat *et al,* 1987).

Immunocytochemistry of ER has also been shown to predict response of breast cancer to endocrine therapy (Pertshuk *et al,* 1985; McClelland *et al,* 1986; Andersen and Poulsen, 1988). Sixty percent of patients having an ER positive tumour will respond to hormonal treatment. Finally, development of immunocytochemical methods for receptor detection on paraffin sections (Poulsen *et al,* 1985; Shimada *et al,* 1985; Shintaku and Said, 1987; Perrot-Applanat *et al,* 1989) now allows the observation of tumours embedded in paraffin some time ago, thus providing opportunities for retrospective studies. However, the assay completed on frozen sections is somewhat more sensitive than the paraffin assay (Perrot-Applanat *et al,* 1989; Andersen *et al,* 1990). A detailed comparison of the different immunocytochemical techniques has also been recently done on frozen specimens, cytological imprints and paraffin embedded tissues (Ozello *et al,* 1991).

### Female Genital Tract

Endometrial carcinoma is the most frequent malignancy of the female genital tract (Silverberg and Lubera, 1987). However, the mechanism underlying its pathogenesis remains obscure. As a step towards understanding hormone action, ER and PR contents were examined. It has first been shown that PR levels were increased in these carcinomas and that quantitation of PR is a useful, but not absolute, predictor of tumour response to progestin therapy. Only 60% of cases positive for PR by biochemical assays will respond to treatment (Erlich *et al,* 1988). It has been suggested that heterogeneity in PR distribution at the cellular level, not detectable by biochemical assay, might contribute to this discrepancy. Precise determination of the extent of heterogeneity required the ability to assess the presence of receptor in individual cells, as well as within the benign stromal and myometrial cells adjacent to the neoplasm.

Immunocytochemical localization of ER and PR has been observed in the endometrial carcinoma, endometrial sarcomas and adenocarcinomas of the fallopian tube (Bergeron *et al,* 1988; Zaino *et al,* 1988; Segreti *et al,* 1989). Some studies have compared both ER and PR content measured biochemically and immunohistochemically with the histological grade of endometrial carcinoma (Bergeron *et al,* 1988b, 1988c; Segreti *et al,* 1989). The best correlation was always found for malignant epithelial cells, a cell type that also showed a high heterogeneity in staining intensity. By contrast, the stroma, myometrium and benign glandular elements tended to be more uniform in receptor immunostaining and both ER and PR content failed to be correlated with the degree of tumour differentiation. These (and other) results suggest that the malignant epithelium is the primary determinant of tumour behaviour. Yet, the role of the surrounding receptor positive tissue elements on tumour growth remains to be elucidated.

The presence of steroid hormone receptors has also been shown in epithelial ovarian cancers (reviewed in Slotman and Rao, 1988). Positivity for PR was associated with a significantly better survival rate, in association with grade and stage of the tumour (Slotman *et al,* 1990). Receptor is found in the nuclei of adenocarcinoma epithelial cells, not in connective tissue or blood vessels (Press *et al,* 1985). A significant correlation between immunocytochemical staining for ER in nuclei and the content of ER in cytosolic tissue extracts determined by steroid binding assay has also been found.

### Other Neoplasms

Several other neoplasms contain substantial amounts of ER and PR, suggesting that these neoplasms are oestrogen and progesterone sensitive tumours.

The distribution of these receptors has been assessed at the cellular level in neoplasms of the gastrointestinal tract (pancreas, Ladanyi *et al*, 1987; stomach, Wu *et al*, 1990), lung (lymphangioleiomyomatosis, Colley *et al*, 1989) and intracranial tumours (Perrot-Applanat M, in press); however, whether the occurrence of ER and PR is related to pathologic status needs additional research.

Prostatic cancer is a hormonally responsive malignancy. Although a correlation between AR concentration in the prostate and clinical response to endocrine therapy has not been shown, AR rich tumours have a better response to treatment (Trachtenberg and Walsh, 1982). Immunocytochemical localization of AR has been completed in the rat prostatic carcinoma (Quarmby *et al*, 1990).

## CONCLUSION

We have summarized the recent developments in immunocytochemical staining for ER and PR and, to a lesser extent, for other steroid receptors. The molecular biology of steroid hormone receptors and other members of nuclear receptor family has progressed strikingly during the past 5 years. The study of the cellular biology of these receptors, for example, sites of synthesis, cellular traffic, is only beginning.

## SUMMARY

Experimental conditions are described for the detection of steroid receptors in tissue sections or cells at the light microscope level. Current knowledge about the ultrastructural distribution of these receptors is summarized; the mechanisms of their nuclear localization are described. Karyophilic signals involved in nuclear translocation are characterized by means of in vitro mutagenesis of steroid receptor cDNAs. Studies analysing the subcellular distribution of various transfected receptor mutants in energy depleted cells together with fusion experiments provide evidence for nucleoplasmic shuttling of progesterone receptors. We conclude that the "nuclear" location of the wild type progesterone receptor reflects a dynamic equilibrium between active nuclear import and outward diffusion. We also describe the use of immunocytochemistry in pathology, especially for the detection of steroid receptors in hormone dependent tumours.

## References

Andersen J and Poulsen HS (1988) Relationship between estrogen receptor status in the primary tumour and its regional and distant metastases. *Acta Oncologica* **27** 761–768

Andersen J, Thorpe SM, King WJ *et al* (1990) The prognostic value of immunohistochemical estrogen receptor analysis in paraffin-embedded and frozen sections versus that of steroid-binding assays. *European Journal of Cancer* **26** 442–449

Antakly T and Eisen HJ (1984) Immunocytochemical localization of glucocorticoid receptor in target cells. *Endocrinology* **115** 1984–1989

Beato M (1989) Gene regulation by steroid hormones. *Cell* **56** 335–344

Berger U, Wilson P, McClelland R, Davidson J and Coombes RC (1987) Correlation of immunocytochemically demonstrated estrogen receptor distribution and histopathology features in primary breast cancer. *Human Pathology* **18** 1263–1267

Berger U, Wilson P, Thethi S, McClelland RA, Greene GL and Coombes RC (1989) Comparison of an immunocytochemical assay for progesterone receptor with biochemical method of measurement and immunocytochemical examination of the relationship between progesterone and estrogen receptors. *Cancer Research* **49** 5176–5179

Bergeron C, Ferenczy A, Toft DO, Schneider W and Shyamala G (1988a) Immunocytochemical study of progesterone receptors in the human endometrium during the menstrual cycle. *Laboratory Investigation* **59** 862–869

Bergeron C, Ferenczy A, Toft DO and Shyamala G (1988b) Immunocytochemical study of progesterone receptors in hyperplastic and neoplastic endometrial tisues. *Cancer Research* **48** 6132–6136

Bergeron C, Ferenczy A and Shyamala G (1988c) Distribution of estrogen receptors in various cell types of normal, hyperplastic, and neoplastic human endometrial tissues. *Laboratory Investigation* **58** 338–345

Blaustein JD, King JC, Toft DO and Turcotte J (1988) Immunocytochemical localization of estrogen-induced progestin receptors in guinea pig brain. *Brain Research* **474** 1–15

Blaustein JD, Olster DH, Delville Y, Nielsen KH, Tetel MJ and Turcotte JC (1990) Hypothalamic sex steroid hormone receptors and female sexual behavior: new insights from immunocytochemical studies, In: Balthazart J (ed). *Hormones, Brain and Behaviour in Vertebrates: Pt. 2: Behavioural Activation in Males and Females—Social Interaction and Reproductive Endocrinology* Comparative Physiology Series, vol. 9, pp 75–90, S Karger, Basel

Borer RA, Lehner CF, Eppenberger HM and Nigg E (1989) Major nucleolar proteins shuttle between nucleus and cytoplasm. *Cell* **56** 379–390

Breeuwer M and Goldfarb DS (1990) Facilitated nuclear transport of histone H1 and other small nucleophilic proteins. *Cell* **60** 999–1008

Brown TJ, McLusky NJ, Leranth C, Shanabrough M and Naftolin F (1990) Progestin receptor-containing cells in guinea pig hypothalamus: afferent connections, morphological characteristics, and neurotransmitter content. *Molecular and Cellular Neurosciences* **1** 58–77

Charpin C, Martin PM, De Victor B *et al* (1988) Multiparametric study (Samba 200) of estrogen receptor immunocytochemical assay in 400 human breast carcinomas: analysis of estrogen receptor distribution heterogeneity in tissues and correlations with dextran coated charcoal assays and morphological data. *Cancer Research* **48** 1578–1586

Charpin C, Andrac L, Habib MC *et al* (1989) Immunodetection in fine-needle aspirates and multiparametric (SAMBA) image analysis. *Cancer* **63** 863–872

Chong MT and Lippman ME (1982) Effects of RNA and ribonuclease on the binding of estrogen and glucocorticoid receptors from MCF7 to DNA cellulose. *Journal of Biological Chemistry* **257** 2996–3002

Clarke CL (1990) Cell-specific regulation of progesterone receptor in the female reproductive system. *Molecular and Cellular Endocrinology* **70** C29–C33

Clarke CL, Zaino RJ, Feil PD *et al* (1987) Monoclonal antibodies to human progesterone receptor: characterization by biochemical and immunohistochemical techniques. *Endocrinology* **121** 1123–1132

Colley MH, Geppert E and Franklin WA (1989) Immunohistochemical detection of steroid receptors in a case of pulmonary lymphangioleiomyomatosis. *American Journal of Surgical Pathology* **13** 803–807

Conneely OM, Dobson ADW, Tsai MJ *et al* (1987) Sequence and expression of a functional chicken progesterone receptor. *Molecular Endocrinology* **1** 517–525

Dworetzky SI, Lanford RE and Feldherr CM (1988) The effects of variations in the number and sequence of targeting signals on nuclear uptake. *Journal of Cell Biology* **107** 1279–1287

Ehrlich CE, Young PCM, Stehman FB, Sutton GP and Alford WM (1988) Steroid receptors and clinical outcome in patients with adenocarcinoma of the endometrium. *American Journal of Obstetrics and Gynaecology* **158** 796–807

Fakan S and Nobis P (1978) Ultrastructural localization of transcription sites and of RNA distribution during the cell cycle of synchronized CHO cells. *Experimental Cell Research* **113** 327–337

Fakan S and Puvion E (1980) The ultrastructural visualization of nucleolar and extranucleolar RNA synthesis and distribution. *International Review of Cytology* **65** 255–299

Fakan S, Puvion E and Spohr G (1976) Localization and characterization of newly synthetized nuclear RNA in isolate rat hepatocytes. *Experimental Cell Research* **99** 155–164

Fakan S, Leser G and Martin T (1986) Immunoelectron microscope visualization of nuclear ribonucleoprotein antigens within spread transcription complexes. *Journal of Cell Biology* **103** 1153–1157

Farman N, Oblin ME, Lombes M *et al* (1991) Immunolocalization of gluco- and mineralocorticoid receptors in rabbit kidney. *American Journal of Physiology* **260** C226–C233

Fuxe K, Wikström AC, Okret S *et al* (1985) Mapping of glucocorticoid receptor immunoreactive neurons in the rat tel- and diencephalon using a monoclonal antibody against rat liver glucocorticoid receptor. *Endocrinology* **117** 1803–1812

Garcia E, Bouchard P, De Brux J *et al* (1988) Use of immunocytochemistry of progesterone and estrogen receptors for endometrial dating. *Journal of Clinical Endocrinology and Metabolism* **67** 80–87

Gasc JM, Renoir JM, Radanyi C, Joab I, Tuohimaa P and Baulieu EE (1984) Progesterone receptor in the chick oviduct: an immunohistochemical study with antibodies to distinct receptor components. *Journal of Cell Biology* **99** 1193–1201

Gasc JM, Ennis BW, Baulieu EE and Stumpf WE (1986) Combined technique of immunohistochemistry and autoradiography for the simultaneous detection of steroid hormone receptors and their ligand. *Journal of Histochemistry and Cytochemistry* **34** 1505–1508

Gasc JM, Renoir JM, Faber LE, Delahaye F and Baulieu EE (1990) Nuclear localization of two steroid receptor-associated proteins, hsp90 and p59. *Experimental Cell Research* **186** 362–367

Gorski J, Toft D, Shyamala G, Smith D and Notides A (1968) Hormone receptors: studies on the interaction of estrogen with the uterus. *Recent Progress in Hormone Research* **24** 45–81

Govindan MV (1980) Immunofluorescence microscopy of the intracellular translocation of glucocorticoid-receptor complexes in rat hepatoma (HTC) cells. *Experimental Cell Research* **127** 293–297

Greene GL, Fitch FW and Jensen EV (1980) Monoclonal antibodies to estrophilin: probes for the study of estrogen receptors. *Proceedings of the National Academy of Sciences of the USA* **77** 157–161

Gronemeyer H, Turcotte B, Quirin-Strickerc *et al* (1987) The chicken progesterone receptor: sequence, expression and functionnal analysis. *EMBO Journal* **6** 3985–3994

Groyer-Picard MT, Vu-Hai MT, Jolivet A, Milgrom E and Perrot-Applanat M (1990) Monoclonal antibodies for immunocytochemistry of progesterone receptors (PR) in various laboratory rodents, livestock, humans and chickens: identification of two epitopes conserved in PR of all these species. *Endocrinology* **126** 1485–1491

Guiochon-Mantel A, Loosfelt H, Lescop P *et al* (1989) Mechanisms of nuclear localization of the progesterone receptor: evidence for interaction between monomers. *Cell* **57** 1147–1154

Guiochon-Mantel A, Lescop P, Christin-Maitre S *et al* (1991) Nucleocytoplasmic shuttle of the progesterone receptor. *EMBO Journal* **10** 3851–3859

Hild-Petito S, Stouffer RL and Brenner RM (1988) Immunocytochemical localization of estradiol and progesterone receptors in the monkey ovary throughout the menstrual cycle. *Endocrinology* **123** 2896–2905

Horwitz KB, Mockus MB, Pike AW, Fennessey PV and Sheridan RL (1983) Progesterone receptor replenishment in T47D human breast cancer cells: roles of protein synthesis and hormone metabolism. *Journal of Biological Chemistry* **258** 7603–7610

Hsu SM, Raine L and Fanger H (1981) The use of antiavidin antibody and avidin-biotin peroxidase complex in immunoperoxidase technics. *American Journal of Clinical Pathology* **75** 816–821

Isola JJ (1987) The effect of progesterone on the localization of progesterone receptors in the nuclei of chick oviduct cells. *Cell and Tissue Research* **249** 317–323

Isola J, Korte JM and Tuohimaa P (1987a) Immunocytochemical localization of progesterone receptor in the chicken ovary. *Endocrinology* **121** 1034–1040

Isola J, Pelto-Huikko M, Ylikomi T and Tuohimaa P (1987b) Immunoelectron microscopic localization of progesterone receptor in the chick oviduct. *Journal of Steroid Biochemistry* **26** 19–23

Isomaa V, Isolato H, Orava M and Janne O (1979) Regulation of cytosol and nuclear progesterone receptors in rabbit uterus by estrogen, antiestrogen and progesterone administration. *Biochimica et Biophysica Acta* **585** 24–33

Iwai T, Nanbu Y, Iwai M, Taii S, Fujii S and Mori T (1990) Immunohistochemical localization of oestrogen receptors and progesterone receptors in the human ovary throughout the menstrual cycle. *Virchows Archiv. A, Pathological Anatomy and Histopathology* **417** 369–375

Jensen T, Suzuki T, Kawashima T, Stumpf WE, Jungblut PW and DeSombre ER (1968) A two-step mechanism for the interaction of oestradiol with the rat uterus. *Proceedings of the National Academy of Sciences of the USA* **59** 632–638

Kalderon D, Roberts BL, Richardson WD and Smith AE (1984) A short aminoacid sequence able to specify nuclear location. *Cell* **39** 499–509

King WJ and Greene GL (1984) Monoclonal antibodies localize oestrogen receptor in the nuclei of target cells. *Nature* **307** 745–747

King WL, DeSombre ER, Jensen EV and Greene GL (1985) Comparision of immunocytochemical and steroid-binding assays for estrogen receptor in human breast tumours. *Cancer Research* **45** 293–304

Korte JM and Isola J (1988) An immunocytochemical study of the progesterone receptor in rabbit ovary. *Molecular and Cellular Endocrinology* **58** 93–101

Kreitmann B, Bugat R and Bayard F (1979) Estrogen and progestin regulation of the progesterone receptor concentration in human endometrium. *Journal of Clinical Endocrinology and Metabolism* **49** 926–929

Kumar V and Chambon P (1988) The estrogen receptor binds tightly to its responsive element as a ligand-induced homodimer. *Cell* **55** 145–156

Ladanyi M, Mulay S, Arseneau J and Bettez P (1987) Estrogen and progesterone receptor determination in the papillary cystic neoplasm of the pancreas. *Cancer* **60** 1604–1611

Lanford RE, Kanda P and Kennedy RC (1986) Induction of nuclear transport with a synthetic peptide homologous to the SV40 antigen transport signal. *Cell* **46** 575–582

Lessey BA, Killam AP, Metzger DA, Haney AF, Greene GL and McCarty KS (1988) Immunohistochemical analysis of human uterine estrogen and progesterone receptors throughout the menstrual cycle. *Journal of Clinical Endocrinology and Metabolism* **67** 334–340

Liao S, Liang T and Tymoczko JL (1973) Ribonucleoprotein binding of steroid-"receptor" complexes. *Nature (New Biology)* **241** 211–213

Liao S, Smythe S, Tymoczko JL, Rossini GP, Chen C and Hiipakka RA (1980) RNA-dependent release of androgen and other steroid receptor complexes from DNA. *Journal of Biological Chemistry* **255** 5545–5551

Lindquist S and Craig EA (1988) The heat-shock proteins. *Annual Review of Genetics* **22** 631–677

Logeat F, Vu Hai MT, Fournier A, Legrain P, Buttin G and Milgrom E (1983) Monoclonal antibodies to rabbit progesterone receptor: crossreaction with other mammalian progesterone receptors. *Proceedings of the National Academy of Sciences of the USA* **80** 6456–6459

Lombes M, Edelman IS and Erlanger BF (1989) Internal image properties of a monoclonal auto-anti-idiotypic antibody and its binding to aldosterone receptors. *Journal of Biological Chemistry* **264** 2528–2536

Lombes M, Farman N, Oblin ME *et al* (1990) Immunohistochemical localization of renal mineralocorticoid receptor by using an anti-idiotypic antibody that is an internal image of aldosterone. *Proceedings of the National Academy of Sciences of the USA* **87** 1086–1088

Loosfelt H, Atger M, Misrahi M *et al* (1986) Cloning and sequence analysis of rabbit progesterone-receptor complementary DNA. *Proceedings of the National Academy of Sciences of the USA* **83** 9045–9049

Lorenzo F, Jolivet A, Loosfelt H *et al* (1988) A rapid method of epitope mapping: application to the study of immunogenic domains and to the characterization of various forms of rabbit progesterone receptor. *European Journal of Biochemistry* **176** 53–60

Mandell RB and Feldherr CM (1990) Identification of two HSP70-related xenopus oocyte proteins that are capable of recycling across the nuclear envelope. *Journal of Cell Biology* **111** 1775–1783

Martins VR, Pratt WB, Terracio L, Hirst MA, Ringold GM and Housley PR (1991) Demonstration by confocal microscopy that unliganded overexpressed glucocorticoid receptors are distributed in a nonrandom manner throughout all planes of the nucleus. *Molecular Endocrinology* **5** 217–225

McClelland R, Berger U, Miller L, Powles T and Coombes RC (1986) Immunocytochemical assay for ER in patients with breast cancer, relationship to a biochemical assay and to outcome of therapy. *Journal of Clinical Oncology* **4** 1171–1176

Milgrom E, Luu Thi MT, Atger M and Baulieu EE (1973) Mechanisms regulating the concentration and the conformation of progesterone receptor(s) in the uterus. *Journal of Biological Chemistry* **248** 6366–6374

Misrahi M, Atger M, D'Auriol L *et al* (1987) Complete amino acid sequence of the human progesterone receptor deduced from cloned cDNA. *Biochemical and Biophysical Research Communications* **143** 740–748

Molinari AM, Medici N, Armetta I, Nigro V, Montcharmont B and Puca GA (1985) Particulate nature of the unoccupied uterine estrogen receptor. *Biochemical and Biophysical Research Communications* **128** 634–642

Newmeyer DD and Forbes DJ (1988) Nuclear import can be separated into distinct steps in vitro: nuclear pore binding and translocation. *Cell* **52** 641–653

Nigg EA (1990) Mechanisms of signal transduction to the cell nucleus. *Advances in Cancer Research* **550** 271–310

Nigg EA, Baeuerle PA and Lührmann R (1991) Nuclear import-export: in search of signals and mechanisms. *Cell* **66** 15–22

Okret S, Wikström AC, Wrange O, Andersson B and Gustafsson JA (1984) Monoclonal antibodies against the rat liver glucocorticoid receptor. *Proceedings of the National of Academy of Sciences of the USA* **81** 1609–1613

Okulicz WC, Savasta AM, Hoberg LM and Longcope C (1989) Immunofluorescent analysis of estrogen induction of progesterone receptor in the rhesus uterus. *Endocrinology* **125** 930–934

Ozzello L, DeRosa C, Habif DV and Greene GL (1991) An immunohistochemical evaluation of progesterone receptor in frozen sections, paraffin sections, and cytologic imprints of breast carcinomas. *Cancer* **67** 455–462

Papamichail M, Tsokos G, Tsawdaroglou N and Sekeris CE (1980) Immunocytochemical demonstration of glucocorticoid receptors in different cell types and their translocation from the cytoplasm to the cell nucleus in the presence of dexamethasone. *Experimental Cell Research* **125** 490–493

Peralta Soler A and Aoki A (1989) Immunocytochelical detection of estrogen receptors in a hormone-unresponsive mammary tumour. *Histochemistry* **91** 351–356

Perrot-Applanat M, Logeat F, Groyer-Picard MT and Milgrom E (1985) Immunocytochemical study of mammalian progesterone receptor using monoclonal antibodies. *Endocrinology* **116** 1473–1484

Perrot-Applanat M, Groyer-Picard MT, Logeat F and Milgrom E (1986) Ultrastructural localization of the progesterone receptor by an immunogold method: effect of hormone administration. *Journal of Cell Biology* **102** 1191–1199

Perrot-Applanat M, Groyer-Picard MT, Lorenzo F *et al* (1987) Immunocytochemical study with monoclonal antibodies to progesterone receptor in human breast tumours. *Cancer Research* **47** 2652–2661

Perrot-Applanat M, Groyer-Picard MT, Garcia E, Lorenzo F and Milgrom E (1988) Immunocytochemical demonstration of estrogen and progesterone receptors in muscle cells of uterine arteries in rabbits and humans. *Endocrinology* **123** 1511–1519

Perrot-Applanat M, Groyer-Picard MT, Vu Hai MT, Pallud C, Spyratos F and Milgrom E (1989) Immunocytochemical staining of progesterone receptor in paraffin sections of human breast cancers. *American Journal of Pathology* **135** 457–468

Perrot-Applanat M, Groyer-Picard MT and Kujas M Immunocytochemical study of progesterone receptor in human meningioma. *Acta Neurochirurgica* (in press)

Pertshuk LP, Eisenberg K, Carter A and Feldman JG (1985) Immunohistologic localization of estrogen receptors in breast cancer with monoclonal antibodies. *Cancer* **55** 1513–1518

Picard D and Yamamoto KR (1987) Two signals mediate hormone-dependent nuclear localization of the glucocorticoid receptor. *EMBO Journal* **6** 3333–3340

Picard D, Salser SJ and Yamamoto KR (1988) A movable and regulable inactivation function within the steroid binding domain of the glucocorticoid receptor. *Cell* **54** 1073–1080

Picard D, Kumar V, Chambon P and Yamamoto KR (1990) Signal transduction by steroid hormones: nuclear localization is differentially regulated in estrogen and glucocorticoid receptors. *Cell Regulation* **1** 291–299

Poulsen HS, Jensen J and Hermansen C (1981) Human breast cancer: heterogeneity of estrogen binding sites. *Cancer* **48** 1791–1793

Poulsen H, Ozzelo L, King WJ, and Greene GL (1985) The use of monoclonal antibodies to estrogen receptors (ER) for immunoperoxidase detection of ER in paraffin sections of human breast cancer tissue. *Journal of Histochemistry and Cytochemistry* **33** 87–92

Press MF and Greene GL (1987) Recent developments in the use of anti-receptor antibodies to study steroid hormone receptors, In: Clark CR (ed). *Steroid hormone receptors: their intracellular localization*, pp 251–275, Ellis Horwood Ltd Publishing, Hertfordshire

Press MF and Greene GL (1988) Localization of progesterone receptor with monoclonal antibodies to the human progestin receptor. *Endocrinology* **122** 1165–1175

Press MF, Holt JA, Herbs AL and Greene GL (1985) Immunocytochemical identification of estrogen receptor in ovarian carcinomas: localization with monoclonal estrophilin antibodies compared with biochemical asays. *Laboratory Investigation* **53** 349–361

Press MF, Udove JA and Greene GL (1988) Progesterone receptor distribution in the human endometrium. *American Journal of Pathology* **131** 112–124

Quarmby VE, Beckman Jr WC, Cooke DB *et al* (1990) Expression and localization of estrogen receptor in the R33-27 dunning rat prostatic carcinoma. *Cancer Research* **50** 735–739

Resnik R (1986) Regulation of uterine blood flow, In: Huszar G (ed). *The Physiology and Biochemistry of the Uterus in Pregnancy and Labor*, pp 25–40, CRC Press, Boca Raton

Richardson WD, Mills AD, Dilworth SM, Laskey RA and Dingwall C (1988) Nuclear protein migration involves two steps: rapid binding at the nuclear envelope followed by slower translocation through nuclear pores. *Cell* **52** 655–664

Ricketts D and Coombes RC (1989) What's new in steroid receptor immunocytochemistry in clinical oncology? *Pathology Research and Practice* **185** 935–941

Robbins J, Dilworth SM, Laskey RA and Dingwall C (1991) Two interdependent basic domains in nucleoplasmin nuclear targeting sequence: identification of a class of bipartite nuclear targeting sequence. *Cell* **64** 615–623

Roberts B (1989) Nuclear location signal-mediated protein transport. *Biochimica et Biophysica Acta* **1008** 263–280

Roth J (1982) The protein A-gold (pAg) technique: a qualitative and quantitative approach for antigen localization on thin sections, In: Bullock GR and Petrusz P (eds). *Techniques in Immunocytochemistry*, vol 1, pp 107–133, Academic Press Inc, New York

Sanchez ER, Hirst M, Scherrer LC *et al* (1990) Hormone-free mouse glucocorticoid receptors overexpressed in chinese hamster ovary cells are localized to the nucleus and are associated with both hsp70 and hsp90. *Journal of Biological Chemistry* **265** 20123–20130

Sar M, Lubahn DB, French FS and Wilson EM (1990) Immunohistochemical localization of the androgen receptor in rat and human tissues. *Endocrinology* **127** 3180–3186

Savouret JF, Bailly A, Misrahi M *et al* (1991) Characterization of the hormone responsive element involved in the regulation of the progesterone receptor gene. *EMBO Journal* **10** 1875–1883

Segreti EM, Novotny DB, Spoer JT, Mutch DG, Creasman WT and McCarty KS (1989) Endometrial cancer: histologic correlates of immunohistochemical localization of progesterone receptor and estrogen receptor. *Obstetrics and Gynecology* **73** 780–784

Shimada A, Kimura S, Abe K *et al* (1985) Immunocytochemical staining of estrogen receptor in paraffin sections of human breast cancer by use of monoclonal antibody: comparison with that in frozen sections. *Proceedings of the National Academy of Sciences of the USA* **82** 4803–4807

Shintaku IP and Said JW (1987) Detection of estrogen receptors with monoclonal antibodies in routinely processed formalin-fixed paraffin sections of breast carcinoma. *American Journal of Clinical Pathology* **87** 161–167

Silver P and Goodson H (1989) Nuclear protein transport. *Critical Review of Biochemical Molecular Biology* **24** 419–435

Silverberg E and Lubera J (1987) Cancer statistics. *CA* **37** 2–19

Slotman BJ and Rao BR (1988) Ovarian cancer (review): etiology, diagnosis, prognosis, surgery, radiotherapy, chemotherapy and endocrine therapy. *Anticancer Research* **8** 417–434

Slotman BJ, Nauta JJP and Rao BR (1990) Survival of patients with ovarian cancer: apart from stage and grade, tumour progesterone receptor contents is a prognostic indicator. *Cancer* **66** 740–744

Sternberger LA, Hardy PH, Cuculis JJ and Meyer HG (1970) The unlabeled antibody enzyme method of immunocytochemistry: preparation and properties of soluble antigen-antibody complex (horseradish peroxidase-antihorseradish peroxidase) and its use in identification of spirochetes. *Journal of Histochemistry and Cytochemistry* **18** 315–333

Sullivan WP, Beito TG, Proper J, Krco CJ and Toft DO (1986) Preparation of monoclonal antibodies to the avian progesterone receptor. *Endocrinology* **119** 1549–1557

Takeda H, Nakamoto T, Kokontis J, Chodak GW and Chang C (1991) Autoregulation of androgen receptor expression in rodent prostate: immunohistochemical and in situ hybridization analysis. *Biochemical and Biophysical Research Communications* **177** 488–496

Tan J, Joseph DR, Quarmby VE *et al* (1988) The rat androgen receptor: primary structure, autoregulation of its messenger ribonucleic acid, and immunocytochemical localization of the receptor protein. *Molecular Endocrinology* **12** 1276–1285

Trachtenberg J and Walsh PC (1982) Correlation of prostatic nuclear androgen receptor content with duration of response and survival following hormonal therapy in advanced prostatic cancer. *Journal of Urology* **127** 466–471

Tsai SY, Carlstedt-Duke J, Weigel NL *et al* (1988) Molecular interactions of steroid hormone receptor with its enhancer element: evidence for receptor dimer formation. *Cell* **55** 361–369

Urda LA, Yen PM, Simons Jr SS and Harmon JM (1989) Region-specific antiglucocorticoid receptor antibodies selectively recognize the activated form of the ligand-occupied receptor and inhibit the binding of activated complexes to deoxyribonucleic acid. *Molecular Endocrinology* **3** 251–256

Vasquez-Nin GH, Echeverria OM, Fakan S, Traish AM, Wotiz HH and Martin TE (1991) Immunoelectron microscopic localization of estrogen receptor on premRNA containing constituents of rat uterine cell nuclei. *Experimental Cell Research* **192** 396–404

Verdi JM and Campagnoni AT (1990) Translational regulation by steroids: identification of steroid modulatory element in the 5′-untranslated region of the myelin basic protein messenger RNA. *Journal of Biological Chemistry* **265** 20314–20320

Vu Hai MT, Jolivet A, Ravet V *et al* (1989) Novel monoclonal antibodies against human uterine progesterone receptor: mapping of receptor immunogenic domains. *Biochemical Journal* **260** 371–376

Walker JJ, Bouzubar N, Robertson J *et al* (1988) Immunocytochemical localization of estrogen receptor in human breast tissue. *Cancer Research* **48** 6517–6522

Walsh PJ, Teasdale J and Cowen PN (1990) Ultrastructural localisation of oestrogen receptor in breast cancer cell nuclei. *Histochemistry* **95** 205–207

Warembourg M, Logeat F and Milgrom E (1986) Immunocytochemical localization of progesterone receptor in the guinea pig central nervous system. *Brain Research* **384** 121–131

Warembourg M, Jolivet A and Milgrom E (1989) Immunohistochemical evidence of the presence of estrogen and progesterone receptors in the same neurons of the guinea pig hypothalamus and preoptic area. *Brain Research* **480** 1–15

Welshons WV, Lieberman ME and Gorski J (1984) Nuclear localization of unoccupied oestrogen receptors. *Nature* **307** 747–749

Wikström AC, Bakke O, Okret S, Brönnegard M and Gustafsson JA (1987) Intracellular localization of the glucocorticoid receptor: evidence for cytoplasmic and nuclear localization. *Endocrinology* **120** 1232–1242

Wu CW, Chi CW, Chang TJ, Lui WY and P'eng FK (1990) Sex hormone receptors in gastric cancer. *Cancer* **65** 1396–1400

Ylikomi T, Gasc JM, Isola J, Baulieu EE and Tuohimaa P (1985) Progesterone receptor in the chick bursa of Fabricius: characterization and immunohistochemical localization. *Endocrinology* **117** 155–160

Zaino RJ, Clarke CL, Mortel R and Satyaswaroop PG (1988) Heterogeneity of progesterone receptor distribution in human endometrial adenocarcinoma. *Cancer Research* **48** 1889–1895

Zaino RJ, Clarke CL, Feil PD and Satyaswaroop PG (1989) Differential distribution of estrogen and progesterone receptors in rabbit uterus detected by dual immunofluorescence. *Endocrinology* **125** 2728–2734

Zegers ND, Claassen E, Neelen C *et al* (1991) Epitope prediction and confirmation for the human androgen receptor: generation of monoclonal antibodies for multi-assay performance following the synthetic peptide strategy. *Biochimica et Biophysica Acta* **1073** 23–32

The authors are responsible for the accuracy of the references.

# The Importance of Normal and Abnormal Oestrogen Receptor in Breast Cancer

**WILLIAM L McGUIRE • GARY C CHAMNESS • SUZANNE A W FUQUA**
*University of Texas Health Science Center, Division of Medical Oncology, 7703 Floyd Curl Drive, San Antonio Texas 78284-7884*



## INTRODUCTION

Oestrogen receptor (ER) is an excellent marker of differentiation. It predicts improved disease free survival in breast cancer and, most important, predicts the likelihood of benefit from tamoxifen therapy.

But there are still many key issues regarding ER to be considered. Firstly, why are some breast tumours ER negative? And secondly, why do some ER positive tumours behave as if they are ER negative (eg fail anti-oestrogen therapy) and some ER negative tumours behave as if they are ER positive, for example synthesize progesterone receptor (PgR)?

With respect to the loss of normal functional ER, there are a number of possibilities that need to be examined. We could have a genomic deletion of the gene itself. We could have mutations or rearrangements of the gene. We could have downregulation of transcription at the promoter level. We could have methylation within the coding domain or the promoter region. And finally, we could have an altered message such as that which occurs with alternative splicing. We must also consider aberrant function, ie outlaw receptors. We should consider particularly the possibility of a dominant positive ER, ie a variant receptor that is active even in the absence of oestrogen. We should also consider the possibility of a dominant negative ER, ie a variant receptor that is not only inactive itself but also prevents the function of normal ER. We will first briefly review published studies from other laboratories on ER DNA, RNA and abnormal function and then turn to our own ER variant studies.

*Cancer Surveys* Volume 14: *Growth Regulation by Nuclear Hormone Receptors*
 0-87969-371-1/92. $3.00 + .00

## PUBLISHED STUDIES ON ER VARIATION

### DNA Studies

In Southern hybridization analysis of DNA from 34 breast cancer patients, Koh found no evidence of ER amplification or rearrangement (Koh *et al,* 1989), whereas Nembrot reported evidence for a 1.6- to 3-fold amplification in 6 of 14 cases (Nembrot *et al,* 1990). Falette found different methylation patterns of the ER gene by Southern analysis in normal breast and adjacent tumour tissue and in ER positive and ER negative tumours, but there was no difference in receptor expression as a function of methylation (Falette *et al,* 1990). Castagnoli found a *Pvu*II restriction fragment length polymorphism (RFLP) in the *ER* gene of 14 of 20 males (Castagnoli *et al,* 1987). Hill studied this same RFLP and found that it correlated with *ER* expression in 188 breast cancer patients (Hill *et al,* 1989). However, Parl found the *Pvu*II RFLP to be correlated with age but not *ER* expression in a smaller number of breast cancer patients (Parl *et al,* 1989). In a follow-up study, Parl located the *Pvu*II RFLP within intron 1; this time no correlation with either age or *ER* expression was seen in 260 breast cancer patients (Yaich *et al,* 1991). And finally, Wanless described a *Hin*dIII RFLP in the *ER* gene in a small percentage of breast cancer patients, a finding that correlated with progesterone receptor expression (Wanless *et al,* 1991).

### Messenger RNA Studies

Barrett-Lee found a good correlation between ER messenger RNA (mRNA), protein and ligand binding (Barrett-Lee *et al,* 1987). Rio, by northern analysis, found no gross structural alterations in *ER* message (Rio *et al,* 1987). Piva found that *ER* mRNA correlated with ER protein (Piva *et al,* 1988), Henry found that *ER* mRNA assays were more sensitive than ligand binding (Henry *et al,* 1988) and May found that a high ratio of ER protein to mRNA correlated with the risk of relapse (May *et al,* 1989). The first RNA variant described was by Garcia *et al,* who used an RNase protection assay and found in 8 of 66 ER positive tumours a nucleotide mismatch in the B coding region which correlated with low ligand binding (Garcia *et al,* 1988). She subsequently found that the mismatch corresponded to a C to T transition at nucleotide 257, resulting in an alanine to valine substitution which removes a *Bbv*I restriction site (Garcia *et al,* 1989). In a rather surprising turn of events, Lehrer found that 50% of breast cancer patients with the B variant had spontaneous abortions compared with only 10% of patients with wild type *ER* (Lehrer *et al,* 1990). Lehrer's group also reported that spontaneous abortions occur only in the B variant ER positive breast cancer patients and not in the ER negative breast cancer patients or patients without breast cancer (Lehrer *et al,* 1991). The variant was also found (in heterozygous form) in about 12% of genomic DNAs, apparently unrelated to ER status of breast tumours or to the presence of breast cancer at all (Schmutzler *et al,* 1991). No explanation for these findings is yet available.

In northern hybridization analyses of breast tumour RNA, Murphy found a number of smaller size *ER* mRNA variants resulting from deletions of the hormone binding domain (Murphy and Dotzlaw, 1989). She prepared a complementary DNA (cDNA) library from one of these breast cancer biopsy specimens and found 84 unique aminoacids introduced at the exon 3 intron boundary (aminoacid 253) that were L-1 repetitive sequences (Dotzlaw and Murphy, 1990). These sequences were followed by a stop codon, resulting in a truncated 37 kDa protein. Dotzlaw has also reported an ER variant with an insertion of six unique aminoacids at the exon 2 intron boundary (aminoacid 214), finally followed by a stop codon for a total of 220 aminoacids (Dotzlaw *et al,* 1991). Benz, using ER gel retardation assays, found that some ER positive tumours either did not bind or bound only weakly to a synthetic oestrogen response element. This decrease in binding was associated with a 50 kDa variant dimer or a 50/67 kDa heterodimer of wild type plus variant receptor (Scott *et al,* 1991).

### Studies of Abnormal Function

Zava in 1977 was one of the first to speculate on the possible existence of biologically active ER without oestrogen (Zava *et al,* 1977). Horwitz suggested that permanently activated ER might explain the persistently high levels of progesterone receptor in T47D tissue culture cells (Horwitz, 1981; Horwitz *et al,* 1982; Graham *et al,* 1990). Sluyser brought a different focus to the problem and hypothesized that mutated or truncated ER without oestrogen might be able to act as an oncogene and stimulate breast cancer growth (Sluyser, 1985). We ourselves have studied ER negative, PgR positive breast cancer patients and have found an exon 5 deletion variant that results in a truncation at aminoacid 370 and constitutively stimulates oestrogen responsive genes (Fuqua *et al,* 1991). In the past few years, much activity has thus been devoted to discovering abnormal ER in clinical breast cancer and to trying to determine whether these receptors are associated with altered function. We should now like to further detail our studies in San Antonio.

## SCREENING FOR ER VARIANTS

Since it is not feasible to sequence the whole ER in both normal and tumour tissue from every breast cancer patient, we decided to develop more selective screening strategies to search for abnormal ER in clinical experiments. We chose to produce polymerase chain reaction (PCR) amplified cDNA from known functional domains of the ER message (Fuqua *et al,* 1990) and to use direct sequencing, chemical mismatch cleavage (CMC) (Cotton *et al,* 1988) and single stranded conformational polymorphism analyses (SSCP) (Orita *et al,* 1989) to detect *ER* RNA variants. Finally, we also used gel retardation assays to detect ER variants with abnormal binding to the oestrogen response elements of oestrogen responsive genes.

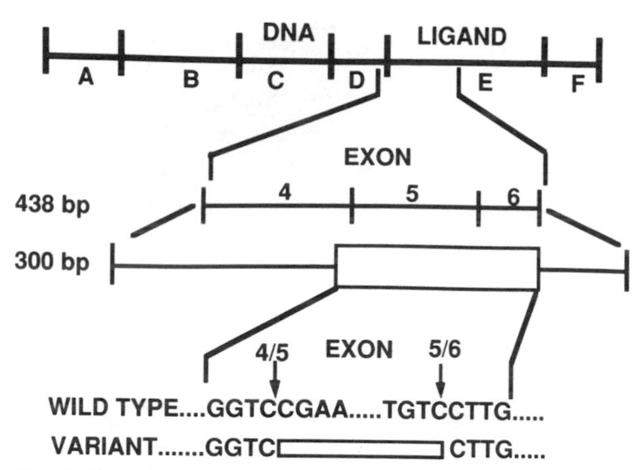


**Fig. 1.** Functional domains of the ER and sequence results of the exon 4/6 border of an exon 5 ER deletion variant derived from PCR amplification of a portion of the ligand binding domain of the ER mRNA in an ER–/PgR+ breast tumour

We first see an application of the direct sequencing procedure. Figure 1 shows the ER structure as detailed by in vitro studies (Kumar *et al,* 1987). There are six domains, designated A to F. The C domain contains the DNA binding portion, and the E domain is associated with ligand binding. We examined ER negative/PgR positive tumours, reasoning that some such tumours might have a variant ER lacking ligand binding but possessing transcriptional activating capability. We used oligonucleotide primers to PCR amplify exons 4, 5 and 6 from the ligand binding region of the *ER* mRNA (Fuqua *et al,* 1990). This normally results in a cDNA of length 438 base pairs. We indeed found this fragment in many of these ER–/PgR+ tumours, although most ER–/PgR– tumours had no *ER* mRNA fragments at all (Fuqua *et al,* 1990). But in some ER–/PgR+ tumours, we also found a 300 base pair PCR fragment, which upon sequencing revealed a precise deletion of exon 5 (Fig. 1). In order to be sure that this finding was not the result of PCR artefacts, we performed an RNase protection assay using total RNA isolated from the tumours, which demonstrated that the exon 5 deletion indeed occurs in clinical breast cancer. It is found in ER+ tumours as well, but the ratio of the variant to wild type ER is two- to threefold higher in some of the apparently ER–/PgR+ tumours. About half of the ER–/PgR+ tumours overexpressed this variant. We will examine its functional capabilities below.

We have also used gel retardation assays (Kumar and Chambon, 1988) to screen for binding of possible variant receptors to a synthetic oestrogen response element (ERE). Figure 2 is an example of a gel retardation assay, showing an extract from a receptor positive tumour retarding the migration of the ERE. We show that the complex does indeed contain ER by further upshifting the complex with antibodies which bind the ER. The results for antibodies D75, H222, and the combination of D75 and H222 are shown. For negative control purposes, we use B39, which is an antibody to progesterone

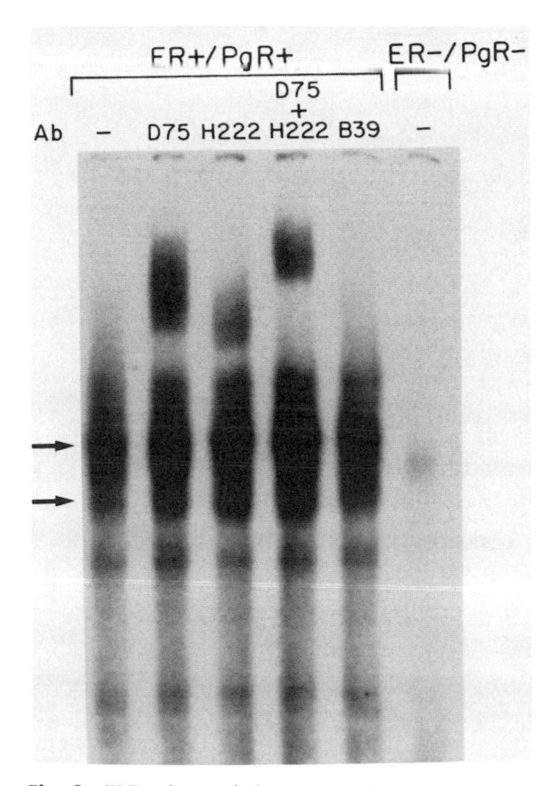


**Fig. 2.** ERE gel retardation assay using protein extracts from a ER+/PgR+ and a ER–/PgR– breast tumour. Specific binding of ER to its ERE is denoted with arrows. ER specific antibodies D75 and H222 were used to confirm the specificity of the complex, and the PgR specific B39 antibody was used as a negative control

receptor. A receptor negative tumour showed no retention of ERE in the gel. Using this assay, we have identified an ER with a 3′ truncation, i.e. it binds ERE and is upshifted by ER antibodies except those with target epitopes in the carboxyterminal region (Fuqua SAW, unpublished). This receptor was cloned and sequenced, and we found that exon 7 was precisely deleted in this tumour.

Figure 3 summarizes some of the variants we have found with these techniques. We have found deletions of exons 3, 5 and 7, an asparagine + arginine insertion in the DNA binding domain and several single aminoacid substitutions in domains D and E. There are also many silent base pair substitutions throughout the gene.

## FUNCTION OF VARIANT RECEPTORS

For ascertaining possible altered function of some of these variant receptors, we established a yeast expression vector function assay (Fuqua *et al,* 1991). Either wild type or variant ER is inserted into a plasmid under the control of a metallothionein promoter. Another plasmid containing the ERE controlling a

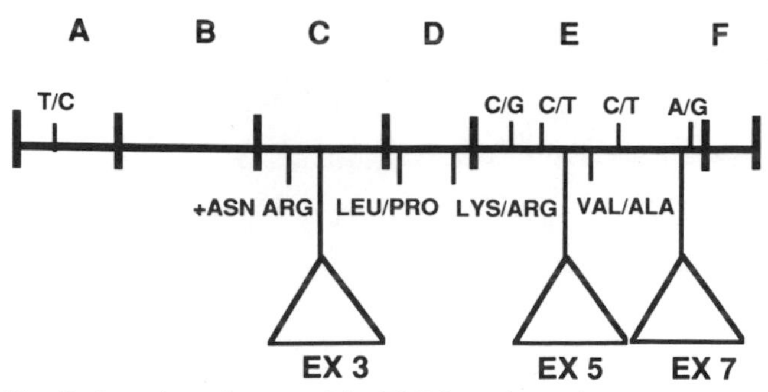


**Fig. 3.** Location of some of the ER RNA variants detected within functional domains of the ER

β-galactosidase reporter gene is also transformed into the same yeast cell. This system thus tests the transcriptional activating ability of wild type and variant ER. In a typical experiment where wild type ER is inserted under oestrogen-free conditions (Table 1), there is no β-galactosidase synthesis. Adding copper to increase the absolute amount of receptor made does not result in any β-galactosidase synthesis, but the addition of oestrogen dramatically increases synthesis. Therefore, this system is exquisitely oestrogen dependent. We also see an example of an exon 3 deletion variant, which is completely inactive in this assay, as might have been expected since it has no DNA binding domain.

We then used this system to examine possible outlaw receptors, those with abnormal function either dominant positive (transcriptionally active in the absence of oestrogen) or dominant negative (transcriptionally inactive but preventing function of normal ER). Examples of both of these are listed in Table 1.

The exon 5 deletion that we first described was cloned from an ER negative, PgR positive tumour, so that a dominant positive variant was suspected. Indeed, in the yeast expression system, there was appreciable β-galactosidase synthesis in the absence of oestrogen. This was increased when the level of the variant receptor protein was increased by copper but was not affected by oestradiol addition. Such a receptor would have the potential of stimulating breast tumour growth in the absence of oestrogen.

Table 1 also shows an example of a dominant negative receptor, the 3′

**TABLE 1. Yeast *trans* activation assay for ER variant function[a]**

| Treatment | β-Galactosidase activity | | | |
|---|---|---|---|---|
| | wild type ER | exon 3 del | exon 5 del | exon 7 del |
| Control | 0 | 0 | 400 | 0 |
| +Copper | 0 | 0 | 800 | 0 |
| +Oestrogen | 2700 | 0 | 800 | 0 |

[a]The ER gene, with type or variant, is present under a copper inducible metallothionein promoter, whereas ER activity is detected by induction of β-galactosidase under control of an oestrogen response element

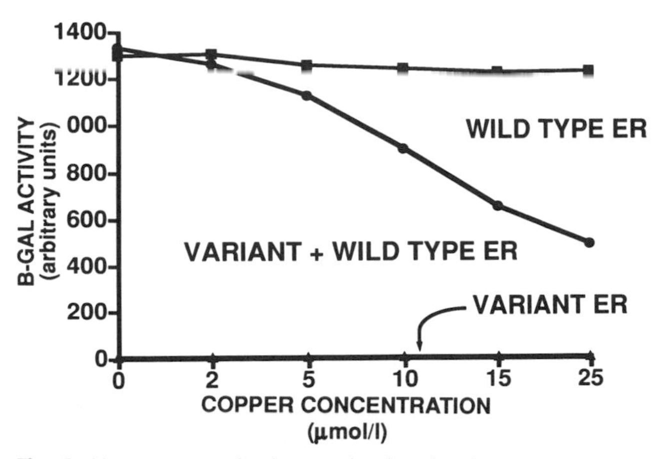


**Fig. 4.** Yeast *trans*-activation results showing the amount in arbitrary units of β-galactosidase (β-GAL) activity as a function of copper concentration in the presence of oestrogen. The exon 7 deletion ER variant and wild type ER were introduced into yeast cells either alone or simultaneously and β-galactosidase levels were determined

truncated exon 7 deletion discovered in our gel retardation studies. In this assay, the exon 7 deletion stimulates no β-galactosidase synthesis under control, copper or oestrogen stimulated conditions. Therefore, it is transcriptionally inactive. But we also questioned whether it could prevent the function of normal ER, ie be dominant negative. Figure 4 illustrates a yeast system assay where wild type and variant are both present in the same cell, the wild type under a constitutive promoter and the variant under an inducible metallothionein promoter. The variant can thus be progressively turned on by copper. The top curve represents wild type ER by itself, and the bottom curve is the exon 7 ER deletion variant by itself. The middle curve represents variant and wild type ER together. It can be seen that as the copper concentration is increased and the variant ER is progressively induced, the variant interferes with the wild type receptor's ability to induce β-galactosidase at the ERE, even though oestradiol is present. It is noteworthy that at the highest copper concentration, western blot analysis revealed no more than equivalent levels of wild type ER and variant ER. Thus, this is a very potent dominant negative receptor variant.

## SUMMARY

We have used the screening techniques of CMC and SSCP of selected PCR fragments, and also gel retardation assays, to discover a number of ER variants in clinical breast cancer tissues. We have found base pair insertions, transitions, and deletions, and deletions of exons 3, 5 and 7. Using a yeast transactivation assay, we have discovered receptors with outlaw function consisting of both dominant positive receptors that were transcriptionally active in the absence of oestrogen and dominant negative receptors that were transcriptionally inactive themselves but prevented normal ER function.

Future efforts should focus in particular on such dominant positive and dominant negative variants. With regard to positive variants, we should like to know whether they stimulate tumour growth and, if so, whether they can be turned off. With regard to dominant negative variants, we should like to determine whether they can inhibit tumour growth and, if so, whether they can be turned on.

## Acknowledgements

We should like to emphasize that our studies summarized here are the result of a collaborative effort. In San Antonio, Shelly Krieg, Richard Elledge and Chye-Ning Weng worked on the gel retardation studies, Saundra Fitzgerald contributed to the cloning of the exon deletion variants and Margaret Benedix performed the SSCP and CMC studies. In Houston, we received invaluable help with the yeast *trans*-activation assays from Donald McDonnell, Zafar Nawaz and Bert O'Malley. Geoffrey Green in Chicago was extremely helpful with the antibody experiments. This work was supported by National Institutes of Health grants CA30195, CA52351 and HD10202. William L McGuire is a clinical research professor of the American Cancer Society.

## References

Barrett-Lee PJ, Travers MT, McClelland RA, Luqmani Y and Coombes RC (1987) Characterization of estrogen receptor messenger RNA in human breast cancer. *Cancer Research* **47** 6653–6659

Castagnoli A, Maestri I, Bernardi F and Del Senno L (1987) PvuII RFLP inside the human estrogen receptor gene. *Nucleic Acids Research* **15** 866

Cotton RGH, Rodrigues NR and Campbell RD (1988) Reactivity of cytosine and thymine in single-base-pair mismatches with hydrozylamine and osmium tetroxide and its application to the study of mutations. *Proceedings of the National Academy of Sciences of the USA* **85** 4397–4401

Dotzlaw H and Murphy C (1990) Cloning and sequencing of a variant sized estrogen receptor (ER) mRNA detected in some human breast cancer biopsies: 13th annual San Antonio breast cancer symposium, San Antonio. *Breast Cancer Research and Treatment* **16** 147 [Abstract]

Dotzlaw H, Alkhalaf M and Murphy LC (1991) Multiple estrogen receptor (ER) like mRNAs in human breast cancer biopsies. In: *The Endocrine Society 1991*, p 173 [Abstract], Endocrine Society, Washington DC

Falette NS, Fuqua SAW, Chamness GC, Cheah MS, Greene GL and McGuire WL (1990) Estrogen receptor gene methylation in human breast tumors. *Cancer Research* **50** 3974–3978

Fuqua SAW, Falette NF and McGuire WL (1990) Sensitive detection of estrogen receptor RNA by polymerase chain reaction assay. *Journal of the National Cancer Institute* **82** 858–861

Fuqua SAW, Fitzgerald SD and Chamness GC (1991) Variant human breast tumor estrogen receptor with constitutive transcriptional activity. *Cancer Research* **51** 105–109

Garcia T, Lehrer S, Bloomer WD and Schachter B (1988) A variant estrogen receptor messenger ribonucleic acid is associated with reduced levels of estrogen binding in human mammary tumors. *Molecular Endocrinology* **2** 785–791

Garcia T, Sanchez M and Cox JL (1989) Identification of a variant form of the human estrogen receptor with an amino acid replacement. *Nucleic Acids Research* **17** 8361

Graham II ML, Krett NL and Miller LA (1990) T47D CO cells, genetically unstable and containing estrogen receptor mutations, are a model for the progression of breast cancers to hormone resistance. *Cancer Research* **50** 6208–6217

Henry JA, Nicholson S, Farndon JR, Westley BR and May FEB (1988) Measurement of oestrogen receptor mRNA levels in human breast tumours. *British Journal of Cancer* **58** 600–605

Hill SM, Fuqua SAW, Chamness GC, Greene GL and McGuire WL (1989) Estrogen receptor expression in human breast cancer associated with an estrogen receptor gene restriction fragment length polymorphism. *Cancer Research* **49** 145–148

Horwitz KB (1981) Is a functional estrogen receptor always required for progesterone receptor induction in breast cancer? *Journal of Steroid Biochemistry* **15** 209–217

Horwitz KB, Mockus MB and Lessey BA (1982) Variant T47D human breast cancer cells with high progesterone-receptor levels despite estrogen and antiestrogen resistance. *Cell* **28** 633–642

Koh EH, Ro J, Wildrick DM, Hortobagyi GN and Blick M (1989) Analysis of the estrogen receptor gene structure in human breast cancer. *Anticancer Research* **9** 1841–1846

Kumar V and Chambon P (1988) The estrogen receptor binds tightly to its responsive element as a ligand-induced homodimer. *Cell* **55** 145–156

Kumar V, Green S, Stack G, Berry M, Jin J-R and Chambon P (1987) Functional domains of the human estrogen receptor. *Cell* **51** 941–951

Lehrer S, Sanchez M and Song HK (1990) Oestrogen receptor B-region polymorphism and spontaneous abortion in women with breast cancer. *Lancet* **335** 622–624

Lehrer S, Schmutzler R and Rabin J (1991) A variant human estrogen receptor (ER) gene is associated with a history of spontaneous abortion in women with ER positive breast cancer, but not in women with ER negative breast cancer or women without breast cancer. In: *The Endocrine Society 1991*, p 38 [Abstract], Endocrine Society, Washington DC

May E, Mouriesse H, May-Levin F, Contesso G and Delarue J-C (1989) A new approach allowing an early prognosis in breast cancer: the ratio of estrogen receptor (ER) ligand binding activity to the ER-specific mRNA level. *Oncogene* **4** 1037–1042

Murphy LC and Dotzlaw H (1989) Variant estrogen receptor mRNA species detected in human breast cancer biopsy samples. *Molecular Endocrinology* **3** 687–693

Nembrot M, Quintana B and Mordoh J (1990) Estrogen receptor gene amplification is found in some estrogen receptor-positive human breast tumors. *Biochemical and Biophysical Research Communications* **166** 601–607

Orita M, Iwahana H, Kanazawa H, Hayashi K and Sekiya T (1989) Detection of polymorphisms of human DNA by gel electrophoresis as single-strand conformation polymorphisms. *Proceedings of the National Academy of Sciences of the USA* **86** 2766–2770

Parl FF, Cavener DR and Dupont WD (1989) Genomic DNA analysis of the estrogen receptor gene in breast cancer. *Breast Cancer Research and Treatment* **14** 57–64

Piva R, Bianchini E, Kumar VL, Chambon P and Del Senno L (1988) Estrogen induced increase of estrogen receptor RNA in human breast cancer cells. *Biochemical and Biophysical Research Communications* **155** 943–949

Rio MC, Bellocq JP and Gairard B (1987) Specific expression of the pS2 gene in subclasses of breast cancers in comparison with expression of the estrogen and progesterone receptors and the oncogene ERBB2. *Proceedings of the National Academy of Sciences of the USA* **84** 9243–9247

Schmutzler RK, Sanchez M and Lehrer S (1991) Incidence of an estrogen receptor polymorphism in breast cancer patients. *Breast Cancer Research and Treatment* **19** 113–120

Scott GK, Kushner P, Vigne J-L and Benz CC (1991) Truncated forms of DNA-binding estrogen receptors in human breast cancer. *Clinical Research* **38** 311

Sluyser M (1985) Oncogenes homologous to steroid receptors? *Nature* **315** 546

Wanless C, Barker S, Puddefoot JR *et al* (1991) Somatic change in the estrogen receptor gene associated with altered expression of the progesterone receptor. *Anticancer Research* **11** 139–142

Yaich LE, Dupont WD, Cavener DR and Parl FF (1991) The estrogen receptor PVU II restriction fragment length polymorphism is not correlated with estrogen receptor content or patient age in 260 breast cancers. In: *The Endocrine Society 1991*, p 175 [Abstract], Washington DC

Zava DT, Chamness GC, Horwitz KB and McGuire WL (1977) Human breast cancer: biologically active estrogen receptor in the absence of estrogen? *Science* **196** 663–664

The authors are responsible for the accuracy of the references.

# Cellular Heterogeneity and Mutant Oestrogen Receptors in Hormone Resistant Breast Cancer

**K B HORWITZ**
*University of Colorado Health Sciences Center, 4200 East 9th Avenue, Denver, Colorado 80262*



## INTRODUCTION

Faithful expression of genetic information is lost in tumour cells owing to the formation of spontaneous cell variants. In breast cancer, this evolution is marked by progression of tumours from hormone dependent, through hormone responsive, to hormone resistant states. Many resistant tumours no longer express oestrogen receptors (ER), and this may be the basis for their oestrogen resistance. However, 50% of all hormone resistant breast cancers are ER positive, yet they too fail to respond to anti-oestrogen therapy. Both the cellular heterogeneity that marks progression of the disease and the hormone resistance that characterizes the end stages of the disease have been longstanding clinical problems that are slowly yielding to basic research focused both on solid tumours taken directly from patients and on breast cancer cell lines derived from such tumours. Studies addressing issues of molecular and cellular heterogeneity and their relation to progression and resistance of breast cancer are reviewed below.

## OESTROGEN RECEPTORS

The molecular biology of oestrogen receptors has been extensively explored in recent years. The complementary DNA (cDNA) was independently cloned

*Cancer Surveys* Volume 14: *Growth Regulation by Nuclear Hormone Receptors*
 0-87969-371-1/92. $3.00 + .00

and sequenced from MCF-7 breast cancer cells by two groups in 1986 (Green *et al,* 1986; Greene *et al,* 1986). The gene for the ER from MCF-7 cells was cloned and analysed 2 years later (Ponglikitmongkol *et al,* 1988). The protein is composed of 595 aminoacids within which Kumar *et al* (1986) distinguished six functional domains identified by the letters A to F. The A/B domains contain regions that regulate the transcriptional function of the proteins. The C domain contains two DNA binding zinc fingers and is the region of the protein that binds to the oestrogen response element (ERE). Mutations in this portion of the protein change its affinity for DNA, resulting in suboptimal, or complete loss of, binding. The hormone binding properties of the receptors reside in region E, as shown by mutagenesis analysis. Since these two functions, DNA binding and hormone binding, are carried out by separate parts of the protein, they are to some extent independent. Thus, it is possible to have variant receptors that can bind to DNA with limited affinity without first binding hormone, and vice versa (Kumar *et al,* 1986; Kumar and Chambon, 1988). Three additional important specialized regions of steroid receptors have also been identified: a nuclear localization signal, a heat shock protein (hsp90) binding region and a dimerization domain. The nuclear localization signal, located downstream from the DNA binding domain, is a region of the protein that must be present for the receptor to remain within the nucleus in the absence of ligand (Guiochon-Mantel *et al,* 1988). It has been identified in progesterone receptors (PR) and is presumed to be similar in ER. The hsp90 appears to bind to regions in the hormone binding domain of some steroid receptors when ligand is absent, and its binding is thought to prevent receptor dimerization and DNA binding (Baulieu, 1987). Ligand activation leads to hsp90 dissociation and monomer dimerization in solution (Kumar and Chambon, 1988). The dimerization domain that mediates this interaction between two ER molecules has been localized to the carboxyterminal end of the hormone binding domain (Fawell *et al,* 1990). A weak dimerization domain may also be present in the second zinc finger of the DNA binding domain (Kumar and Chambon, 1988). Additional sites for heterologous protein interactions may also be located in the hormone binding domain (Adler *et al,* 1988), and covalent modifications by phosphorylation (Washburn *et al,* 1991) further enhance the complexity of this protein molecule.

## MOLECULAR HETEROGENEITY

Several reports of naturally occurring mutant or variant ER forms have recently appeared (Murphy, 1990, and references therein). In addition, polymorphic forms of the ER gene have been described (Castagnoli *et al,* 1987; Hill *et al,* 1989; Parl *et al,* 1989). The majority of these genetic changes are found in introns, which do not directly encode the mRNA or, in turn, the protein. Of interest is the recent report by Keaveney *et al* (1991) identifying an alternative oestrogen receptor messenger RNA (mRNA), which appears to be the primary transcript present in the human uterus as opposed to the breast cancer line MCF-7. This newly identified transcript is alternatively spliced in the 5′ untranslated region and has an additional exon with two small open reading frames upstream of the alternative splice site. Although the receptor proteins encoded by these two types of messages are identical, the nucleotide sequences that flank the translated regions are different and are likely to lead to differential regulation of the protein, depending on which type of message predominates in the tissue in question. Equally interesting is a truncated ER message specific to pituitary cells (Shupnik *et al*, 1989). This deletion involves the translated region and presumably encodes a variant receptor, although expression of the protein has not yet been documented. Thus, in normal cells, the regulation of ER gene transcription, and even ER protein structure, may be tissue specific. These issues have received little attention.

## Mutant Oestrogen Receptors in Solid Tumours

Turning to malignant cells, there is now mounting evidence that in addition to silent mutations and regulatory heterogeneity, mutations in ER exons exist that would influence protein structure and protein function. Garcia *et al* (1988, 1989) identified a polymorphic variant in the B region of ER mRNA in some human breast cancer biopsy specimens. This variant has since been correlated with lower than normal levels of hormone binding activity, and preliminary evidence suggests that women who are heterozygous for this variant have a higher proportion of spontaneous abortions than those who are homozygous at the same locus (Lehrer *et al*, 1990).

Wild type ER mRNAs from several normal and malignant tissues and species are reported to be approximately 6.2 kb in size. However, Murphy and Dotzlaw (1989) have identified truncated ER like mRNAs in human breast cancer biopsy samples by northern blotting. These messages appear to lack significant portions of the 3′ region, including the hormone binding domain. By polymerase chain reaction (PCR) amplification of mRNA from breast tumour specimens, Fuqua *et al* (1991) have also identified mutant forms of ER missing part of the hormone binding domain because of deletion of exon 5. These mutants are an alternatively spliced form capable of constitutively activating transcription of a normally oestrogen dependent gene construct in yeast cells. PCR amplification was also used to identify a mutation in the D domain of *ER* mRNA expressed in a murine transformed Leydig cell line, B-1 F (Hirose *et al*, 1991). The functional importance of these mutations has yet to be fully explored, but they clearly suggest mechanisms by which mutant receptor forms can subvert the activity of wild type forms, when both are expressed in the same tumour cell.

The weakness in all these analyses is the assumption that message variants reflect protein variants. Although this may indeed be the case, no mutant proteins had been detected until recently. This may now have been rectified by two studies in which gel shift assays were used to examine the ability of tumour ERs to bind an ERE (Foster *et al,* 1991; Scott *et al,* 1991). The studies show that some tumours containing abundant immunoreactive ER failed to demonstrate DNA binding ER, or the DNA binding ER forms appeared to be truncated or they were immunologically ER negative but positive by the mobility shift assay. On the basis of these preliminary data, the prevalence of non-DNA binding ER forms or of truncated ER forms among ER positive or PR positive tumours may exceed 50%—a significant number whose structural analysis may become a critically important prognostic tool.

## Mutant Oestrogen Receptors in Breast Cancer Cell Lines

Oestrogen receptors have a critical role in the development, progression and hormone responsiveness of breast cancers. Their structural analysis, with methods like those described above, can be used to generate functional predictions. Alternatively, a product of ER action can be monitored, and PR have served this role for many years (Horwitz *et al,* 1975). In all oestrogen/progesterone target tissues, oestradiol is required for PR induction. This relationship holds true for breast cancers (Horwitz and McGuire, 1978) and led us to propose that the presence of PR could be used to predict the hormone dependence of human breast tumours. Thus, a tumour that contains PR would, of necessity, have a functional ER. This idea has in general been borne out by studies which show that ER positive tumours that also have PR are much more likely (75%) to respond to hormone treatment than tumours that are ER positive but PR negative (35%) (Horwitz *et al,* 1985, and references therein). These studies also identify a small group of puzzling tumours that are ER negative but PR positive and have a higher response rate than is usually expected of ER negative tumours. They are puzzling because according to dogma, such tumours should not exist. Thus, either PR synthesis in these tumours is entirely independent of ER action (in defiance of a long accepted principle) or a variant or other unmeasured form of ER is stimulating PR synthesis.

In 1978, while measuring the steroid receptor content of a series of cultured human breast cancer cells, we found one cell line, T47D, that had no soluble ER by sucrose density gradient analysis yet had the highest PR levels of any cell line surveyed (Horwitz *et al,* 1978; Keydar *et al,* 1979). These cells seemed to be ideally suited to study this ER negative but PR positive paradox.

We found that a subline, $T47D_{co}$, did have ER, but they were in a permanently activated state in the nucleus. The ER were not sensitive to the action of oestrogens, suggesting that the oestrogen regulatory mechanism was defective at a step beyond the initial interaction of the steroid-receptor complex with DNA. PR were also insensitive to oestradiol or to anti-oestrogens but were synthesized in extraordinary amounts and were functional. Additional studies suggested that the PR retained characteristics of inducible proteins. Thus, we suggested that persistent nuclear ER were constitutively stimulating PR, even in the absence of exogenous oestradiol (Horwitz, 1981; Horwitz *et al,* 1982). Recently, the tools became available to test this conjecture. Two cDNA

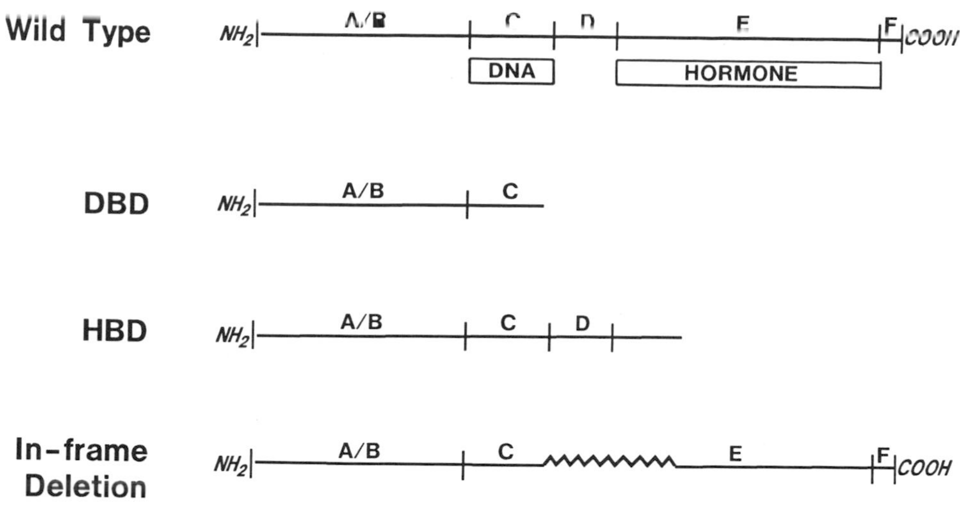


**Fig. 1.** Schematic representation of functional domains in wild type ER and the proteins encoded by mutant ER cDNAs of $T47D_{co}$ cells. Functional regions of wild type ER (595 aa) include the aminoterminal domain A/B, the DNA binding domain C, the hinge region D, the hormone binding domain E and the carboxyterminus F (34). The predicted structures of the mutant proteins are also shown including the DBD mutant truncated at aa 250, the HBD mutant truncated at aa 417 and the 153 aa in frame deletion (jagged line) mutant (Graham *et al*, 1990)

libraries constructed from $T47D_{co}$ cells (Graham *et al*, 1990) have yielded mutant ER cDNA clones encoding the putative ER proteins shown in Fig. 1. Library I, made from total poly(A)+ mRNA, yielded two clones having wild type ER sequences and one clone that encodes part of ER exon 2 and all of exon 3. At the precise junction between exons 3 and 4, the ER sequence homology ends and the adjacent sequence is not found in ER cDNA, in the reported ER intron sequences or in the DNA data base. Compared with wild type ER, the region immediately upstream of the exon 3/intron 3 junction has two inserted T residues. This leads to a disruption of the protein reading frame by generating a TGA termination codon. There are also two point mutations just distal to the insertions. Together, these mutations may account for the abnormal splicing reaction observed. This cDNA would encode an ER truncated at aminoacid (aa) 250 in the DBD, just beyond the last cysteine (aa 245) of the second DNA binding finger. The putative "DBD" protein (Fig. 1) would lack the nuclear localization signal and hormone binding domains of ER and epitopes for most anti-ER antibodies.

Library I also yielded an ER clone that appears to be an RNA processing intermediate or splicing error and contains approximately 1 kb of intron 5 linked upstream of exon 6 and three clones with an insertion in exon 5. The insert in the three clones contains at least two blocks of direct repeats of about 130 nucleotides terminating in A residues that are 70–85% homologous to the human *Alu* family.

To ensure that the first library did not have an overrepresentation of unprocessed nuclear mRNAs, a second library was made from cytoplasmic mRNA. In addition to clones consistent with wild type ER, library II yielded mutant cDNAs that would encode proteins with potentially important biological activity. One clone has a point deletion in the hormone binding domain just upstream of the end of exon 5. This leads to a frameshift and a translation termination seven codons later. This mutant cDNA would encode an ER truncated in the middle of the HBD at aa 417, with a unique 7 aa carboxyterminal end. The putative "HBD" protein (Fig. 1) would be unable to bind hormone or the anti-ER antibody H222.

Library II also yielded two independent clones having an identical in-frame deletion. These clones contain a wild type upstream sequence from the 5′ untranslated region to the first four codons of exon 4. There follows a loss of 460 nucleotides, including most of exons 4 and 5. The last codon of exon 5 is preserved, as are downstream exon 6 sequences. The nucleotide sequences at the borders of the deletion are identical (GAC) and could represent either nucleotides 1004–1006 (aa 258) of exon 4 or nucleotides 1463–1465 (aa 411) of exon 5. This cDNA would encode a mutant ER of 442 aa instead of the normal 595 aa, having a 153 aa deletion from the end of the DNA binding domain C, through the hinge region D to the mid hormone binding domain E (Fig. 1). The deletion originates in the sequence encoding the putative nuclear localization signal (aa 256-263; R-K-D-R-R-G-G-R). However, the aminoacid sequence encoded by the deletion mutant (R-K-D-R-N-Q-G-K) preserves four of the five basic aminoacid residues of the wild type sequence.

We do not know whether the abnormal proteins are expressed. Gel mobility shift analyses of $T47D_{co}$ nuclear extracts show considerable amounts of specific ERE binding proteins, which neither co-migrate with wild type receptors nor are supershifted by antibodies H222 or H226. The identity of these proteins is still under investigation. However, on the basis of deletion mutagenesis analyses (Kumar and Chambon, 1988), we can begin to predict the consequence to the cells of such mutant ER. Especially in $T47D_{co}$ sublines with hypertetraploid subpopulations (see further below), which contain four or five alleles of the *ER* gene (Graham *et al,* 1989b), cells having a mixture of wild type and mutant receptors could co-exist. Heterodimers of the wild type and mutant monomers, having dominant positive or dominant negative activity (Forman *et al,* 1989), could override the oestrogen requirement of the wild type receptors. This would result in ER positive but oestrogen resistant cells, a phenotype that describes 50% of hormone resistant breast cancers.

## CONSEQUENCES OF MUTANT OESTROGEN RECEPTORS: CELLULAR HETEROGENEITY?

The consequences of this newly discovered molecular diversity in ER may reach beyond issues of hormone dependence to the broader problems of tumour progression and cellular heterogeneity that also characterize advanced breast cancer. Cellular heterogeneity has usually been assumed to exist within tumours but has been difficult to demonstrate. The concept is, however, important, since it means that in practice, the clinician must treat not just one tumour but a variety of possibly heterogeneous subtumours. Is it possible that heterogeneity of ER among cells can lead to heterogeneity of cells among tumours? Although the analyses of ER described above have led to the discovery of variant receptor forms, the methods cannot answer a fundamental question: do all or only some of the cells carry the variants? Moreover, wild type ER are always present together with the variants. Are wild type ER present alone in some cells of the tumour or are they always co-expressed with the variants in any one cell? We postulated that the genetic diversity of ER would be reflected in heterogeneity of other molecular markers and set out to develop an assay that could simultaneously measure DNA content and PR heterogeneity in subpopulations of tumour cells (Graham *et al,* 1989a,b). We have used this immunological, dual parameter flow cytometry based assay to demonstrate and quantitate a remarkable heterogeneity in PR content, DNA ploidy and mitotic indices among subpopulations of breast cancer cells (Graham *et al,* 1992).

## Progesterone Receptor Heterogeneity

Progesterone receptor heterogeneity is illustrated in Fig. 2 by three cell lines derived from $T47D_{co,}$ in which only the first, $T47D_v$ (panel A), has a PR phenotype that most current receptor measurement methods assume, namely that cells are positive at a level greater than measured background. However, even $T47D_v$ cells have PR levels that range extensively as shown by the width of the receptor peak on the log scale. Panels B and C illustrate entirely different PR positive patterns: two cell lines (V22 and V26) that have more than one PR positive population despite the fact that they were derived as single cell clones from $T47D_v$.

To quantitate PR in the subpopulations, we have developed a computer program, 1-Par. Table 1 shows calculations made with this software, namely that 12.8% of cells in V22 and 23.2% of cells in V26 are PR negative and that each cell line also contains two distinctly different PR positive subpopulations.

Does a two population model adequately describe cells such as those depicted in panels B and C of Fig. 2? Probably not. Bimodality of a single variable like PR hints at still greater numbers of subpopulations when a second variable is analysed simultaneously. The simultaneous analysis of PR and DNA indices shows that V26 is a mixture of 47.2% hyperdiploid (HD) cells and 52.8% hypertetraploid (HT) cells. The HD cells, with 24.3% of cells in S and $G_2$/M, grow slightly faster than the HT cells, which have 17.0% of cells in the proliferating fraction. Combining the PR and DNA data shows that there are two distinct HD subpopulations: one contains cells with low PR levels and the other cells with high PR levels. In addition, the HT cells also contain subpopulations with low and high PR levels. Thus, there are at least four subpopulations in this cell line, each having a different combination of PR, DNA content and mitotic indices. V22 cells are similarly heterogeneous.

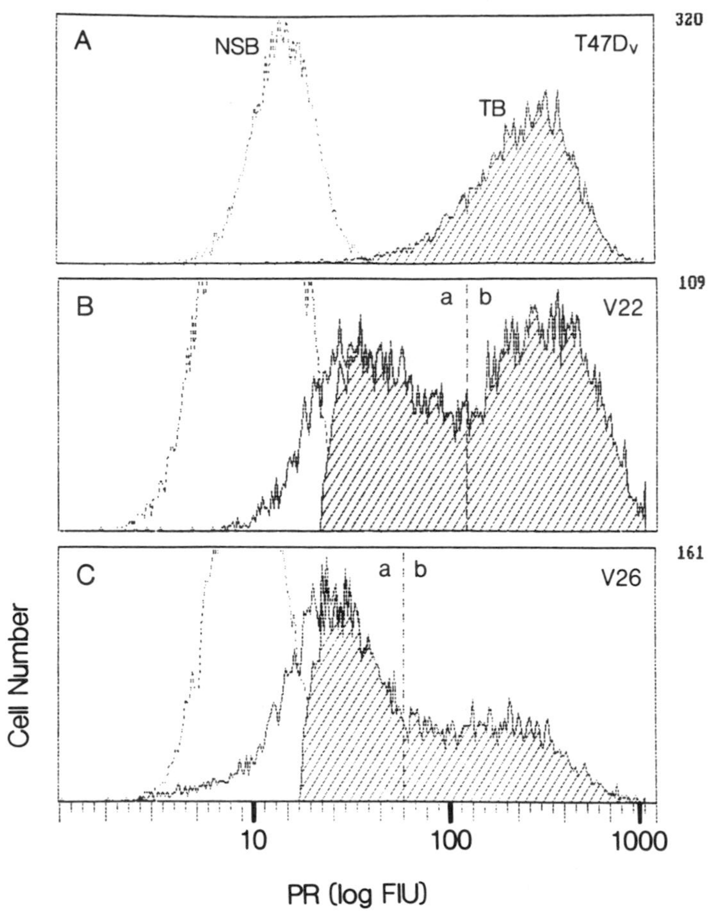


**Fig. 2.** Heterogeneity of PR levels in three T47D cell lines. Total (TB) and non-specific (NSB) PR were measured by dual parameter flow cytometry in $T47D_V$ cells and two clonal sublines, V22 and V26. Fluorescence intensity unit levels in 10 000 cells of each set were analysed and plotted. Specific binding (SB) of PR positive cells, indicated by the shaded areas, was calculated by curve subtraction. Cells with TB levels falling under the NSB curve are considered to be PR negative. Using the 1-Par program, subpopulations with low (a) and high (b) PR levels were gated and analysed as described in Table 1 and the text. The vertical scales of the panels are different, as shown by the number on the right. PR levels are plotted on a log scale (Graham *et al*, 1992). In panels B and C, the peak of the NSB curve has been cut off in order to expand the SB curves

## Tumour Cell "Remodelling" by Tamoxifen

The practical consequence of this PR heterogeneity in breast cancer cells is illustrated by an experiment in which the $T47D_V$ cell line was treated for 8 weeks with or without 1 μmol/l tamoxifen. Tamoxifen at 1 μmol/l generally suppresses growth and PR in oestrogen target tissues (Horwitz and McGuire, 1978) that carry a normal ER and is the major endocrine therapeutic drug used in breast cancer (Lerner and Jordan, 1990). But what is the effect of tamoxifen in cells that carry not only normal but also variant ER?

Figure 3 shows a univariate PR analysis of control cells (panel A), 1 μmol/l

**TABLE 1. Quantitation of PR heterogeneity in $T47D_v$ and its two clonal sublines shown in Fig. 2[a]**

| | | PR negative | PR positive | |
|---|---|---|---|---|
| **Cell line** | **Population** | **% of cells[b]** | **% of cells[b]** | **Log mean PR (FIU)** |
| $T47D_v$ | total | 1.0 | 99.0 | 269.7 |
| V22 | total | 12.8 | 87.2 | 224.0 |
| | pop "a" | — | 36.1 | 55.7 |
| | pop "b" | — | 51.1 | 343.1 |
| V26 | total | 23.2 | 76.8 | 113.4 |
| | pop "a" | — | 41.8 | 33.6 |
| | pop "b" | — | 35.0 | 208.2 |

[a]Parallel sets of cells were fixed and permeabilized. One set was treated with two anti-PR monoclonal antibodies to measure TB; the other set was treated with an isotypic control first antibody to measure NSB. PR in all sets were labelled with a fluorescein isothiocyanate conjugated secondary antibody, and DNA was simultaneously labelled with propidium iodide. Cells were analyzed for fluorescence by flow cytometry as described in Graham *et al* (1989). Specific PR levels and PR subpopulations were analysed by curve subtraction using 1-Par software

[b]Percent cells of 10 000 analysed (Graham *et al* 1992)

tamoxifen treated cells (panel B) and a comparison of the two (panel C) to demonstrate the shifts in PR patterns after 8 weeks under the influence of the drug. Cell growth was suppressed by 40% (not shown), and there was a marked shift in the PR pattern—mostly to the left, reflecting a complete loss or decrease in PR, as shown in panel B, and by the large shadowed area in panel C, but with an unexpected small subpopulation reproducibly shifted to the right in which PR levels have apparently been induced by tamoxifen. This subpopulation represents 5.2% of the cells in this experiment and contains an average PR of 571.6 fluorescence intensity units (FIU), or greater than one million PR molecules/cell—levels that none of the untreated cells attain. Thus, tamoxifen, although decreasing PR levels in a majority of cells, appears paradoxically to increase PR levels in a selected subset of cells. The ominous consequence of tumour cell populations that may be stimulated by tamoxifen requires little comment.

In addition, analysis of the DNA indices (not shown) demonstrates that tamoxifen has a dual effect on proliferation. Firstly, for the same number of cells, fewer tamoxifen treated cells are in mitosis. Secondly, the populations that are in mitosis under tamoxifen differ from the controls. Thus, although the overall growth of the tamoxifen treated cells lags behind that of the control cells, the DNA data show what we term the remodelling influence of the drug Thus, at least two new subpopulations of cells emerge that are not present in controls: a PR negative or low PR HD subset and an ultra high PR HT subset. If the biological behaviour of this cell line mimics the pattern seen in patients

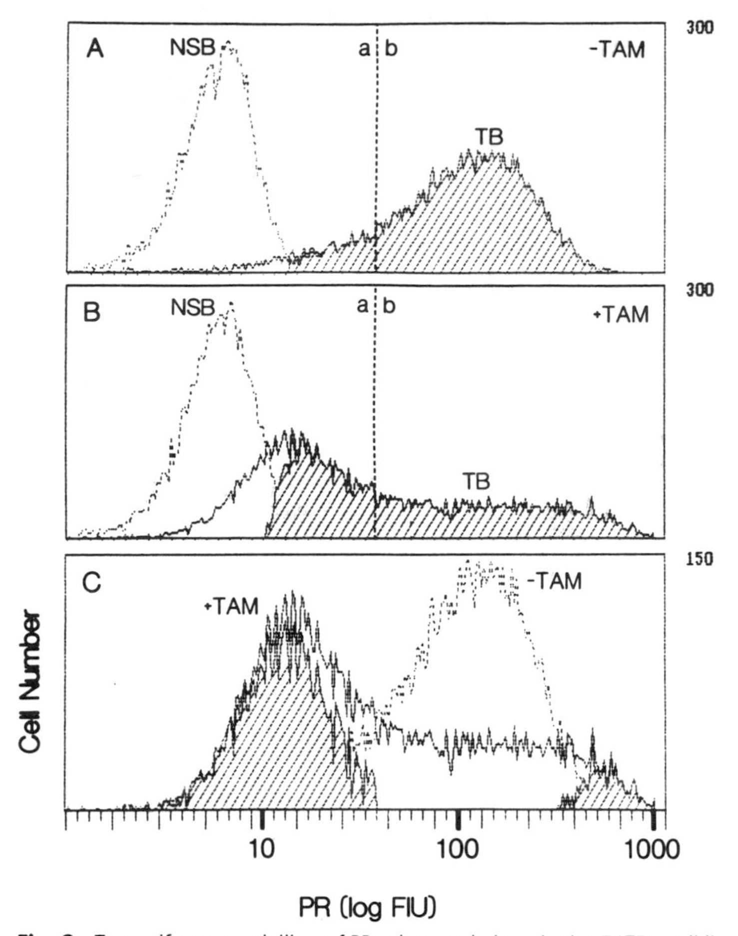


**Fig. 3.** Tamoxifen remodelling of PR subpopulations in the $T47D_v$ cell line. Total (TB) and non-specific (NSB) PR signals and simultaneous DNA signals were measured by flow cytometry in $T47D_v$ after 8 weeks of growth in control med (-TAM) or in 1 μmol/l tamoxifen (+TAM). PR fluorescence intensity unit levels in 10 000 cells of each set were analysed and are shown. In panels A and B, specific binding was calculated by curve subtraction and is shown by the shaded area. Using the 1-Par program, subpopulations with low (a) and high (b) PR levels were gated and analysed as described in Graham *et al* (1992). A, PR in control cells, showing TB and NSB. B, PR in cells treated with 1 μmol/l tamoxifen showing TB and NSB. C, TB curves from control and tamoxifen treated cells superimposed; PR subpopulations present in +TAM but excluded in -TAM were calculated by curve subtraction and are shaded, note change in the vertical scale (Graham *et al*, 1992)

with metastatic breast cancer who have an initial growth inhibitory response to tamoxifen, it may be these emerging subpopulations that lead to later clinical tumour progression and our present impression of recurrent breast cancer as an incurable disease.

A variety of mechanisms have been proposed for development of the acquired resistance to tamoxifen that arises in animal model systems (Osborne *et al*, 1987; Gottardis and Jordan, 1988) and in virtually all patients (Lippman, 1984; McGuire *et al*, 1987) undergoing hormone therapy. Genetic mechanisms include the variant and mutant forms of ER described above, which may exert dominant controls over oestrogen and anti-oestrogen regulated growth. Additionally, heterogeneity of ER and mutant ER may in part explain the extreme PR heterogeneity documented here. Epigenetic mechanisms center on pharmacokinetic issues related to drug absorption, distribution and metabolism. Whereas some of the metabolites of tamoxifen are more potent anti-oestrogens than the parent compound (Lerner and Jordan, 1990), other metabolites may be oestrogenic (Gottardis *et al,* 1989). Recent data indicate that tamoxifen and its anti-oestrogenic metabolite, *trans*-4-hydroxytamoxifen, may be selectively excluded from tamoxifen resistant breast cancers or be further metabolized to relatively inactive forms (Osborne *et al,* in press).

Although one or both general mechanisms of resistance may become operative in different tumours and different cells, we propose that tumour progression to the resistant state includes the selection and expansion of cell subpopulations, some of which remain strongly influenced by tamoxifen. That hormone treatment may itself provide the selective remodelling pressure is suggested by the studies described here and by studies showing that human breast cancer cells change significantly in response to hormone deprivation (Katzenellenbogen *et al,* 1987; Daly *et al,* 1990) or stimulation (Gottardis and Jordan, 1988; Gottardis *et al,* 1989).

Our data suggest that subsets of cells may actually be stimulated by tamoxifen. Little is known about the mechanisms underlying these "agonist" actions of some anti-oestrogens. It is possible that binding of tamoxifen to specific types of ER mutants establishes a transcriptionally productive receptor complex. The agonist activities of tamoxifen are usually expressed at low doses (Horwitz and McGuire, 1978), but they may also be tissue specific (Turner *et al,* 1987). Although tamoxifen at high doses suppresses PR, it induces PR at low doses (Horwitz and McGuire, 1978). The tumour "flare" that occurs during initiation of tamoxifen therapy (Henderson *et al,* 1989) and the withdrawal response that occurs when the drug is stopped after tumours become resistant (Legault-Poisson *et al,* 1979) may also be explained by this property. Additionally, we have previously shown that pretreatment of cells with an anti-oestrogen can sensitize them to a subsequent challenge with oestrogens. In this state, cells respond more rapidly and more extensively to oestrogens, for example superinduction of PRs is observed (Horwitz *et al,* 1981). It is possible that anti-oestrogen pretreatment can sensitize tumour cells to low levels of oestrogens or to weak oestrogens, to which in other settings they would be unresponsive. The molecular mechanisms underlying the phenomenon of superinduction remain unknown.

## SUMMARY

In summary, we propose that the molecular heterogeneity of ER in breast tumour cells, characterized by the presence of mutant receptor forms, generates the cellular heterogeneity evident when PR or DNA ploidy are analysed in cell subpopulations. Furthermore, it is likely that cellular heterogeneity leads to the lack of uniformity in response to tamoxifen that we have described. We find that heterogeneity of PR and DNA ploidy reflects the existence of mixed subpopulations of breast cancer cells that are substantially remodelled under the influence of tamoxifen. It appears likely that different subsets of cells, rather than being "resistant", can be inhibited or stimulated by tamoxifen and that their suppression or outgrowth alters the phenotype of the tumour. PR heterogeneity in solid tumours of patients may predict a mixed, and potentially dangerous, response to anti-oestrogen treatment. As we learn more about the heterogeneity of PR, ER and other proteins in tumours, we may be able to recognize such lethal subpopulations, which the flow cytometry immunoassay can simply and rapidly measure. Our data also suggest that the use of tamoxifen as a chemopreventant in women at high risk of developing breast cancer (Kiang, 1991) should be viewed with caution.

## Acknowledgements

The substantive contributions of my colleagues and collaborators are gratefully acknowledged. I am especially indebted to Dr Louise Miller, Dr Kimberly Leslie and Dr Mark Graham. I am also grateful to the National Cancer Institute, NIH, and the American Cancer Society for their financial support.

## References

Adler S, Waterman ML, He X and Rosenfeld MG (1988) Steroid-receptor mediated inhibition of rat prolactin gene expression does not require the receptor DNA-binding domain. *Cell* **52** 685–695

Baulieu E-E (1987) Steroid hormone antagonists at the receptor level: a role for the heat-shock protein MW 90 000 (hsp 90). *Journal of Cellular Biochemistry* **35** 161–174

Castagnoli A, Maestri I, Bernardi F and Del Senno L (1987) PvuII RFLP inside the human estrogen receptor gene. *Nucleic Acids Research* **15** 866

Daly RJ, King RJB and Darbre PD (1990) Interaction of growth factors during progression towards steroid independence in T47D human breast cancer cells. *Journal of Cellular Biochemistry* **43** 199–211

Fawell SE, Lees JA, White R and Parker MG (1990) Characterization and colocalization of steroid binding and dimerization activities in the mouse estrogen receptor. *Cell* **60** 953–962

Forman BM, Yang C, Au M, Casanova J, Ghysdael J and Samuels HH (1989) A domain containing leucine zipper-like motifs mediate novel *in vivo* interactions between the thyroid hormone and retinoic acid receptors. *Molecular Endocrinology* **3** 1610–1626

Foster BD, Cavener DR and Parl FF (1991) Binding analysis of the estrogen receptor to its specific DNA target site in human breast cancer. *Cancer Research* **51** 3405–3410

Fuqua SAW, Fitzgerald SD, Chamness GC et al (1991) Variant human breast tumor estrogen receptor with constitutive transcriptional activity. *Cancer Research* **51** 105–109

Garcia T, Lehrer S, Bloomer W and Schachter B (1988) A variant estrogen receptor messenger ribonucleic acid is associated with reduced levels of estrogen binding in human mammary tumors. *Molecular Endocrinology* **2** 785–791

Garcia T, Sanchez M, Cox JL et al (1989) Identification of a variant form of the human estrogen receptor with an amino acid replacement. *Nucleic Acids Research* **17** 8364

Gottardis MM and Jordan VC (1988) Development of tamoxifen-stimulated growth of MCF-7 tumors in athymic mouse after long-term tamoxifen administration. *Cancer Research* **48** 5183–5188

Gottardis MM, Jiang S-Y, Jeng M-H and Jordan VC (1989) Inhibition of tamoxifen stimulated growth of an MCF-7 tumor variant in athymic mice by novel steroidal antiestrogens. *Cancer Research* **49** 4090–4093

Graham II ML, Bunn Jr PA, Jewett PB, Gonzalez-Aller C and Horwitz KB (1989a) Simultaneous measurement of progesterone receptors and DNA indices by flow cytometry: characterization of an assay in breast cancer cell lines. *Cancer Research* **49** 3934–3942

Graham II ML, Dalquist KE and Horwitz KB (1989b) Simultaneous measurement of progesterone receptors and DNA indices by flow cytometry: analyses of breast cancer cell mixtures and genetic instability of the T47D line. *Cancer Research* **49** 3943–3949

Graham II ML, Krett NL, Miller LA et al (1990) $T47D_{co}$ cells, genetically unstable and containing estrogen receptor mutations, are a model for the progression of breast cancers to hormone resistance. *Cancer Research* **50** 6208–6217

Graham II ML, Smith JA, Jewett PB and Horwitz KB (1992) Heterogeneity of progesterone receptor content and remodeling by tamoxifen characterize subpopulations of cultured human breast cancer cells: analysis by quantitative dual parameter flow cytometry. *Cancer Research* **52** 593–602

Green S, Walter P, Kumar V et al (1986) Human oestrogen receptor cDNA: sequence, expression and homology to v-*erb*-A. *Nature* **320** 134–139

Greene GL, Gilna P, Waterfield M, Baker A, Hort Y and Shine J (1986) Sequence and expression of human estrogen receptor complementary DNA. *Science* **231** 1150–1154

Guiochon-Mantel A, Loosfelt H, Ragot T et al (1988) Receptors bound to antiprogestins form abortive complexes with hormone responsive elements. *Nature* **336** 695–698

Henderson IC, Harris JR, Kinne DW and Hellman S (1989) Cancer of the breast, In: VT DeVita Jr, S Hellman and SA Rosenberg (eds). *Cancer: Principles and Practice of Oncology*, p 1252, JB Lippincott Co, Philadelphia Hill SM, Fuqua SAW, Chamness GC, Greene GL and McGuire WL (1989) Estrogen receptor expression in human breast cancer associated with an estrogen receptor gene restriction fragment length polymorphism. *Cancer Research* **49** 145–148

Hirose T, Koga M, Matsumoto K and Sato B (1991) A single nucleotide substitution in the D domain of estrogen receptor cDNA causes amino acid alteration from Glu-279 to Lys-279 in a murine transformed Leydig cell line (B-1 F). *Journal of Steroid Biochemistry and Molecular Biology* **39** 1–4

Horwitz KB (1981) Is a functional estrogen receptor always required for progesterone receptor induction in breast cancer? *Journal of Steroid Biochemistry* **15** 209–217

Horwitz KB and McGuire WL (1978) Estrogen control of progesterone receptor in human breast cancer: correlation with nuclear processing of estrogen receptor. *Journal of Biological Chemistry* **253** 2223–2228

Horwitz KB, McGuire WL, Pearson OH and Segaloff A (1975) Predicting response to endocrine therapy in human breast cancer: a hypothesis. *Science* **189** 726–727

Horwitz KB, Zava DT, Thiligar AK, Jensen EM and McGuire WL (1978) Steroid receptor analyses of nine human breast cancer cell lines. *Cancer Research* **38** 2434–2437

Horwitz KB, Aiginger P, Kuttenn F and McGuire WL (1981) Nuclear estrogen receptor release from antiestrogen suppression: amplified induction of progesterone receptor in MCF-7 human breast cancer cells. *Endocrinology* **108** 1703–1709

Horwitz KB, Mockus MB and Lessey BA (1982) Variant T47D human breast cancer cells with high progesterone receptor levels despite estrogen and antiestrogen resistance. *Cell* **28** 633–642

Horwitz KB, Wei LL, Sedlacek S and d'Arville CN (1985) Progestin action and progesterone receptor structure in human breast cancer: a review. *Recent Progress in Hormone Research*

**41** 249–316

Katzenellenbogen BS, Kendra KL, Normal MJ and Berthois Y (1987) Proliferation, hormonal responsiveness, and estrogen receptor content of MCF-7 human breast cancer cells grown in the short-term and long-term absence of estrogens. *Cancer Research* **47** 4355–4360

Keaveney M, Klug J, Dawson MT et al (1991) Evidence for a previously unidentified upstream exon in the human oestrogen receptor gene. *Journal of Molecular Endocrinology* **6** 111–115

Keydar I, Chen L, Karbey S et al (1979) Establishment and characterization of a cell line of human breast carcinoma origin. *European Journal of Cancer* **15** 659–670

Kiang DT (1991) Chemoprevention for breast cancer: are we ready? *Journal of the National Cancer Institute* **83** 462–463

Kumar V and Chambon P (1988) The estrogen receptor binds tightly to its responsive element as a ligand-induced homodimer. *Cell* **55** 145–156

Kumar V, Green S, Staub A and Chambon P (1986) Localisation of the oestradiol-binding and putative DNA-binding domains of the human oestrogen receptor. *EMBO Journal* **5** 2231–2236

Legault-Poisson S, Jolivet J, Poisson R, Beretta-Piccoli M and Band PR (1979) Tamoxifen-induced tumor stimulation and withdrawal response. *Cancer Treatment Reports* **63** 1839–1841

Lehrer S, Sanchez M, Song HK et al (1990) Oestrogen receptor B-region polymorphism and spontaneous abortion in women with breast cancer. *Lancet* **335** 622–624

Lerner LJ and Jordan VC (1990) Development of antiestrogens and their use in breast cancer. *Cancer Research* **50** 4177–4189

Lippman ME (1984) An assessment of current achievements in the systemic management of breast cancer. *Breast Cancer Research and Treatment* **4** 69–77

McGuire WL, Lippman ME, Osborne CK and Thompson EB (1987) Resistance to endocrine therapy. *Breast Cancer Research and Treatment* **9** 165–173

Murphy LC (1990) Estrogen receptor variants in human breast cancer. *Molecular and Cellular Endocrinology* **74** C83–C86

Murphy LC and Dotzlaw H (1989) Variant estrogen receptor mRNA species detected in human breast cancer biopsy samples. *Molecular Endocrinology* **3** 687–693

Osborne CK, Coronado EB and Robinson JR (1987) Human breast cancer in the athymic nude mouse: cytostatic effects of long-term antiestrogen therapy. *European Journal of Cancer and Clinical Oncology* **23** 1189–1196

Osborne CK, Coronado EB, Wiebe V and DeGregorio M Acquired tamoxifen resistance correlates with reduced breast tumor levels of tamoxifen and isomerization of trans-4-hydroxytamoxifen. *Journal of the National Cancer Institute* (in press)

Parl FF, Cavener DR and Dupont WD (1989) Genomic DNA analysis of the estrogen receptor gene in breast cancer. *Breast Cancer Research and Treatment* **14** 57–64

Ponglikitmongkol M, Green S and Chambon P (1988) Genomic organization of the human oestrogen receptor gene. *EMBO Journal* **7** 3385–3388

Scott GK, Kushner P, Vigne J-L and Benz CC (1991) Truncated forms of DNA-binding estrogen receptors in breast cancer. *Journal of Clinical Investigation* **88** 700–706

Shupnik MA, Gordon MS and Chin WW (1989) Tissue-specific regulation of rat estrogen receptor mRNAs. *Molecular Endocrinology* **3** 660–665

Turner RT, Wakely GK, Hannum KS and Bell NH (1987) Tamoxifen prevents the skeletal effects of ovarian deficiency in rats. *Journal of Bone and Mineral Research* **2** 449–456

Washburn T, Hocutt A, Brautigan DL and Korach KA (1991) Uterine estrogen receptor *in vivo*: phosphorylation of nuclear-specific forms on serine residues. *Molecular Endocrinology* **5** 235–242

The author is responsible for the accuracy of the references.

# A Molecular Strategy to Control Tamoxifen Resistant Breast Cancer

**S-Y JIANG • V C JORDAN**

*Department of Human Oncology, University of Wisconsin Comprehensive Cancer Center, Madison, Wisconsin 53792*



## INTRODUCTION: TREATMENT STRATEGIES

Tamoxifen is a non-steroidal anti-oestrogen that was originally evaluated for the palliative treatment of advanced breast cancer in postmenopausal women. The drug is now available to treat selected patients at all stages of breast cancer. Adjuvant therapy with tamoxifen appears to confer a survival advantage in node positive postmenopausal women (Early Breast Cancer Trialists' Collaborative Group, 1988) especially if the primary tumour was originally oestrogen receptor (ER) positive. Similarly, patients with ER positive node negative breast cancer benefit from adjuvant therapy with tamoxifen (Fisher *et al*, 1989). Indeed, these promising clinical results, and supporting laboratory studies (Jordan, 1991), have encouraged the evaluation of tamoxifen to prevent (or rather chemosuppress) the appearance of primary breast cancer (Fig. 1) (Powles *et al*, 1990). Should this strategy prove to be successful, hundreds of thousands (perhaps millions) of women will be taking tamoxifen therapy. It is estimated that breast cancer incidence will be reduced by 50% in women receiving tamoxifen.

Adjuvant therapy with tamoxifen has also changed in the past decade. Early studies employed a treatment duration of 1–2 years (Early Breast Cancer Trialists' Collaborative Group, 1988). But on the basis of laboratory studies (Jordan, 1983) and the finding that tamoxifen produces few serious side effects, the duration of treatment has been extended to more than 5 years. Unfortunately, tamoxifen is unlikely to be effective indefinitely, and therapeutic failure will result with the development of drug resistance.

Several mechanisms of drug resistance are possible for the ER positive

*Cancer Surveys* Volume 14: *Growth Regulation by Nuclear Hormone Receptors*
 0-87969-371-1/92. $3.00 + .00

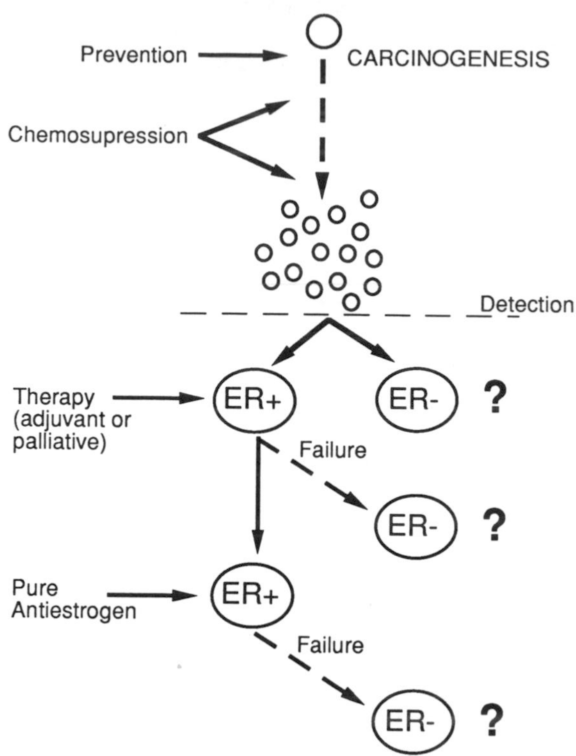


**Fig. 1.** Strategies designed to control the growth of breast cancer cells. A normal mammary epithelial cell is converted to a cancer cell through carcinogenesis, which could be prevented by avoiding carcinogens. Once cancer cells are formed, the expansion of the cell population to a detectable tumour can be stopped by using chemicals that will block cell replication (chemosuppression). Breast tumours that express the ER can be controlled by anti-oestrogens. However, no targeted hormone therapy is available for the treatment of ER negative breast cancer

tumour. One possibility is that the oestrogen-like properties of tamoxifen will eventually encourage tumours to grow in response to the therapy. A new pure anti-oestrogen (Wakeling *et al*, 1991) is to be tested as a second line therapy following the failure of tamoxifen treatment.

Overall, several treatment strategies (Fig. 1) with tamoxifen are in place to reduce the number of primary tumours and to control the recurrence of the disease following mastectomy. Additionally, there are plans to use pure anti-oestrogens after the failure of long term tamoxifen therapy. However, it is the relentless development of ER negative disease that has confounded all attempts to develop a successful therapy for breast cancer.

Since oestrogen regulated growth is thought to be mediated via the ER, we have taken the strategic step to determine whether the ER can reassert control of growth in an ER negative breast cancer cell. Our goal is to establish a biological basis that could lead to the development of a targeted "gene therapy" for cancer. If steroid receptors can reassert control within a cancer cell, then either a targeted vector could be devised to introduce a gene that constitutively produces the appropriate receptor or perhaps techniques could be developed to reactivate pre-existing steroid receptor genes.

## OESTROGEN RECEPTOR TRANSFECTION

Several reports describe the stable transfection of the ER complementary DNA (cDNA) into cells in culture (Watts *et al,* 1989; Kushner *et al,* 1990; Touitou *et al,* 1990; Maminta *et al,* 1991). However, a mutant cDNA for the human ER was used in these studies. We have constructed a vector containing the cDNAs for the ER driven by the cytomegalovirus promoter that will produce a polycistronic messenger RNA (mRNA) for the ER and aminoglycoside phosphotransferase to confer resistance to G418 (Jiang and Jordan, in press). The vector was transfected (Felgner *et al,* 1987) into the ER negative breast cancer cell line MDA-MB-231 clone 10A. A number of transfectants have been obtained and characterized (Fig. 2).

### Mutant and Wild Type Receptors

Two cDNAs for the human ER were available: one that produces the wild type receptor and the other a mutant form that produces a protein with valine substituted for glycine at position 400 in the steroid binding domain. Comparison of the effect of oestrogen on the growth rate of the transfectants demonstrated a strong inhibition by oestradiol, but the transfectants with the mutant receptor were less sensitive to oestradiol (Fig. 3). The mutant receptor is, however, known to be less effective (Tora *et al,* 1989), and the affinity for oestradiol is reduced (Jiang and Jordan, in press).

One observation that merits further investigation is that although the levels of receptor in wild type and mutant transfectants are equivalent by ligand binding, the monoclonal antibodies used in enzyme immunoassay (H222/D547) appear to have a reduced ability to detect mutant receptor (Fig. 2). We are following up this finding because it implies that a critical epitope is no longer exposed in the mutant receptor for antibody interaction.

The growth inhibitory effect of oestradiol is in complete contrast to the recognized action of oestrogen to stimulate growth in ER positive cells. However, further examination of the mechanism(s) of the receptor mediated action in the transfectants might provide some clues to the perplexing efficacy of high dose oestrogen (DES) therapy to inhibit receptor positive breast cancer growth. As a hypothesis, it is possible that oestrogen initially causes an increase in receptors that activate inappropriate genes to cause the demise of the cancer cells. Alternatively, high dose oestrogen may cause a profound negative feedback inhibition of receptor resynthesis, which ultimately results in receptor deprivation (equivalent to oestrogen withdrawal).

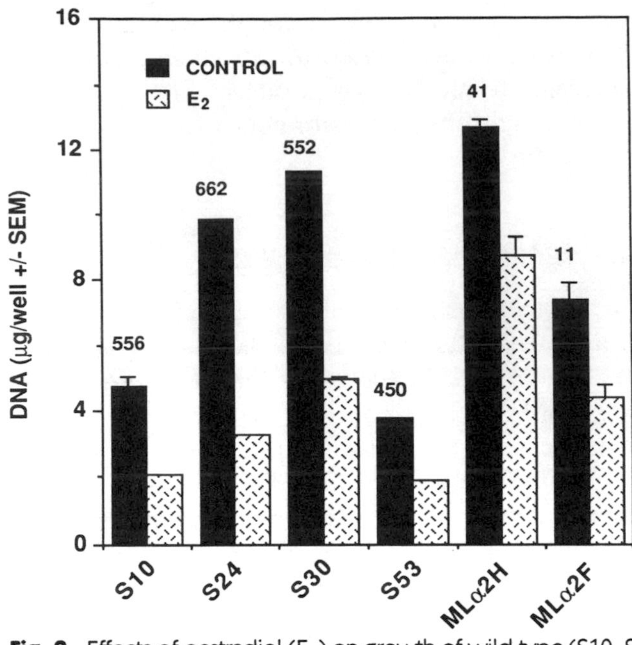


**Fig. 2.** Effects of oestradiol ($E_2$) on growth of wild type (S10, S24, S30 and S53) and mutant (MLα2H and MLα2F) ER transfectants. Cells were plated in 24-well plates in oestrogen free media and then cultured in media containing $E_2$ ($10^{-9}$ mol/l for S10, S24, S30 and S53 cells and $10^{-8}$ mol/l for MLα2H and MLα2F cells) or ethanol vehicle (0.1%) (control) for 6 days. Histograms were combined from different experiments with different seeding densities (~40 000 cells per well). The levels of ER (expressed as fmol/mg protein determined by enzyme immunoassay) are indicated at the top of the control histogram of each transfectant. The S30 and MLα2H cells expressed similar levels of ER as determined by ligand binding method (437 and 458 fmol/mg protein, respectively)

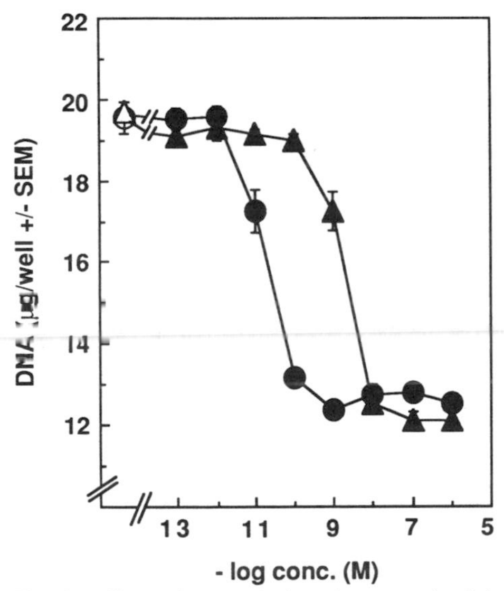


**Fig. 3.** Effect of oestradiol on the growth of the wild type ER transfectant S30 (circle) and the mutant ER transfectant MLα2H (triangle). Cells were plated at a density of 40 000 cells per well in oestrogen free media for 2 days and then cultured in media containing $E_2$ (solid symbols) or control media (open symbols) for 8 days with media change every 2 days. Experiments were conducted in triplicate and data were presented as mean +/- standard error of mean

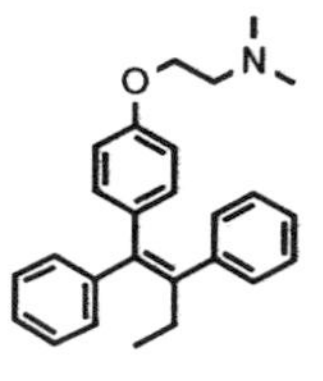

tamoxifen

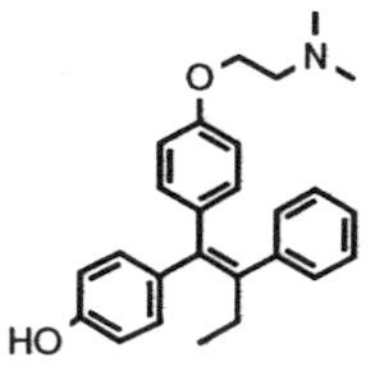

4 hydroxytamoxifen

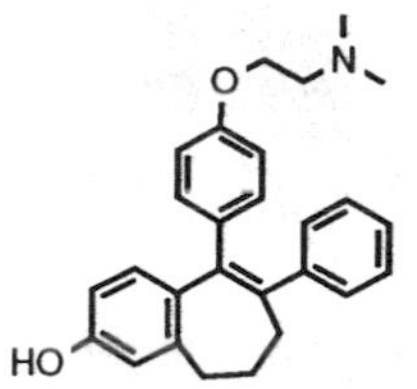

fixed ring 4-hydroxytamoxifen

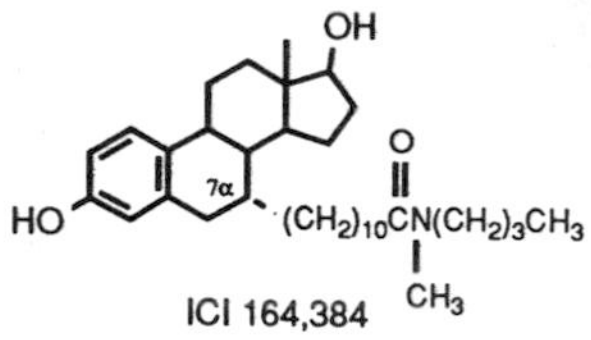

ICI 164,384

**Fig. 4.** Structure of anti-oestrogens

### Anti-oestrogen Action

Non-steroidal anti-oestrogens display a remarkable spectrum of oestrogenic and anti-oestrogenic properties depending on the structure of the compound and the test system employed (Jordan, 1984; Jordan and Murphy, 1990). Recently, however, a series of steroidal compounds has been reported that exhibit only pure anti-oestrogenic properties (Wakeling and Bowler, 1987). We have compared and contrasted the action of a number of compounds (Fig. 4) to determine whether the effects of oestrogen in inhibiting the growth of transfectants can be blocked. In general, pure anti-oestrogens can effectively block oestrogen action in both mutant and wild type transfectants. However, triphenylethylene compounds such as 4-hydroxytamoxifen or its fixed ring derivative appear to be unable to block oestrogen action in the mutant transfectants compared with their activity in wild type transfectants. The compound expresses much more oestrogenic activity on cell growth in the mutant transfectant. The contrasting actions of fixed ring 4-hydroxytamoxifen in wild type and mutant transfectants are shown in Fig. 5. The pure anti-oestrogen ICI 164 384 is an anti-oestrogen in both cell types.

These, and additional observations in our laboratory, have led us to devise a model system to dissect some of the components of anti-oestrogen action that could result in a change in the pharmacology of tamoxifen from an antagonist to an agonist.

## A MODEL SYSTEM FOR STUDYING ANTI-OESTROGEN PHARMACOLOGY

Breast cancer cells stably transfected with constitutive vectors for the ER can be transiently transfected with a construct containing an oestrogen response

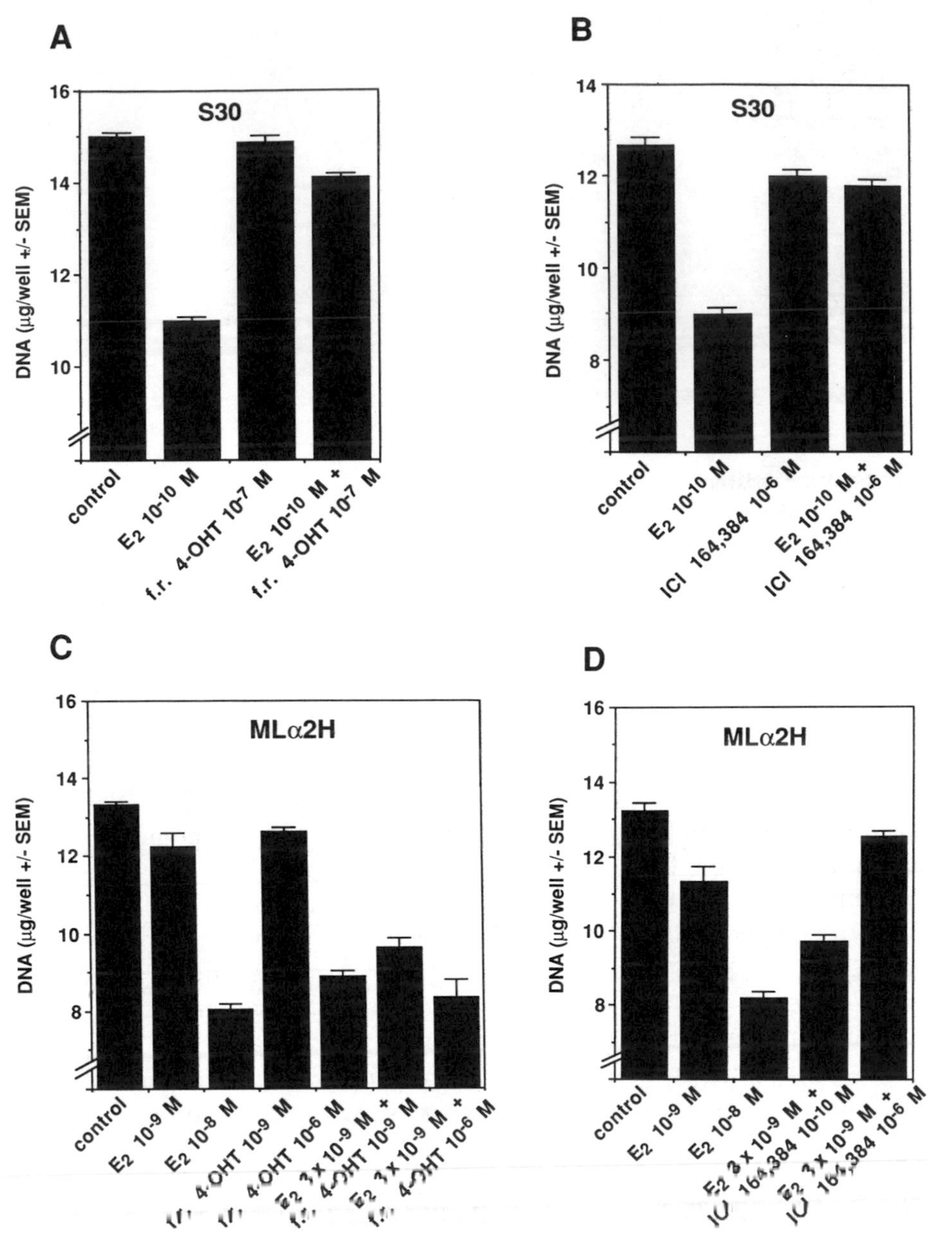


**Fig. 5.** Effects of fixed ring 4-hydroxytamoxifen (f.r. 4-OHT) and ICI 164 384 on the growth of wild type (A,B) and mutant (C,D) ER transfectants S30 and MLα2H, respectively. Wells were plated at a density of 50 000 cells in oestrogen free media for 2 days and then cultured in media containing $E_2$, f.r. 4-OHT (A,C), ICI 164 384 alone (B,D) at the indicated concentration or $E_2$ in combination with anti-oestrogens at the indicated concentrations for 6 days

element (ERE) linked to a chloramphenicol acetyltransferase (CAT) reporter gene and appropriate promoter.

Receptor mediated CAT activity is dependent on the cellular environment and the promoter located 5′ to the CAT gene in a given construct (Berry *et al*,

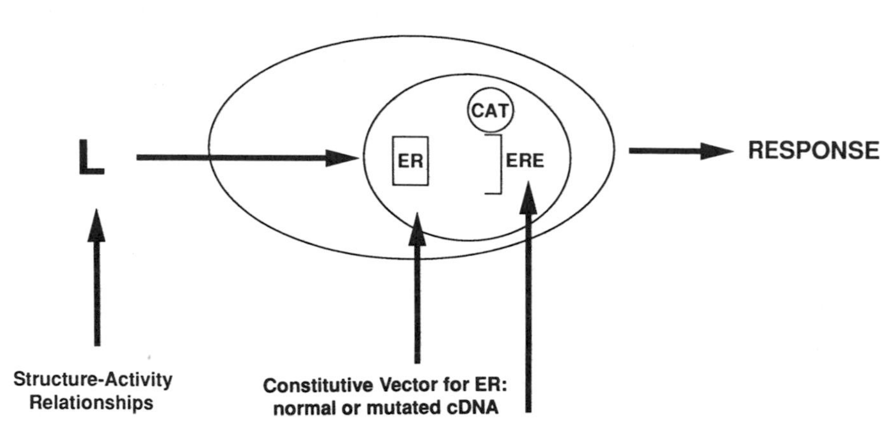


**Fig. 6.** A model system to study anti-oestrogen pharmacology in ER negative breast cancer cell MDA-MB-231 clone 10A transfected with a cDNA for ER

1990). However, the relative aspects of drug receptor pharmacology can be dissected out by using a single clone of breast cancer cells for study. The implication of the pharmacology of an oestrogenic or anti-oestrogenic compound can be studied by altering the ERE, the promoter or the ER cDNA. Similarly, the ligand can be altered to study the impact of structure-activity relationships (Fig. 6). Oestradiol will effectively activate the CAT reporter gene (*vitERE-TK-CAT*) in both mutant and wild type transfectants. The wild type ER appears to confer high potency, but the mutant receptor appears to cause a greater induction in CAT activity (Tora *et al*, 1989; Jiang and Jordan, 1991). There are, however, differences in the basal CAT activity (which is not affected by pure anti-oestrogens in our oestrogen free system) that might complicate direct comparison. The wild type receptor confers higher basal CAT activity, so the increase in CAT activity appears to be reduced.

The imperfect palindromic ERE from the *pS2* gene is less efficient than the ERE from the vitellogenin gene (Berry *et al*, 1989). However, by altering the four discriminating variables (two ER cDNAs and two reporter plasmid constructs), we show that a change occurs in the pharmacology of a single compound fixed ring 4-hydroxytamoxifen in transfectants. The results for the CAT assays are summarized in Table 1. We show that the pharmacology of this model compound is changed from an anti-oestrogen in the wild type transfectant to a partial oestrogen in the mutant transfectant* using the vitERE-TK-CAT construct. For the pS2-CAT construct, fixed ring 4-hydroxytamoxifen is an oestrogen in both types of transfectants. The full agonist activity of fixed ring 4-hydroxytamoxifen, which is due to the presence of different promoter or the imperfect palindromic ERE, is not yet known. We are currently constructing plasmids that contain the pS2ERE and TK promoter in order to carry out this analysis.

**TABLE 1. Change in the pharmacology of fixed ring 4 hydroxytamoxifen with changes in the cDNA for the ER or ERE for the CAT reporter gene transfected into MDA-MB-231 breast cancer cells**

| ER | ERE | Pharmacology[a] |
|---|---|---|
| Wild type | vitellogenin[b] | antagonist |
| Mutant | vitellogenin[b] | partial agonist |
| Mutant | *pS2*[c] | agonist |
| Wild type | *pS2*[c] | agonist |

[a]This describes whether the compound produces an action that is of the same magnitude as oestradiol (agonist) or blocks the action of oestradiol (antagonist)
[b]The reporter plasmid contains a perfect palindromic ERE derived from vitellogenin and thymidine kinase promoter derived from herpes simplex virus
[c]The reporter plasmid contains an imperfect palindromic ERE and promoter derived from *pS2* gene

This model is currently being developed to make a systematic study of different EREs and will be valuable for describing the impact of site directed mutagenesis on the ER cDNA. However, these laboratory data might also provide an insight into mechanisms of the development of tamoxifen resistance during breast cancer therapy.

## POTENTIAL MECHANISMS FOR RESISTANCE

The pharmacological requirement for an anti-oestrogen such as tamoxifen to block oestrogen stimulated growth of ER positive breast cancer cells is firstly to bind to the receptor and to cause an inappropriate conformational change, so that the target gene cannot be activated even though the anti-ER complex may bind to the ERE. Some of the steps to be considered when an anti-oestrogenic drug is taken by the patient for breast cancer are illustrated in Fig. 7. The drug can be metabolized, and each of the metabolites might be unstable and spontaneously change its structures. The drug and each of the metabolites will compete with each other and with oestradiol (and its metabolites)

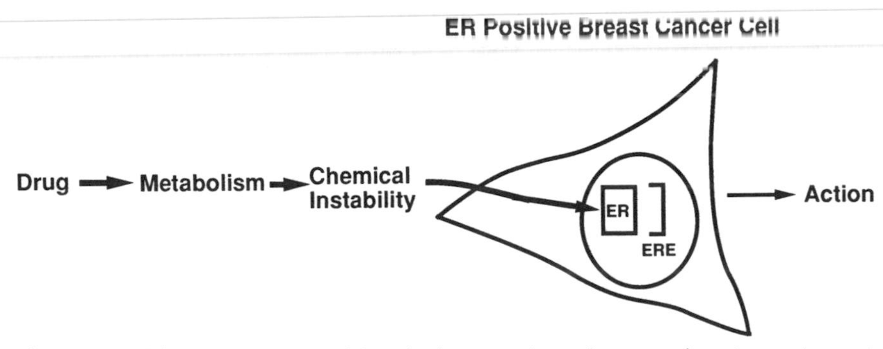


**Fig. 7.** Possible steps to be considered when a patient takes an anti-oestrogen to control the growth of ER positive breast cancer

to occupy the ER. Successful binding to the receptor depends on the local molar concentration and the relative binding affinities for the receptor. In circumstances when there are non-saturating concentrations of an anti-oestrogen in the target tissues, either oestradiol-ER or anti-ER complexes will compete for ERE. As saturating concentrations of an anti-oestrogen are achieved, the action observed in the cell will depend on the intrinsic activity of the anti-ER complex alone. The triphenylethylene anti-oestrogens exhibit some oestrogen like responses, but the pure anti-oestrogens do not. The preceding events can be summarized as follows:

oestrogen→ERC→ERE→growth
anti-oestrogen→ERC*→ERE→blocks growth

Oestrogen thus forms an oestrogen receptor complex (ERC), which can interact with as yet undetermined response elements (ERE) to initiate growth. The anti-oestrogen blocks growth by forming an inappropriate receptor complex (ERC*).

The development of resistance to an anti-oestrogen could occur through the induction of enzyme systems that convert the anti-oestrogen to oestrogenic metabolites. However, on the basis of our studies with mutant receptors and different reporter plasmids, we suggest that the anti-oestrogenic ligand could be translated into a stimulatory complex by binding to a mutant receptor. The two hypotheses, which are not mutually exclusive, would be as follows:

1. anti-oestrogen→oestrogenic metabolite→ERC→ERE→growth
2. anti-oestrogen→$ER_mC$→$ERE_m$/ERE→growth

The anti-oestrogen may bind to a mutant receptor ($ER_m$) and the complex could bind to either a normal or mutant ($ERE_m$) response element to initiate growth. Therefore, a scheme for resistance to tamoxifen that produces an alteration of the pharmacology from an antagonist to an agonist could be represented as follows:

metabolism→mutated receptors→mutated response elements→altered response

This scheme can now be used in laboratory models to make a systematic study of tamoxifen resistance. Tamoxifen stimulated tumour growth has been described and could provide valuable information to understand clinic resistance. At present, though, most research efforts have focused on the potential of metabolic transformation to be the cause of tamoxifen resistance.

Initial studies focused on the unusual pharmacology of tamoxifen in different species: tamoxifen is an oestrogen in the mouse, a partial agonist in the rat and a complete antagonist in the chick (reviewed in Furr and Jordan, 1984). The species differences could be caused by differential metabolism to oestrogens in mice, whereas the metabolic changes would not occur in the chick. Studies in vivo and in vitro demonstrate that the principal metabolite in the chick and the mouse is 4-hydroxytamoxifen (Lyman and Jordan, 1985;

Robinson *et al,* 1991). A potential oestrogenic metabolite, metabolite E (tamoxifen without the dimethylaminoethane side chain) has not been detected in these species (Lyman and Jordan, 1985), although the metabolite has been detected in bile in the dog, a species in which tamoxifen has oestrogenic properties (Furr and Jordan, 1984).

Studies on the pharmacology of tamoxifen in the mouse have, however, demonstrated that the oestrogenic properties in the uterus and vagina are transient, and long term therapy causes the uterus and vagina to become refractory to subsequent oestrogen treatment (Jordan, 1975; Jordan *et al,* 1990). Tamoxifen also inhibits the development of spontaneous mammary tumours in mice (Jordan *et al,* 1991), so the development of metabolic tolerance through long term therapy is unlikely. A clinical evaluation of tamoxifen metabolism in patients during 10 years of therapy has not demonstrated any dramatic changes in the circulating blood levels of either tamoxifen or its anti-oestrogenic metabolites (Langan-Fahey *et al,* 1990). Metabolite E has a low binding affinity for the ER (Lieberman *et al,* 1983) and has only a low potency as an oestrogen in vivo (Jordan *et al,* 1983; Lyman and Jordan, 1985). If its presence is to contribute to tamoxifen resistance, considerable levels (easily detectable) of metabolite E would need to be present to reverse the action of tamoxifen and its metabolites.

An alternative hypothesis would be that metabolism and chemical instability occur in the tumour. The tumour would then be unique, and all events would occur intracellularly. Recently, Osborne and co-workers (1991) have proposed that metabolic instability results in the local production of oestrogenic compounds that support the growth of MCF-7 breast tumours, which grow, after long term tamoxifen therapy (Osborne *et al,* 1987; Gottardis and Jordan, 1988), in athymic mice. Increased levels of the *cis* geometric isomer of 4-hydroxytamoxifen are noted in the nuclear fraction of tamoxifen stimulated tumours, and metabolite E has been detected. Although these observations are certainly of interest, they must, however, be considered in the context of the model and the work that has been conducted to date.

There are two model systems of tamoxifen stimulated tumour growth: the MCF-7 breast tumours that grow during long term tamoxifen therapy (MCF-7 TAM tumour) (Osborne *et al,* 1987; Gottardis and Jordan, 1988) and the human endometrial carcinoma EnCa101, which is a primary tumour maintained in tamoxifen treated athymic mice (Satyaswaroop *et al,* 1984). The MCF-7 TAM tumours grow only in response to tamoxifen in vivo (Gottardis *et al,* 1989b), which would support the position that an intact tumour (cells and stroma) is necessary to produce the required metabolites. MCF-7 TAM tumours also grow in athymic rats (Gottardis *et al,* 1989), in which metabolism is altered compared with that of athymic mice (Robinson *et al,* 1991), but again, this would support the view that the tumour is unique and the type of host is of secondary importance. A pure anti-oestrogen ICI 164 384 (Fig. 4) will inhibit tamoxifen stimulated growth of both MCF-7 TAM and EnCa101 tumours (Gottardis *et al,* 1989a, 1990), so the position could be taken that the tamoxifen metabolites were being competed out so that the result could have been predicted.

Superficially then, the models support the position that local tumour metabolism is the sole determinant of tumour growth. In fact, the target site specific actions of tamoxifen certainly rule out whole body metabolism of tamoxifen to oestrogens in the mouse. Doubly transplanted athymic animals with an MCF-7 tumour on one side and an EnCa101 tumour on the other side show growth of the endometrial carcinoma during oestradiol and tamoxifen treatment but growth inhibition of the MCF-7 tumour during tamoxifen treatment (Gottardis *et al,* 1988). Also in support of the concept of tumour specific effects of tamoxifen is the observation that during long term tamoxifen therapy, the uterine wet weight of athymic animals is reduced and refractory to oestrogen (Gottardis and Jordan, 1988), but the MCF-7 TAM tumours continue to grow in the same tamoxifen environment. Thus, local tumour metabolism must be the focus for investigation.

The hypothesis (Osborne *et al,* 1991) proposes that tamoxifen is converted to oestrogenic metabolites that preferentially support tumour growth. Clearly, only potent oestrogen will be able to subvert control of tumour growth from an environment saturated by tamoxifen and its high potency (Jordan *et al,* 1977) metabolite 4-hydroxytamoxifen.

The *cis* geometric isomer of tamoxifen ICI 47 699 is a weak oestrogen in vivo and in vitro (Harper and Walpole, 1966; Jordan *et al,* 1981; Lieberman *et al,* 1983). However, there is no evidence that tamoxifen is converted to its oestrogenic isomer in vivo. By analogy, it has been proposed that 4-hydroxytamoxifen is converted to its *cis* isomer. The pharmacology of *cis* 4-hydroxytamoxifen is difficult to study because it is very unstable and readily converts to the *trans* isomer (Fig. 8) (Katzenellenbogen *et al,* 1984). Nevertheless the fixed ring derivative of *cis* 4-hydroxytamoxifen has been studied in detail (Jordan *et al,* 1988; Murphy *et al,* 1990). The compound is a very weak anti-oestrogen with approximately the same potency as tamoxifen in vitro. The compound is a partial agonist, weak antagonist in vivo (McCague *et al,* 1988). Thus, very high concentration of the *cis* isomer of 4-hydroxytamoxifen would need to be present to reduce the inhibitory activity of tamoxifen and its other metabolites. The mechanism of resistance would depend on the conversion of a potent anti-oestrogen (*trans* 4-hydroxytamoxifen) to a weak anti-oestrogen (*cis* 4-hydroxytamoxifen) with some increase in agonist actions.

The source of agonist metabolites could be derived from metabolite E. This compound is only weakly active as an oestrogen in vivo (Jordan *et al,* 1983; Lyman and Jordan, 1985), but it can convert to a more potent isomer in vitro (Fig. 8) (Murphy *et al,* 1990). The detection of increased local concentration of the *cis* isomer of metabolite E would be essential to support the hypothesis of local metabolism as the cause of tamoxifen resistance. However, the isomers are so unstable that it may be unclear whether the compounds were present in vivo or the isomerization has occurred during storage, tissue handling, extraction and assay. Isomerization must occur to produce adequate

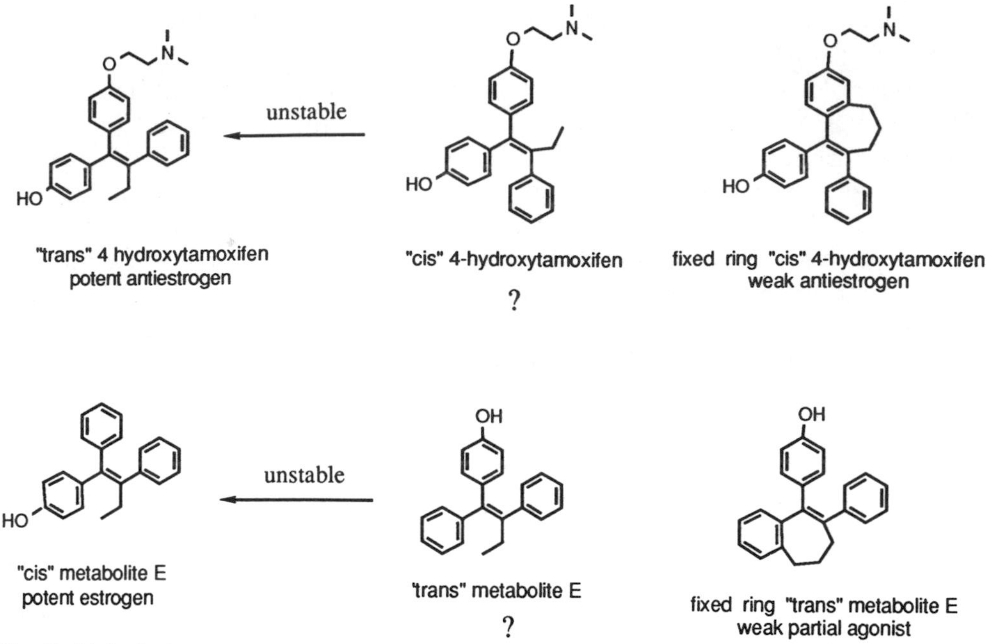

unstable

"trans" 4 hydroxytamoxifen
potent antiestrogen

"cis" 4-hydroxytamoxifen
?

fixed ring "cis" 4-hydroxytamoxifen
weak antiestrogen

unstable

"cis" metabolite E
potent estrogen

"trans" metabolite E
?

fixed ring "trans" metabolite E
weak partial agonist

**Fig. 8.** Biological activities and isomerization of metabolites derived from tamoxifen. Because of the physical instability of "cis" 4-hydroxytamoxifen and "trans" metabolite E (middle), the true biological activities of these compounds are unknown. However, their activities can be studied by using the non-isomerizable fixed ring compounds (right)

local concentrations of oestrogens, because the fixed ring derivatives of metabolite E (Fig. 8) form only a weakly oestrogenic partial agonist (Murphy *et al*, 1990). Finally it is possible that the bisphenolic triphenylethylene (4-hydroxytamoxifen without the dimethylaminoethane side chain) is produced. This compound is a partial agonist in vitro (Jordan *et al*, 1984; Murphy *et al*, 1990). Overall, it is clear that the whole concept of local drug metabolism and ligand stability is rather speculative at present. An alternate approach would be to rely on structure-activity relationships in designing compounds that cannot isomerize but still cause tumour growth.

Although local metabolism might occur and produce oestrogenic metabolites, it is possible that the steroid regulatory systems are responding to the inherent agonist activity of the drug molecule. The concept that a drug molecule can interact with any receptor to exhibit partial agonist/antagonist actions is an established fact throughout the whole of pharmacology. The intrinsic efficacy of the ligand-receptor complex is a basic tenet on which pharmacology seeks to modulate physiological processes. One can therefore pose the questions: "Are there anti-oestrogenic drug molecules with similar potencies and similar biological properties to the triphenylethylene type anti-oestrogens that are metabolically and chemically stable? Do these compounds stimulate tumour growth just like tamoxifen?" Indeed, compounds that cannot isomerize, for example nafoxidine and trioxifene, can support the growth of

EnCa101 tumours in athymic mice (Gottardis *et al*, 1990). Similarly, the fixed ring version of tamoxifen, which cannot form high potency oestrogenic metabolites, will support the growth of MCF-7 TAM tumours in athymic mice (Wolf D, unpublished results from our laboratory).

Further structure-activity relationship studies will clearly help to delineate the requirements for tamoxifen stimulated growth in these laboratory models. However, our observations in the transfectants with wild type and mutant cDNAs for the ER suggest that study of the sequence of the ER in tamoxifen stimulated tumours may be valuable. We are currently exploring this research strategy.

With regard to additional information concerning mutated ERE, we are currently ignorant about the precise mechanism of oestrogen regulated growth, so it may be premature to study ERE in random genes with the hope of discovering mutants. It is possible that the key oestrogen regulated genes that orchestrate growth have not yet been discovered or that the receptor may act in concert with additional transcription factors that in turn are the key to oestrogen regulated growth. The possibilities may be infinite, but with developing techniques in molecular biology and the mapping of the human genome the isolated pieces of information we currently possess may soon converge to provide the answers to at least one form of resistance to anti-oestrogen therapy.

## SUMMARY

Our research goal is to develop possible strategies that could have therapeutic implications for the control of breast cancer. Although tamoxifen therapy is successful for some patients, it does not provide adequate benefit for the majority, who have ER negative disease. There is also both laboratory and clinical evidence to support the position that initially responsive tumours will eventually develop resistance to tamoxifen therapy. Since the ER mechanism is the key to the successful control of tumour growth with anti-oestrogens, we have taken the strategic step of determining whether the ER will reassert growth control in breast tumour cells. We have demonstrated that this is feasible, and it might now be appropriate to plan a "gene therapy" approach to cancer control that is based on reactivation of the ER or the development of a targeted vector. Since the concept of growth control in refractory breast cancer has become a reality, there can now be enthusiasm about developing a means to achieve this therapeutic goal. Indeed, the implications for cancer therapy could be enormous. There is every reason to suppose that other types of cancer cells transfected with steroid receptor genes will respond with growth suppression to the appropriate ligand.

We have further broadened our studies of anti-oestrogens to describe a laboratory model to dissect the molecular pharmacology of hormone and antihormone action. In practical terms, the model has provided an insight into tamoxifen stimulated growth as a mechanism of tamoxifen resistance. The pharmacology of a model compound, fixed ring 4-hydroxytamoxifen, was changed by a mutant ER. It is possible to envisage the clonal selection of cells with mutated ER or ERE that will thrive on the partial agonist actions of tamoxifen. These cells would, as is observed in the MCF-7 TAM and EnCa101 laboratory models, also respond to oestradiol stimulation. Future studies of mutant receptors in the laboratory and clinic might provide support for a novel mechanism of tamoxifen resistance. What is most encouraging, though, is the finding that the pharmacology of new pure anti-oestrogens is not affected by the mutation in the ER and that the pure anti-oestrogens can control tamoxifen stimulated growth. These observations provide additional support for the development of pure anti-oestrogens as a therapy for breast cancer. A clinically acceptable compound could be used as a therapy after tamoxifen failure or perhaps the compound may prevent the development of receptor mutants if it is used as a first line therapy.

## References


Berry M, Nunez A-M and Chambon P (1989) Estrogen-responsive element of the human pS2 gene is an imperfectly palindromic sequence. *Proceedings of the National Academic Sciences of the USA* **86** 1218–1222

Berry M, Metzger D and Chambon P (1990) Role of the two activating domains of the oestrogen receptor in the cell-type and promoter contact dependent agonist activity of the antioestrogen 4-hydroxytamoxifen. *EMBO Journal* **9** 2811–2818

Early Breast Cancer Trialists' Collaborative Group (1988) Effects of adjuvant tamoxifen and of cytotoxic therapy on mortality in early breast cancer. *New England Journal of Medicine* **319** 1681–1692

Felgner PL, Gadek TR, Holm M *et al* (1987) Lipofection: a highly efficient, lipid-mediated DNA-transfection procedure. *Proceedings of the National Academy of Sciences of the USA* **84** 7413–7417

Fisher B, Costantino J, Redmond C *et al* (1989) A randomized clinical trial evaluating tamoxifen in the treatment of patients with node negative breast cancer who have estrogen receptor positive tumors. *New England Journal of Medicine* **320** 479–484

Furr BJA and Jordan VC (1984) The pharmacology and clinical uses of tamoxifen. *Pharmacology and Therapeutics* **25** 127–205

Gottardis MM and Jordan VC (1988) Development of tamoxifen-stimulated growth of MCF-7 tumors in athymic mice after long-term antiestrogen administration. *Cancer Research* **48** 5183–5187

Gottardis MM, Robinson SP, Satyaswaroop PG and Jordan VC (1988) Contrasting actions of tamoxifen on endometrial and breast tumor growth in the athymic mouse. *Cancer Research* **48** 812–815

Gottardis MM, Jiang SY, Jeng MH and Jordan VC (1989a) Inhibition of tamoxifen-stimulated growth of an MCF-7 tumor variant in athymic mice by novel steroidal antiestrogens. *Cancer Research* **49** 4090–4093

Gottardis MM, Wagner RJ, Borden EC and Jordan VC (1989b) Differential ability of antiestrogens to stimulate breast cancer cell (MCF-7) growth *in vivo* and *in vitro*. *Cancer Research* **49** 4765–4769

Gottardis MM, Ricchio M, Satyaswaroop PG and Jordan VC (1990) Effect of steroidal and non-steroidal antiestrogens on the growth of a tamoxifen-stimulated human endometrial carcinoma (EnCa 101) in athymic mice. *Cancer Research* **50** 3189–3192

Harper MJK and Walpole AL (1966) Contrasting endocrine activities of *cis* and *trans* isomers in a series of substituted triphenylethylenes. *Nature* **212** 87

Jiang SY and Jordan VC Growth regulation of estrogen receptor negative breast cancer cells transfected with cDNA's for estrogen receptor. *Journal of the National Cancer Institute* (in press)

Jordan VC (1975) Prolonged antioestrogenic activity of ICI 46 474 in the ovariectomized mouse. *Journal of Reproduction and Fertility* **42** 251–258

Jordan VC (1983) Laboratory studies to develop general principles for the adjuvant treatment of breast cancer with antiestrogens: problems and potential for future clinical applications. *Breast Cancer Research and Treatment* **3** (Supplement 1) 73–86

Jordan VC (1984) The biochemical pharmacology of antiestrogen action. *Pharmacological Reviews* **36** 245–276

Jordan VC (1991) Chemosuppression of breast cancer by long-term tamoxifen therapy. *Cancer Prevention* **20** 3–14

Jordan VC and Murphy CS (1990) Endocrine pharmacology of antiestrogens as antitumor agents. *Endocrine Reviews* **11** 578–610

Jordan VC, Collins MM, Rowsby L and Prestwich G (1977) A monohydroxylated metabolite of tamoxifen with potent antioestrogenic activity. *Journal of Endocrinology* **75** 305–316

Jordan VC, Haldemann B and Allen KE (1981) Geometric isomers of substituted triphenylethylenes and antiestrogen action. *Endocrinology* **108** 1353–1361

Jordan VC, Bain RR, Brown RR, Gosden B and Santos MA (1983) Determination and pharmacology of a new hydroxylated metabolite of tamoxifen observed in patient sera during therapy for advanced breast cancer. *Cancer Research* **43** 1446–1450

Jordan VC, Lieberman ME, Cormier E, Koch R, Bagley JR and Ruenitz PC (1984) Structural requirements for the pharmacological activity of nonsteroidal antiestrogens *in vitro*. *Molecular Pharmacology* **26** 272–278

Jordan VC, Koch R, Langan SM and McCague R (1988) Ligand interactions at the estrogen receptor to program antiestrogen action: a study with non-steroidal compounds *in vitro*. *Endocrinology* **122** 1449–1454

Jordan VC, Lababidi MK and Mirecki DM (1990) The antioestrogenic and antitumour properties of prolonged tamoxifen therapy in C3H/OUJ mice. *European Journal of Cancer* **26** 718–721

Jordan VC, Lababidi MK and Langan-Fahey SM (1991) The suppression of mouse mammary tumorigenesis by long-term tamoxifen therapy. *Journal of the National Cancer Institute* **83** 492–496

Katzenellenbogen BS, Norman MJ, Eckert RL, Peltz SW and Mangel WF (1984) Bioactivities, estrogen receptor interactions and plasminogen activator activities of tamoxifen and hydroxytamoxifen isomers in MCF-7 human breast cancer cells. *Cancer Research* **44** 112–119

Kushner PG, Hart E, Shine J *et al* (1990) Construction of cell lines that express high levels of the human estrogen receptor and are killed by estrogens. *Molecular Endocrinology* **4** 1465–1473

Langan-Fahey SM, Tormey DC and Jordan VC (1990) Tamoxifen metabolites in patients on long-term adjuvant therapy for breast cancer. *European Journal of Cancer* **26** 883–888

Lieberman ME, Gorski J and Jordan VC (1983) An estrogen receptor model to describe the regulation of prolactin synthesis by antiestrogens *in vitro*. *Journal of Biological Chemistry* **258** 4741–4745

Lyman SD and Jordan VC (1985) Metabolism of tamoxifen and its uterotrophic activity. *Biochemical Pharmacology* **34** 2787–2794

Maminta MLD, Molteni A and Rosen ST (1991) Stable expression of the human estrogen receptor in HeLa cells by infection: effect of estrogen on cell proliferation and c-myc expression. *Molecular and Cellular Endocrinology* **78** 61–69

McCague R, LeClerq G and Jordan VC (1988) Non-isomerizable analogues of (Z) and (E)-4 hydroxytamoxifen: synthesis and endocrinological properties of substituted diphenylbenzocycloheptenes. *Journal of Medicinal Chemistry* **31** 1285–1290

Murphy CS, Langan-Fahey S, McCague R and Jordan VC (1990) Structure-function relationships of hydroxylated metabolites of tamoxifen that control the proliferation of estrogen responsive T47D breast cancer cells *in vitro*. *Molecular Pharmacology* **39** 421–428

Osborne CK, Coronado EB and Robinson JP (1987) Human breast cancer in the athymic nude mouse: cytostatic effects of long-term antiestrogen therapy. *European Journal of Cancer and Clinical Oncology* **23** 1189–1195

Osborne CK, Coronado E, Wiebe V and DeGregoria M (1991) Acquired tamoxifen (TAM) resistance in breast cancer correlates with reduced tumor accumulation of TAM and trans-4-hydroxyTAM (4HT). *Proceedings of the 27th annual meeting of American Society of Clinical Oncology (Houston)* [Abstract 59]

Powles TJ, Tillyer CR, Jones AL *et al* (1990) Prevention of breast cancer with tamoxifen—an update on the Royal Marsden Hospital pilot programme. *European Journal of Cancer* **26** 680–684

Robinson SP, Langan-Fahey SM, Johnson DA and Jordan VC (1991) Metabolites, pharmacodynamics and pharmacokinetics of tamoxifen in rats and mice compared to the breast cancer patient. *Drug Metabolism and Disposition* **19** 36–43

Satyaswaroop PG, Zaino RJ and Mortel R (1984) Estrogen-like effects of tamoxifen on human endometrial carcinoma transplanted into nude mice. *Cancer Research* **44** 4006

Tora L, Mullick A, Metzger D *et al* (1989) The cloned human oestrogen receptor contains a mutation which alters its hormone binding properties. *EMBO Journal* **8** 1981–1986

Touitou I, Mathieu M and Rochefort H (1990) Stable transfection of the estrogen receptor cDNA into HeLa cells induces estrogen responsiveness of endogenous cathepsin D gene but not of cell growth. *Biochemistry and Biophysics Research Communications* **169** 109–115

Wakeling AE and Bowler J (1987) Steroidal pure antioestrogens. *Journal of Endocrinology* **112** R7–R10

Wakeling AE, Dukes M and Bowler J (1991) A potent specific pure antiestrogen with clinical potential. *Cancer Research* **51** 3867–3873

Watts CKW, Parker MG and King RJB (1989) Stable transfection of the oestrogen receptor gene into a human osteosarcoma cell line. *Journal of Steroid Biochemistry* **34** 483–490

The authors are responsible for the accuracy of the references.

# Steroid Antagonists as Nuclear Receptor Blockers

**A E WAKELING**
*Bioscience I, ICI Pharmaceuticals, Alderley Park, Macclesfield, Cheshire SK10 4TG*



## INTRODUCTION

I shall try in this review to convey a flavour of the excitement offered by the convergence of knowledge about steroid antagonists from several areas of research, including clinical oncology, endocrine physiology, receptor biochemistry and molecular biology, that promises new insights into the treatment of "hormone-dependent" tumours. I shall also try to illustrate why these insights into the mode of action of steroid antagonists may indicate novel approaches to the discovery of agents for the treatment of cancer and other diseases.

The discovery of steroid hormone antagonists more than 30 years ago occurred somewhat unexpectedly in the course of testing synthetic agents in physiological assays for sex steroid like activity (reviewed in Neumann, 1987; Lerner and Jordan, 1990). The primary impetus for the pharmaceutical interest in novel sex hormone antagonists was their potential application to contraception. Although with the possible exception of progesterone antagonists (Baulieu, 1989), such hopes were not fulfilled, an alternative obvious major therapeutic use has evolved, namely in the treatment of breast and prostate cancer (Shalet, 1990). The precedent for the use of agents that ablate the action of sex hormones in these tumours was also an historical one, based on surgical removal of the gonadal source(s) of sex steroids. Beatson (1896) showed that removal of the ovaries provided effective palliation of breast cancer, and Huggins and Hodges (1941) demonstrated the effectiveness of orchiectomy in prostate cancer.

Although classical endocrine pharmacology assays, for example oestrogen induced vaginal cornification or uterine growth, facilitated the discovery of

*Cancer Surveys* Volume 14: *Growth Regulation by Nuclear Hormone Receptors*

anti-oestrogens (Lerner *et al,* 1958; Harper and Walpole, 1967), such studies also revealed a paradox in their biological effects. Thus, among the first described anti-oestrogens, MER 25, clomiphene and tamoxifen each produced differing degrees of stimulatory (partial oestrogen agonist) activity. This varied depending on the test species and organ response measured, as well as on the more conventional pharmacodynamic determinants such as dose and duration of treatment (Martin, 1980; Jordan, 1984). More recent studies have confirmed and extended this pharmacological diversity, illustrating a changing balance between stimulatory (oestrogen like) and inhibitory (anti-oestrogenic) activity at the organ, cellular and gene level (Wakeling, 1987; Jordan and Murphy, 1990). The central question posed by these observations, which have similar but less extensively investigated parallels among androgen and progesterone antagonists, is, "what is the molecular mechanism underlying such extreme pharmacological variations?" As we shall see, the answer to this question, which is beginning to emerge from the application of molecular genetic analysis, may have profound implications for future drug discovery.

In the remainder of this discussion I shall focus specifically on my area of personal interest, the anti-oestrogens, but will cite supportive data for other classes of antagonists where available.

## RECEPTOR BIOCHEMISTRY: THE LEAD TO NEW COMPOUNDS

It was not until the discovery of the oestrogen receptor in the early 1960s that the beginnings of an understanding of the biochemical mode of action of steroids (and thus of their antagonists) began to emerge (reviewed in Jensen and Jacobson, 1962). Further work led Jensen *et al* (1968) to propose a two step mechanism of oestrogen action in vivo, in which the hormone binds with high affinity and specificity to a cognate receptor in the cytoplasm and the receptor hormone complex migrates to the cell nucleus where it is retained in association with chromatin. Anti-oestrogens were shown to compete effectively with oestradiol for receptor binding in vitro and to block effectively the receptor mediated nuclear accumulation of oestradiol in vivo (Terenius, 1970; Rochefort *et al,* 1972; Jensen and DeSombre, 1973). Although the advent of receptor specific antibodies showed that the ligand free receptor for oestrogens (and other steroid hormones) is normally resident in the cell nucleus (King and Greene, 1984; Welshons *et al,* 1984), and precipitated a revised view of the original two step hypothesis, it is still useful conceptually and in a practical sense to distinguish two separate steps in the early phase of oestrogen action.

The initial high affinity binding reaction releases sufficient energy to promote a secondary transformation of the receptor complex, and it is assumed that this structural change is a necessary precursor of DNA binding and transcriptional activation. Although the molecular changes involved remain to be described, the qualitative differences between different classes of ligands in terms of receptor activation have been analysed (Franklin, 1980, 1986). A parallel was drawn between the thermodynamics of receptor activation and enzyme catalysis: in receptor activation, binding energy may be utilized in a non-productive or productive manner, and in enzyme-substrate interactions, binding energy is used both to lower the free energy of the transition state and to increase the rate of reaction. Maximum productive use of binding energy would be provided by full agonists in transforming receptors to an "excited" state capable of generating biologically relevant "signals", whereas with partial agonists, some binding energy would be used non-productively, for example in stabilizing the ligand-receptor complex. Pure antagonists would generate only non-productive interactions. Thus, anti-oestrogens such as tamoxifen would be less effective than oestradiol in promoting the formation of productive receptor complexes or, because of their lower affinity for the receptor, be incapable of full activation of the receptor. I shall reconsider these ideas later but first describe how the two step hypothesis provided a practical view of what might be required to identify a novel class of anti-oestrogens.

## PURE ANTI-OESTROGENS

### Rationale

By the end of the 1970s, it was clear that tamoxifen provides highly effective treatment of breast cancer (Litherland and Jackson, 1988). It was equally clear that a further improvement in treatment was medically attractive, because of the long period that had elapsed since the discovery of tamoxifen, but that a similarly successful replacement agent would be difficult to discover. We recognized that a novel anti-oestrogen might also be useful in treating a range of non-malignant oestrogen dependent diseases such as endometriosis and uterine fibroids, to which tamoxifen is not applied (Wakeling, 1985, 1987). We reasoned that a compound which would bind the oestrogen receptor with a high affinity without promoting either nuclear accumulation or activation of the receptor would be distinctively different from tamoxifen and should display the pharmacological characteristics of a pure antagonist in all tissues responsive to oestrogens (Wakeling, 1991). Such compounds should allow complete blockade of oestrogen action in breast cancer patients—an effect not achieved by any therapy available currently. Pure anti-oestrogens may provide more effective palliation of breast cancer, and this hypothesis is nearing clinical evaluation (Wakeling *et al*, 1991).

We were encouraged by precedents for the synthesis of non-translocating oestrogen receptor ligands. For example, 4-mercuri-17β-oestradiol did not promote nuclear accumulation of the oestradiol receptor (Muldoon, 1971) or receptor interaction with DNA (Muldoon, 1980), although the compound was, paradoxically, a potent oestrogen in vivo. Similarly, the anti-mineralocorticoid spirolactone, SC-26304, did not form nuclear receptor complexes (Marver *et al*, 1974).

## Chemistry

As a starting point for chemistry, we chose to explore the activity of analogues of oestradiol bearing an alkyl chain at the C7 position of the steroid nucleus. This choice was dictated by two considerations—firstly, that we and others had been unsuccessful in separating completely agonist and antagonist activity in non-steroidal molecules based on tamoxifen and, secondly, that high affinity for the receptor was deemed essential. Literature review revealed that 7α-oestradiol analogues used in receptor purification were optimum in retaining affinity for the receptor (Bucourt *et al*, 1978). An initial series of compounds bearing a ten carbon methylene bridge and a variety of terminal functional groups produced molecules with mixed agonist and antagonist activity but also, for the first time, agents devoid of trophic activity, which were also able to block the effects of oestrogens completely (Wakeling and Bowler, 1987; Bowler *et al*, 1989).

The pure anti-oestrogen ICI 164 384, which emerged from this work, has been studied extensively and proved to fulfil in all respects the pharmacological criteria defined for a pure antagonist (reviewed in Wakeling, 1991).

Recently, a new compound, ICI 182 780, with substantially increased intrinsic (receptor binding) activity and in vivo potency has been described (Wakeling *et al*, 1991). In this compound, the amide function is replaced by a sulphoxide, and the terminal chain is fluorinated to reduce metabolic degradation. Others have also described new non-steroidal pure anti-oestrogens (von Angerer *et al*, 1990).

## Physiology

The key feature that distinguishes pure anti-oestrogens from partial agonists is the complete absence of oestrogenic activity. Perhaps the clearest demonstration of this is provided by assays demonstrating that compounds such as ICI 164 384 can block the trophic actions of partial agonists such as tamoxifen. This was demonstrated originally in the rat uterus, where ICI 164 384 abolished tamoxifen induced weight gain in a dose dependent manner (Wakeling and Bowler, 1987) and has subsequently been confirmed in studies with breast cancer cells in vitro (Wakeling *et al*, 1989) and the rat mammary gland in vivo (Nicholson *et al*, 1988). Such studies provide compelling evidence that the physiological effects of both pure and partial agonist anti-oestrogens are mediated through the oestrogen receptor.

In the context of tumour therapy, model studies with human breast tumours in nude mice have shown that the growth of tumours that escape the inhibitory effect of tamoxifen after long term treatment (Gottardis and Jordan, 1988) can also be inhibited with ICI 164 384 (Gottardis *et al*, 1989). This finding suggests that the tumours which are initially inhibited by tamoxifen but which subsequently resume growth are not resistant to anti-oestrogen treatment in the classical sense but have adapted to a dependence on the oestrogenic component of tamoxifen's action(s). These experiments appear to model the progression of breast cancer in patients, since the majority of tumours that respond initially to tamoxifen eventually resume growth despite continuous therapy. A substantial proportion of such patients may be expected to show a further response to treatment with a pure anti-oestrogen (Howell *et al,* 1990). Since a similar mechanism for the development of resistance is unlikely to apply to the pure anti-oestrogens, Howell *et al* (1990) have argued that initial treatment with a pure anti-oestrogen should lead to a more complete and longer lasting tumour response. This hypothesis remains to be tested clinically.

Another interesting feature to emerge from in vivo studies with pure anti-oestrogens is their selectivity of action. For example, at doses that cause an oophorectomy like regression of the uterus, the hypothalamic-pituitary-ovarian axis continues to function normally (Wakeling and Bowler, 1988; Wakeling *et al,* 1991). The potential clinical consequences of such selectivity are profound, because it may prove possible to correct uterine dysfunction without other endocrine sequelae associated with blockade of receptors in the brain. The blood-brain barrier may play an important part in this apparent selectivity, but there is also evidence that different peripheral oestrogen responsive organs are also differentially sensitive to pure anti-oestrogens. For example, in a study of ICI 164 384 in intact rats, the uterus regressed similarly in both anti-oestrogen treated and oophorectomized animals, but the loss of bone observed in the latter group was not seen in the anti-oestrogen treated rats (Dukes M and Wakeling AE, unpublished). Thus, the target specific differential activity of pure antagonists parallels previously described differential effects in organ responses to full and partial agonists (Martin, 1980; Kelner *et al,* 1982; Curtis and Korach, 1991). Any proposed mechanism of hormone action must recognize and account for these physiological differences.

## MODE OF ACTION OF OESTROGEN RECEPTOR LIGANDS

The key to explaining steroid action lies in defining the mechanisms that are interposed between the initial ligand recognition event and the activation of the transcription of hormone responsive genes. Identification of the genes encoding each of the steroid receptors, together with the description of specific sequences of DNA termed hormone response elements (HREs) that bind the hormone receptor(s) and control the efficiency of transcription (Beato, 1989; Tsai *et al,* 1991), has allowed the application of molecular genetic techniques to this problem. Original studies, which defined three functionally important domains in the oestrogen receptor (Kumar *et al,* 1986), have proved to reflect a general consensus (Green and Chambon, 1988). Each receptor has a centrally located DNA binding domain, which shows a high degree of sequence homology across receptor classes, a less well conserved carboxyterminal ligand binding domain and an aminoterminal domain widely variant in size and sequence between receptors. In terms of the efficiency of transcriptional activation induced by ligand-receptor binding to HREs, specific sequences termed transcriptional activation functions (TAFs) are thought to be important (Webster *et al,* 1988; Lees *et al,* 1989; Tora *et al,* 1989). Transcriptional activation function 1 (TAF-1), located in the aminoterminus, can function independently of the ligand binding domain, and hence of TAF-2 or ligand. The activity of TAF-1 is variable in different cells, and its efficiency in activating transcription also varies between promoters. Transcriptional activation function 2 (TAF-2) is thought to be active only for ligand-receptor complexes containing the natural hormone (Green and Chambon, 1988).

The role of TAFs in transmitting ligand-receptor "signals" following binding of different anti-oestrogens has been analysed (Berry *et al,* 1990). In tamoxifen-receptor complexes, TAF-2 was disabled but TAF-1 retained functional activity. Thus, one hypothesis to explain the remarkable variations of biological responses arising from tamoxifen-receptor complexes attributes the diversity to the continued activity of TAF-1. Variations in the strength of the oestrogen like signals at the organ and cellular levels thus attributed beg the question of what controls the efficiency of TAF-1. It is assumed that this requires the concurrent presence of other cell specific transcription factors, the identity of which remains to be elucidated. Similarly, differences in the efficiency with which specific oestrogen responsive genes respond to tamoxifen in individual cells presumably reflect the strength of interaction between TAF-1 and particular promoters (Green and Chambon, 1988). The effects of ICI 164 384-receptor complexes on transcriptional activation are the subject of controversy. Berry *et al* (1990) showed that, in contrast to tamoxifen, the ICI 164 384-receptor complexes failed to promote transcription. The failure of ICI 164 384 to inhibit the activity of TAF-1 implies that the receptor complexed with ICI 164 384 did not bind to DNA (Berry *et al,* 1990). These studies are in agreement with previous work showing that ICI 164 384 did not promote receptor binding to DNA (Fawell *et al,* 1990b; Wilson *et al,* 1990). If this "poisoning" of the receptor's ability to recognize DNA proves correct, it would conform to the mechanistic objective in our original definition of pure anti-oestrogens. Other studies, however, do indicate that ICI 164 384 can promote receptor binding to DNA (Lees *et al,* 1989; Martinez and Wahli, 1989, Pham *et al,* 1991; Sabbah *et al,* 1991) and are supported by observations that in some experimental conditions, very weak agonist effects of ICI 164 384 can be detected (Weaver *et al,* 1988, 1989; Pasqualini *et al,* 1990; Jamil *et al,* 1991).

A necessary prerequisite for DNA recognition is the ligand induced dissociation of a 90 kDa heat shock protein (hsp90) (Denis *et al,* 1988), which is a component common to the native ligand free steroid receptors (Joab *et al,* 1984). Dissociation of hsp90 is thought to allow receptors to dimerize (Miller *et al,* 1985). The homodimeric receptor complex binds tightly to DNA (Kumar and Chambon, 1988). Further analysis of the oestrogen receptor has shown that domains controlling hsp90 binding and receptor dimerization overlap in the hormone binding region of the receptor (Chambraud *et al,* 1990; Fawell et 1990a). It was suggested that hormone antagonists fail to promote dissociation of hsp90 and would therefore prevent dimerization and DNA recognition (Baulieu, 1989). Recent studies by Fawell *et al* (1990b) with ICI 164 384 support this hypothesis. Dimerization of the oestrogen receptor was severely impaired by the pure antagonist. Other investigators have reached opposite conclusions (Pham *et al,* 1991; Sabbah *et al,* 1991). These apparently conflicting conclusions may be reconciled by assuming that the dimerization function is not absolutely required for DNA binding but merely enhances HRE binding affinity (Pham *et al,* 1991). Whatever the molecular mechanism, there is no dispute that the net result of pure anti-oestrogen binding is a receptor complex with severely attenuated transcriptional activity.

Although there remains some uncertainty about whether there is an absolute distinction between pure and partial agonist anti-oestrogens in terms of their biochemical mode of action, it is clear that the former compounds are much less efficient than the latter in promoting the changes in receptor conformation necessary for DNA binding and, subsequently, efficient activation of transcription. Binding energy is assumed, therefore, to be dissipated non-productively by antagonist binding: how might this occur?

Since the affinity of pure anti-oestrogens for the receptor is comparable to that of oestradiol and substantially greater than that of tamoxifen (Wakeling and Bowler, 1987), how is the high binding energy dissipated in a non-productive manner? The first clue to this puzzle has emerged from studies of the fate of the receptor in cells after treatment with ICI 164 384. ICI 164 384 dramatically reduced receptor concentration of the mouse uterus to 5% of control levels 4 hours after treatment (Gibson *et al,* 1991). Since cycloheximide did not affect this response and no change in receptor messenger RNA was detected, it appears that the receptor-pure anti-oestrogen complex becomes more "fragile" and perhaps more susceptible to the normal processes involved in receptor degradation. Whether the binding energy released is actively involved in enhancing receptor degradation remains to be seen, but the half life of the ICI 164 384-receptor complex appears to be substantially less than the half lives of the oestradiol-receptor and tamoxifen-receptor complexes (Eckert *et al,* 1984). In the case of the tamoxifen-receptor complex, the lower energy available from the binding reaction may simply reduce the efficiency of activation and be insufficient to promote more rapid receptor degradation.

## OTHER NUCLEAR RECEPTOR ANTAGONISTS

Other than anti-oestrogens, the ligands studied most intensively at the molecular level are the glucocorticoid and progesterone antagonists. Of particular interest is the compound RU 486, which has a high affinity for both classes of receptor and is an antagonist of both glucocorticoid and progesterone action (Sakiz, 1987). As in the case of anti-oestrogens, there has been disagreement about whether RU 486 can induce receptor dimerization and

DNA recognition (Groyer *et al,* 1987; Guiochon-Mantel *et al,* 1988; Baulieu, 1989; Chao *et al,* 1991). Recent functional studies of progesterone receptor in whole cells have revealed close parallels between the actions of RU 486 and tamoxifen. The RU 486-receptor complex recognizes progesterone response elements and blocks the transcriptional activation function of TAF-2 but not that of TAF-1, permitting the cell and promoter specific expression of partial agonist activity (Meyer *et al,* 1990).

Another series of progesterone antagonists, structurally related to RU 486 (ie 11β-aryl analogues) were synthesized in a successful attempt to reduce affinity for the glucocorticoid receptor (Henderson, 1987). Analysis of the capacity of these compounds to activate transcription suggests that a further differential exists among compounds with very similar structures. The two progesterone antagonists ZK 98734 and ZK 98299, which differ principally in the configuration of the C/D steroid ring junction (Henderson, 1987), are markedly different in their capacity to promote DNA binding (Klein-Hitpass *et al,* 1991). ZK 98299, which has the unnatural *cis* C/D structure, did not provoke DNA binding and inhibited induction of progesterone receptor-DNA binding by progesterone or ZK 98734 and RU 486. It was suggested that progesterone antagonists can be classified as either type I (molecules such as ZK 98299 that interfere with the formation of stable receptor dimers) or type II (compounds such as RU 486 that do not). The apparent parallel with "type I" (pure antagonist) and "type II" (partial agonist) anti-oestrogens needs further confirmation in terms of the "signalling" capacity of the different progesterone antagonist-progesterone receptor complexes. However, Klein-Hitpass *et al* (1991) have shown for the first time that an in vitro transcription assay, using cell free nuclear extracts of appropriate indicator cells, can readily distinguish type I and type II progesterone antagonists. Such assays could provide a basis for rapid screening of novel chemicals for pro- or anti-hormonal activity with reasonable confidence that such data would reflect in vivo activity.

Molecular mode of action studies with anti-androgens are less well advanced, largely because the cloning of the receptor proved difficult, being first reported in 1988 (Chang *et al,* 1988; Lubahn *et al,* 1988), and there are few androgen responsive cell lines to provide suitable in vitro indicators of hormone action. Recent studies with the human prostate carcinoma derived, androgen responsive LNCaP cell line have shown that, with one exception, all anti-androgens actually stimulate cell growth (Veldscholte *et al,* 1992). This result implies that the anti-androgens can form transcription competent complexes with the androgen receptor. The exception to this general conclusion was the compound ICI 176 334 (casodex) (Freeman *et al,* 1989), which failed to cause dissociation of the androgen receptor-heat shock protein complex, induce DNA binding or stimulate transcription. Casodex blocked the growth stimulatory action of androgens and of other anti-androgens, suggesting that androgen antagonists can also be of two different types. However, Veldscholte *et al* (1992) also showed that the apparent transcriptional activation function of antagonists other than Casodex could be attributed to the presence in LNCaP receptors of a previously described single point mutation in the ligand binding domain (residue 868, Thr to Ala). The fact that a single point mutation can produce such a dramatic shift in signalling activity has a number of implications. For example, could the outgrowth of tamoxifen resistant tumours in animals and humans be due to such a point mutation? The fact that such tumours in nude mice are susceptible to treatment with ICI 164 384 (Gottardis *et al,* 1989) suggests that it would be worth while analysing receptors in those tumours for similar mutations. Similarly, could such point mutations account for the apparent agonist activity of ICI 164 384 in the Ishikawa human endometrial carcinoma cell line (Jamil *et al,* 1991)?

In summary, it can be concluded that the most effective hormone antagonists are likely to be found among those compounds that display a type I profile of activity on receptor dimerization/activation. The availability of the receptor genes, the capacity to engineer functional mutations in such genes and the availability of in vitro transcription systems that faithfully mimic in vivo hormone effects (Allan *et al,* 1991; Klein-Hitpass, 1991) offer practical means to search for such compounds.

## THE STEROID RECEPTOR SUPERFAMILY

The isolation of DNA sequences complementary to the steroid receptors revealed the existence of a large family of proteins that bind ligands with structures widely divergent from those of the classical adrenal and sex steroids. They include receptors for thyroid hormones, retinoids and "orphan" receptors with no known ligand (Evans, 1988; Wang *et al,* 1989; O'Malley, 1991). Much remains to be learned about the functional significance of these receptors, but, on the basis of the steroid receptor antagonist precedent and existing knowledge of the importance of retinoid and thyroid ligands in growth, development, differentiation and metabolism, it seems reasonable to assume that many new targets for therapeutic intervention in disease processes will emerge in the future.

Many incidental spin-offs from this rapidly developing field can already be glimpsed, for example the recent description of a steroid receptor like protein that is transcriptionally activated by agents that induce peroxisome proliferation and are also liver carcinogens (Issemann and Green, 1991), and the activation of an "orphan" receptor, COUP-TF, by dopamine (Power *et al,* 1991). The latter observation is of particular interest because it provides a clear example of a mechanism by which "cross talk" between apparently unrelated physiological signals may occur (Power *et al,* 1991). This represents an important extension of previous work describing mechanisms for "cross talk" between different members of the steroid receptor family (Meyer *et al,* 1989) and between the steroid receptors and other nuclear transcription factors including *jun* and *fos* (Diamond *et al,* 1990; Gaub *et al,* 1990; Jonat *et al,* 1990; Beato, 1991).

In view of the intense interest in the role of growth factors in tumour cell proliferation (Cross and Dexter, 1991), the potential for "cross talk" between growth factor and steroid signalling pathways merits closer examination. From a practical standpoint, it has already been shown that anti-oestrogens can attenuate the stimulatory effects of growth factors on human breast cancer cells (Vignon *et al,* 1987; Koga *et al,* 1989; Wakeling *et al,* 1989). That these effects occur only in cells expressing oestrogen receptors suggests that some interaction between this receptor and the growth factor stimulated receptor tyrosine kinase pathway (Ullrich and Schlessinger, 1990) influences cell growth. Since growth factors stimulate the transcription of *fos/jun* (Cantley *et al,* 1991), "cross talk" could reflect attenuation of productive interactions between the two classes of transcription factor. Biochemically, the common link between growth factor and steroid signalling pathways might be at the level of control of phosphorylation of the cognate receptors (Migliaccio *et al,* 1989; Denner *et al,* 1990; Freiss *et al,* 1990). The combination of inhibitors designed to attenuate more than one of the interacting growth signalling pathways simultaneously might represent a prototype for novel approaches to cancer therapy.

## SUMMARY

Molecular analysis of sex steroid receptors, together with the availability of a range of ligands for these receptors that affect differentially the physiological actions of the ligand-receptor complex in the cell nucleus, is providing new insights into the processes that control transcriptional activation leading to cell growth. These studies serve to focus attention on the importance of other cell specific factors that are important in determining the transcriptional efficiency of the hormone-receptor complex. Elucidation of the nature and role of such factors may lead ultimately to hormonal or anti-hormonal agents targeted to modulate selected hormone mediated events in specific cells or organs.

The sex steroid receptors are prototypes of a much larger family of proteins that control transcription, including proteins that bind to ligands known to be important physiologically, such as thyroid hormones and retinoids, as well as "orphan" receptors for which no ligands have been described. The established value of steroid antagonists in cancer therapy suggests that further developments in the understanding of new members of the steroid receptor superfamily may lead to the development of novel anti-cancer agents. The developing understanding of "cross talk" between steroid-receptor and growth factor-receptor mediated tumour growth stimulatory pathways may also provide previously unrecognized opportunities for tumour therapy.

## References

Allan GF, Tsai SY, O'Malley BW and Tsai M-J (1991) Steroid hormone receptors and in vitro transcription. *BioEssays* **13** 73–78

Baulieu E-E (1989) Contragestation and other clinical applications of RU 486, an anti-progesterone at the receptor. *Science* **245** 1352–1357

Beato M (1989) Gene regulation by steroid hormones. *Cell* **56** 335–344
Beato M (1991) Transcriptional control by nuclear receptors. *FASEB Journal* **5** 2044–2051
Beatson GT (1896) On the treatment of inoperable cases of carcinoma mamma: suggestions for a new method of treatment with illustrative cases. *Lancet* **ii** 104–107
Berry M, Metzger, D and Chambon P (1990) Role of two trans activating domains in the cell-type and promotor-context dependent agonistic activity of the anti-oestrogen 4-hydroxytamoxifen. *EMBO Journal* **9** 2811–2818
Bowler J, Lilley TJ, Pittam JD and Wakeling AE (1989) Novel steroidal pure antiestrogens. *Steroids* **54** 71–99
Bucourt R, Vignau M, Torelli V *et al* (1978) New biospecific adsorbents for the purification of estradiol receptor. *Journal of Biological Chemistry* **253** 8221–8228
Cantley LC, Augur KR, Carpenter C *et al* (1991) Oncogenes and signal transduction. *Cell* **64** 281–302
Chambraud B, Berry M, Redeuilh G, Chambon P and Baulieu E-E (1990) Several regions of human estrogen receptor are involved in the formation of receptor-heat shock protein 90 complexes. *Journal of Biological Chmistry* **265** 20686–20691
Chang C, Kokontis J and Liao S (1988) Molecular cloning of the human and rat cDNA encoding androgen receptors. *Science* **240** 324–326
Chao C-C, Thomas TJ, Ebede P, Gallo MA and Thomas T (1991) Ionic and ligand-specific effects on the DNA binding of progesterone receptor bound to the synthetic progestin R5020 and the antiprogestin RU 486. *Cancer Research* **51** 3938–3945
Cross M and Dexter TM (1991) Growth factors in development, transformation and tumorigenesis. *Cell* **64** 271–280
Curtis SW and Korach KS (1991) Uterine estrogen receptor-DNA complexes: effects of different ERE sequences, ligands, and receptor forms. *Molecular Endocrinology* **5** 959–966
Denis M, Poellinger L, Wikstrom A-C and Gustafsson J-A (1988) Requirement of hormone for thermal conversion of the glucocorticoid receptor to a DNA-binding state. *Nature* **333** 686–688
Denner LA, Weigel NL, Maxwell BL, Schrader WT and O'Malley BW (1990) Regulation of progesterone-mediated transcription by phosphorylation. *Science* **250** 1740–1743
Diamond MI, Miner JN, Yoshinaga SK and Yamamoto KR (1990) Transcription factor interactions: selectors of positive or negative regulation from a single DNA element. *Science* **249** 1266–1272
Eckert RL, Mullick A, Rorke EA and Katzenellenbogen BS (1984) Estrogen receptor synthesis and turnover in MCF-7 breast cancer cells measured by a density shift technique. *Endocrinology* **114** 629–637
Evans RM (1988) The steroid and thyroid hormone receptor superfamily. *Science* **240** 889–895
Fawell SE, Lees JA, White R and Parker MG (1990a) Characterization and colocalization of steroid binding and dimerization activities in the mouse estrogen receptor. *Cell* **60** 953–962
Fawell SE, White R, Hoare S, Sydenham M, Page M and Parker MG (1990b) Inhibition of estrogen receptor-DNA binding by the pure antiestrogen ICI 164 384 appears to be mediated by impaired receptor dimerization. *Proceedings of the National Academy of Sciences of the USA* **87** 6883–6887
Franklin TJ (1980) Binding energy and the activation of hormone receptors. *Biochemical Pharmacology* **29** 853–856
Franklin TJ (1986) Binding energy and the stimulation of hormone receptors, In: Van Binst G (ed). *Design and Synthesis of Organic Molecules Based on Molecular Recognition,* pp 103–116, Springer-Verlag, Berlin
Freeman SN, Mainwaring WIP and Furr BJA (1989) A possible explanation for the peripheral selectivity of a novel non-steroidal pure antiandrogen, Casodex (ICI 176 334). *British Journal of Cancer* **60** 664–668
Freiss G, Rochefort H and Vignon F (1990) Mechanisms of 4-hydroxytamoxifen anti-growth factor activity in breast cancer cells: alterations of growth factor receptor binding sites and

tyrosine kinase activity. *Biochemical and Biophysical Research Communications* **173** 919–926

Gaub M-P, Bellard M, Scheuer I, Chambon P and Sassone-Corsi P (1990) Activation of the ovalbumin gene by the estrogen receptor involves the fos-jun complex. *Cell* **63** 1267–1276

Gibson MK, Nemmers LA, Beckman WC Jr, Davis VL, Curtis SW and Korach KS (1991) The mechanism of ICI 164 384 antiestrogenicity involves rapid loss of estrogen receptor in uterine tissue. *Endocrinology* **129** 2000–2010

Gottardis MM and Jordan VC (1988) Development of tamoxifen-stimulated growth of MCF-7 tumors in athymic mice after long-term antiestrogen administration. *Cancer Research* **48** 5183–5187

Gottardis MM, Jiang S-Y, Jeng M-H and Jordan VC (1989) Inhibition of tamoxifen-stimulated growth of an MCF-7 variant in athymic mice by novel steroidal antiestrogens. *Cancer Research* **49** 4090–4093

Green S and Chambon P (1988) Nuclear receptors enhance our understanding of transcription regulation. *Trends in Genetics* **4** 309–314

Groyer A, Schweizer-Goyer G, Cadepond F, Mariller M and Baulieu E-E (1987) Antiglucocorticoid effects suggest why steroid hormone is required for receptors to bind DNA in vivo but not in vitro. *Nature* **328** 624–626

Guiochon-Mantel A, Loosfelt H, Ragot T *et al* (1988) Receptors bound to antiprogestin form abortive complexes with hormone responsive elements. *Nature* **336** 695–698

Harper MJK and Walpole A (1967) A new derivative of triphenylethylene: effect on implantation and mode of action in rats. *Journal of Reproduction and Fertility* **13** 101–119

Henderson D (1987) Antiprogestational and antiglucocorticoid activities of some novel 11β-aryl substituted steroids, In: Furr BJA and Wakeling AE (eds). *Pharmacology and Clinical Uses of Inhibitors of Hormone Secretion and Action*, pp 184–211, Bailliere Tindall, London

Howell A, Dodwell DJ, Laidlaw I, Anderson H and Anderson E (1990) Tamoxifen as an agonist for metastatic breast cancer, In: Goldhirsch A (ed). *Endocrine Therapy of Breast Cancer IV*, pp 49–58, Springer-Verlag, Berlin

Huggins C and Hodges CF (1941) Studies on prostatic cancer: I The effect of castration, of oestrogen and androgen injection on serum phosphatases in metastastic carcinoma of the prostate. *Cancer Research* **1** 292–297

Issemann I and Green S (1991) Activation of a member of the steroid receptor superfamily by peroxisome proliferators. *Nature* **347** 645–650

Jamil A, Croxtall JD and White JO (1991) The effect of antioestrogens on cell growth and progesterone receptor concentration in human endometrial cancer cells (Ishikawa). *Journal of Molecular Endocrinology* **6** 215–221

Jensen EV and Jacobson HI (1962) Basic guides to the mechanisms of estrogen action. *Recent Progress in Hormone Research* **18** 387–414

Jensen EV and DeSombre ER (1973) Estrogen-receptor interaction. *Science* **182** 126–131

Jensen EV, Suzuki T, Kawashima T, Stumpf WE, Jungblut PW and DeSombre ER (1968) A two-step mechanism for the interaction of estradiol with rat uterus. *Proceedings of the National Academy of Sciences of the USA* **59** 632–638

Joab I, Radanyi C, Renoir M *et al* (1984) Common non-hormone binding component in non-transformed chick oviduct receptors of four steroid hormones. *Nature* **308** 850–853

Jonat C, Rahmsdorf HJ, Park K-K *et al* (1990) Antitumor and antiinflammation: down-modulation of AP-1 (fos/jun) activity by glucocorticoid hormone. *Cell* **62** 1189–1204

Jordan VC (1984) Biochemical pharmacology of antiestrogen action. *Pharmacological Reviews* **36** 245–276

Jordan VC and Murphy CS (1990) Endocrine pharmacology of antiestrogens as antitumor agents. *Endocrine Reviews* **11** 578–610

Kelner KL, Kirchick HJ and Peck EJ Jr (1982) Differential sensitivity of estrogen target tissues: the role of the receptor. *Endocrinology* **111** 1986–1995

King WJ and Greene GL (1984) Monoclonal antibodies localize oestrogen receptor in the nuclei of target cells. *Nature* **307** 745–747

Klein-Hitpass L, Cato ACB, Henderson D and Ryffel G (1991) Two types of antiprogestins identified by their differential action in transcriptionally active extracts from T47D cells. *Nucleic Acids Research* **19** 1227–1234

Koga M, Musgrove EA and Sutherland RL (1989) Modulation of the growth-inhibitory effects of progestins and the antiestrogen hydroxyclomiphene on human breast cancer cells by epidermal growth factor and insulin. *Cancer Research* **49** 112–116

Kumar V and Chambon P (1988) The estrogen receptor binds tightly to its responsive element as a ligand-induced dimer. *Cell* **55** 145–156

Kumar V, Green S, Staub A and Chambon P (1986) Localisation of the oestradiol-binding and putative DNA-binding domains of the human oestrogen receptor. *EMBO Journal* **5** 2231–2236

Lees JA, Fawell SE and Parker MG (1989) Identification of two transactivation domains in the mouse oestrogen receptor. *Nucleic Acids Research* **17** 5477–5488

Lerner LJ and Jordan VC (1990) Development of antiestrogens and their use in breast cancer: eighth Cain memorial award lecture. *Cancer Research* **50** 4177–4189

Lerner LJ, Holthaus FJ Jr and Thompson CR (1958) A non-steroidal estrogen antagonist 1-(p-2-diethylaminoethoxyphenyl)-1-phenyl-2-p-methoxyphenyl ethanol. *Endocrinology* **63** 295–318

Litherland S and Jackson IM (1988) Antioestrogens in the management of hormone-dependent cancer. *Cancer Treatment Reviews* **15** 183–194

Lubahn DB, Joseph DR, Sullivan PM, Willard HF, French FS and Wilson EM (1988) Cloning of the human androgen receptor cDNA and localization to the X-chromosome. *Science* **240** 327–330

Martin L (1980) Estrogens, anti-estrogen and the regulation of cell proliferation in the female reproductive tract in vivo, In: McLachlan JA(ed). *Estrogens in the Environment*, pp 103–130, Elsevier North Holland Inc, New York

Martinez E and Wahli W (1989) Cooperative binding of estrogen receptor to imperfect estrogen-responsive DNA elements correlates with their synergistic hormone-depenedent enhancer activity. *EMBO Journal* **8** 3781–3791

Marver D, Stewart J, Funder JW, Feldman D and Edelman IS (1974) Renal aldosterone receptors: studies with [$^3$H]aldosterone and the anti-mineralocorticoid [$^3$H]spirolactone (SC-26304). *Proceedings of the National Academy of Sciences of the USA* **71** 1431–1435

Meyer M-E, Gronemeyer H, Turcotte B, Bocquel M-T, Tasset D and Chambon P (1989) Steroid hormone receptors compete for factors that mediate their enhancer function. *Cell* **57** 433–442

Meyer M-E, Pornon A, Ji J, Bocquel M-T, Chambon P and Gonemeyer H (1990) Agonistic and antagonistic activities of RU486 on the functions of the human progesterone receptor. *EMBO Journal* **9** 3923–3932

Migliaccio A, Di Domenico M, Green S *et al* (1989) Phosphorylation on tyrosine of in vitro synthesized human estrogen receptor activates its hormone binding. *Molecular Endocrinology* **3** 1061–1069

Miller MA, Mullick A, Greene GL and Katzenellenbogen BS (1985) Characterization of the subunit nature of nuclear estrogen receptors by the chemical cross-linking and dense amino acid labeling. *Endocrinology* **117** 515–522

Muldoon TG (1971) Characterization of steroid-binding sites by affinity labeling: further studies of the interaction between 4-mercuri-17β-estradiol and specific estrogen-binding proteins. *Biochemistry* **10** 3780–3784

Muldoon TG (1980) Molecular and functional anomalies in the mechanism of the estrogenic action of 4-mercuri-17β-estradiol. *Journal of Biological Chemistry* **255** 1358–1366

Neumann F (1987) Pharmacology and clinical uses of cyproterone acetate, In: Furr BJA and Wakeling AE (eds). *Pharmacology and Clinical Uses of Inhibitors of Hormone Secretion and Action*, pp 132–159, Bailliere Tindall, London

Nicholson RI, Gotting KE, Gee J and Walker KJ (1988) Actions of oestrogens and antioestrogens on rat mammary gland development: relevance to breast cancer prevention. *Journal of Steroid Biochemistry* **30** 95–103

O'Malley BW (1991) Steroid hormones as transactivators of gene expression. *Breast Cancer Research and Treatment* **18** 67–71

Pasqualini JR, Giambiagi N, Gelly C and Chetrite G (1990) Antiestrogen action in mammary cancer and in fetal cells. *Journal of Steroid Biochemistry* **37** 343–348

Pham TA, Elliston JF, Nawaz Z, McDonnell DP, Tsai M-J and O'Malley BW (1991) Antiestrogen can establish nonproductive receptor complexes and alter chromatin structure at target enhancers. *Proceedings of the National Academy of Sciences of the USA* **88** 3125–3129

Power RF, Lydon JP, Conneely OM and O'Malley BW (1991) Dopamine activation of an orphan of the steroid receptor superfamily. *Science* **252** 1546–1548

Rochefort H, Lignon F and Capony F (1972) Formation of estrogen nuclear receptor in uterus: effect of androgens, estrone and nafoxidine. *Biochemical and Biophysical Research Communications* **47** 662–670

Sabbah M, Gouilleux F, Sola B, Redeuilh G and Baulieu E-E (1991) Structural differences between the hormone and antihormone estrogen receptor complexes bound to the hormone response element. *Proceedings of the National Academy of Sciences of the USA* **88** 390–394

Sakiz E (1987) Mifepristone (RU486): a new antiprogestin, In: Furr BJA and Wakeling AE (eds). *Pharmacology and Clinical Uses of Inhibitors of Hormone Secretion and Action*, pp 170–183, Bailliere Tindall, London

Shalet SM (ed) (1990) *Endocrine Aspects of Malignancy* Bailliere Tindall, London

Terenius L (1970) Two modes of interaction between oestrogen and antioestrogen. *Acta Endocrinologica* **64** 47–58

Tora L, White J, Brou C *et al* (1989) The human estrogen receptor has two independent nonacidic transcriptional activation functions. *Cell* **59** 477–487

Tsai SY, Tsai M-Y and O'Malley BW (1991) The steroid receptor superfamily: transactivators of gene expression, In: Parker MG (ed). *Nuclear Hormone Receptors*, pp 103–124, Academic Press, London

Ullrich A and Schlessinger J (1990) Signal transduction by receptors with tyrosine kinase activity. *Cell* **61** 203–212

Veldscholte J, Berrevoets CA, Brinkmann AO, Grootegoed JA and Mulder E (1992) Antiandrogens and the mutated androgen receptor of LNCaP cells: differential effects on binding affinity, heat shock protein interaction and transcription activation. *Biochemistry* **31** 2393–2399

Vignon F, Bouton M-M and Rochefort H (1987) Antiestrogens inhibit the mitogenic effect of growth factors on breast cancer cells in the total absence of estrogens. *Biochemical and Biophysical Research Communications* **140** 1502–1508

von Angerer E, Knebel N, Kager M and Ganss B (1990) 1-(aminoalkyl)-2-phenylindoles as novel pure estrogen antagonists. *Journal of Medicinal Chemistry* **33** 2635–2640

Wakeling AE (1985) Anti-oestrogens in oncology: past, present and prospects, In: Pannuti F (ed). *Current Clinical Practice Series* No. 31, pp 43–53, Excerpta Medica, Amsterdam

Wakeling AE (1987) Pharmacology of antioestrogens, In: Furr BJA and Wakeling AE (eds). *Pharmacology and Clinical Uses of Inhibitors of Hormone Secretion and Action,* pp 1–19, Bailliere Tindall, London

Wakeling AE (1991) Steroidal pure antioestrogens, In: Lippman ME and Dickson RB (eds). *Regulatory Mechanisms in Breast Cancer*, pp 239–257, Kluwer, Boston

Wakeling AE and Bowler J (1987) Steroidal pure antioestrogens. *Journal of Endocrinology* **112** R7–R10

Wakeling AE and Bowler J (1988) Biology and mode of action of pure antioestrogens. *Journal of Steroid Biochemistry* **30** 141–147

Wakeling AE, Newboult E and Peters SW (1989) Effects of antioestrogens on the proliferation of MCF-7 human breast cancer cells. *Journal of Molecular Endocrinology* **2** 225–234

Wakeling AE, Dukes M and Bowler J (1991) A potent specific pure antiestrogen with clinical potential. *Cancer Research* **51** 3867–3873

Wang L-H, Tsai SY, Cook RG, Beattie WG, Tsai M-J and O'Malley BW (1989) COUP transcription factor is a member of the steroid receptor superfamily. *Nature* **340** 163–166

Weaver CA, Springer PA and Katzenellenbogen BS (1988) Regulation of pS2 gene expression by affinity labeling and reversibly binding estrogens and antiestrogens: comparison of effects on the native gene and on pS2-chloramphenicol acetyltransferase fusion genes transfected into MCF-7 human breast cancer cells. *Journal of Molecular Endocrinology* **2** 936–945

Webster NJC, Green S, Jin JR and Chambon P (1988) The hormone-binding domains of the estrogen and glucocorticoid receptors contain an inducible transcription activation function. *Cell* **54** 199–207

Welshons WV, Lieberman ME and Gorski J (1984) Nuclear localization of unoccupied oestrogen receptors. *Nature* **307** 747–749

Wilson APM, Weatherill PJ, Nicholson RI, Davies P and Wakeling AE (1990) A comparative study of the interaction of oestradiol and the steroidal pure antioestrogen, ICI 164 384, with the molybdate-stabilized oestrogen receptor. *Journal of Steroid Biochemistry* **35** 421–428

The author is responsible for the accuracy of the references.

# Perspectives on Oestrogen Receptor Function: Past and Future

**GERALD C MUELLER**
*McArdle Laboratory, University of Wisconsin, 1400 University Avenue, Madison, Wisconsin 53706*



## INTRODUCTION: A LITTLE EARLY HISTORY

In 1950, as a newly appointed Assistant Professor of Oncology at the McArdle Laboratory, I surveyed the fields of cancer research and endocrinology for possible entries into an understanding of the role of hormones in cancer induction and cancer growth. It was already abundantly apparent that the oestrogens were the most interesting, since there had been many reports of their involvement in mammary, uterine, pituitary and lymphoid cancers. At that time, however, there were practically no insights into the mechanism by which oestrogens functioned—or for that matter knowledge of the basic mechanisms that operate in carcinogenesis. I was impressed, however, that the speed of action and the high biological activity of oestrogens made this class of hormones prime candidates for investigation. In addition, the advent of $^{14}C$ isotope technology made possible many metabolic inquiries that had defied earlier investigators.

Accordingly, I launched a series of studies to look for oestrogen effects on basic metabolic pathways—with experiments that bridged from responses in living animals to responses that were elicited in surviving uteri. With the able assistance of early colleagues, including Jack Gorski, Hiroshi Ui, Barbara Vonderhaar, Erich Hecker and Wolfram Zillig, it was shown that oestradiol, not a metabolite, produced the early hormonal response via the activation of gene expression mechanisms (ie processes leading to the induced synthesis of RNA and protein). These early studies were summarized first in 1958 (Mueller *et al*, 1958) and expanded upon in 1972 (Mueller *et al*, 1972). In a number of ways, these articles anticipated significantly the directions in which research on the mechanisms of steroid action was going to develop. Specific references to individual contributions by other investigators during these first 20 years are presented in these publications.


 0-87969-371-1/92. $3.00 + .00

An early fascination for me was the vanishingly small level of oestradiol that was required to produce easily measurable responses (ie changes that occurred within an hour). A steroid concentrating mechanism in target tissues was looked for in the early 1950s, but we only had $^{14}C$ labelled oestradiol to work with, and our resolution was thus limited to cells exhibiting concentrations of steroid above 30 000 molecules per cell. As it turns out, this was just at the limits of receptor associated oestradiol in an organ such as the uterus. Instead of finding the receptor system that was expected, we were forced to publish on the formation of protein-bound metabolites of oestradiol in non-target tissues (Riegel and Mueller, 1954). It remained for Elwood Jensen and Herbert Jacobson, who were able to synthesize $^{3}H$ labelled oestradiol of high specific activity to provide evidence for the presence of the expected oestrogen receptor mechanism (Jensen and Jacobson, 1960). Their findings launched the study of steroid receptors.

In passing, I recall that Jack Gorski called to ask me what I thought of Jensen's new findings. After acknowledging their discovery as a major achievement, I remember telling Jack, "It's Elwood's finding—let's allow him to tie it down to a definite protein before we jump in. I am sure, however, that we will be able to make good use of these new insights". Fortunately for the field, Jack did not heed my comment but proceeded instead to make many meritorious findings in this area. Over 30 years later, new insights into receptor function are still emerging.

## SOME LESS WELL KNOWN PROPERTIES OF OESTROGEN RECEPTORS

One of the most impressive properties of oestrogen receptors as they reside in cytosols or in partially purified states is their high sensitivity to temperature, local anaesthetics, arachidonic acid plus potassium chloride, heparin, A/B ring analogues of oestradiol and phosphorylation state (Mueller *et al,* 1987). In addition, fractionation efforts have revealed the remarkable propensity of oestrogen receptors to associate with other proteins. In fact, oestrogen receptors that have been purified with affinity chromatography (Van Oosbree *et al,* 1984) and thereby separated from many of their associating proteins have a very low affinity for oestradiol until recombined with certain cytosolic proteins or with specific histones (Bhattacharyya *et al,* 1987). In the case of the histones, the formation of sedimentable aggregates of the receptor with the histones can be correlated with the recovery of oestradiol binding activity. Two proteins, of 43 kDa and 70 kDa, isolated from rabbit uterine cytosol restore binding when added in only a 50-fold excess over the number of receptor molecules (Olsen MR and Mueller GC, unpublished).

A very interesting feature of the reactivation response is the fact that specific peptide fragments from these proteins can effect the reactivation. In the case of histone H2B, a major reactivation was obtained with the carboxyterminal half of the H2B molecule derived by means of cyanogen bromide

(CNBr) cleavage. The aminoterminal half was H2B was less than half as effective (Bhattacharyya *et al,* 1987).

This finding is accepted as evidence that a rather specific topographical interaction between the reactivating peptide and the oestrogen receptor is involved, ie only 50 times as many reactivating molecules as receptor implies a significant heterocomplimentarity between the two structures. Because the reactivating peptide is not likely to be functioning enzymatically, it follows that the observed reactivation of the oestrogen receptor may reflect refolding of the oestrogen receptor in a manner that is very probably guided by the interaction event.

A second parameter of ligand binding that points again to the oestrogen receptor molecule as being highly responsive to its environment is the finding (Ahrens H, Gorski J and Mueller GC, unpublished) that phosphatase treatment of the receptor eliminates affinity for oestradiol. In this case, rephosphorylation by protein kinase A re-establishes binding, whereas rephosphorylation by protein kinase C fails to do so. Protein kinase A addresses two sites close to the hinge region of the molecule that are distinct from a protein kinase C phosphorylation site near by. Experiments are in progress to assess the role of phosphate groups at the specific sites in receptor function, but it seems likely that the introduction of a phosphate group in the structure mediates specific directive effects on the folding of the receptor protein and thus on receptor function. For a general review of receptor phosphorylation, see Auricchio (1989).

These observations have prompted a view of oestrogen receptors as "molecular transducers for mediating the interaction of specific macromolecules in the regulation of genetic expression or the catalytic function of multi-enzyme systems" (Kim *et al,* 1982). This view of oestrogen receptors as highly flexible and adaptive molecules has been clearly supported by recent molecular genetic characterization studies (Green *et al,* 1986; Kumar *et al,* 1987). Five distinct interactive domains in the receptor have been demonstrated amid an aminoacid sequence that is highly conserved in nature. These domains have been shown to interact both sequentially and selectively with the steroid or analogue and with other components of the cytosol or nuclear chromatin. Each encounter appears to impart a specific conformational change in the receptor, which is relayed in turn and reciprocally to other members of the assembly. Site specific effects appear to be the order of the day, with the end result depending on the particular components and their post-translationally modified states.

## A CURRENT VIEW OF HOW OESTROGEN RECEPTORS FUNCTION

As insights into the mechanisms that operate in the regulated transcription of specific genes have emerged, it has become clear that the expression (or suppression) of a gene requires the co-operative interactions of a number of protein factors in addition to the RNA polymerase that does the transcribing. The molecular genetic dissection of oestrogen and other steroid receptors has clearly defined the domains of the receptor molecule that interact preferentially with a recognition sequence in a target gene, the activating ligand, the dimerization site of a like receptor and a number of gene specific proteins (ie enhancing elements). Accordingly, it was reasonable to postulate that a ligand dependent expression of a specific gene also involved distinct physical association of the different components at a regulatory site in order to be effective. Using the 5′ regulatory region from both the rat prolactin and human PS2 genes as targets, our laboratory has now demonstrated the expected phenomenon using a photo-crosslinking technology. In this approach, which is a modification of an earlier technology (Blanco *et al*, 1987) that was used in the study of chromosomal DNA replication, DNA containing $^{32}P$ deoxycytidylic acid and bromodeoxyuridine (BrdU) as a substitute for thymidine is combined with nuclear and cytosolic proteins in the presence and absence of a ligand for the oestrogen receptor. Irradiation with 313 nm light promotes a crosslinking of the BrdU sensitized DNA with adjacent proteins, resulting in a covalent attachment of the $^{32}P$ label of DNA to specific proteins—proteins that are clustered within the crosslinking range (ie a few angstroms). In practice, the DNA that has been crosslinked to adjacent proteins is degraded by nucleases to leave short radioactive tags (ie approximately 3–5 nucleotides long), which on the autoradiographs of the proteins separated by SDS gel electrophoresis give discrete bands corresponding to the molecular size of the protein plus the DNA tag (ie about a 2000 unit increase in apparent molecular weight). When this process is carried out in the presence of diluting levels of unlabelled DNA as a dilutor for regions of little interest, the marking and potential identification of proteins assembled around specific gene segments is revealed (Schuh, 1990; Schuh TJ and Mueller GC, unpublished).

These studies have clearly shown that the assembly of a distinct set of proteins around the 5′ regulatory regions of both PS2 and prolactin genes is guided in an oestrogen receptor and ligand dependent process. Using specific antibodies, the 68 kDa oestrogen receptor has been shown to be one of the components of the assembled proteins. When the anti-oestrogen ICI 164 384 is used as the ligand, the influence of the receptor is negated (Schuh TJ and Mueller GC, unpublished). In addition, hydroxytamoxifen yields some differences in proteins crosslinked to the regulatory DNA sequence compared with the representation assembled under the influence of oestradiol.

An interesting finding from these experiments is that the antibodies against the oestrogen receptor, immobilized on beads, retrieve the same group of proteins—even after the complexes have been treated with nucleases to digest any co-ordinating excess of DNA. Accordingly, we have been led to the conclusion that the oestrogen receptor operates in a ligand dependent manner to guide the assembly of a cluster of specific proteins around the regulatory sites for these genes and, very notably, all target genes. Since each assembly contains the oestrogen receptor itself, we have termed these assemblies "receptorsomes". Although the other components of such assemblies have not yet been absolutely identified, preliminary immunological testing indicates the presence of c-*fos* and c-*jun* among the crosslinkable components.

The significance of the formation of these assemblies in the staging of an oestrogen response is supported by the observations that the formation of such assemblies around regulatory regions in oestrogen responsive CAT gene constructs correlates well with their responsiveness to oestrogenic ligands.

In addition, it has been found that the expression of such constructs is cell specific. For example, PS2 based constructs show an oestradiol dependent expression in MCF-7 cells, whereas the prolactin based CAT constructs function in pituitary cells but exhibit an oestradiol mediated suppression. Thus, oestrogen receptors can have both a positive and negative role in the expression of specific genes (depending on the spectrum of helping protein units that are available in the target cell) and participate in the formation of the "receptorsome" structures (Schuh TJ and Mueller GC, unpublished).

A summary of oestrogen receptors in action is presented in Fig. 1 as determined by studies of receptors and the applications of the photo-crosslinking technology.

## FUTURE POTENTIALS OF RECEPTOR MEDIATED BIOLOGY

The message of this chapter is that oestrogen receptors, and probably other members of the rapidly expanding family of steroid receptors (Wahli and Martinez, 1991) and fatty acid responsive receptors as well function as highly adaptive and flexible mediators of the assembly of multiprotein complexes around the regulatory regions of receptor-ligand responsive genes. As indicated, their effects may be either positive or negative with respect to the ensuing transcription events. The formation of active biological assemblies appears to reflect in particular the availability of the receptor and the participating proteins. In all cases, the physical state of each entity appears to be very important. Such factors as the substitution of antagonist for an agonist play an important part in the character and function of such assemblies. In addition, the different target genes respond individually, probably reflecting the role of the base sequence specificity of the interactions.

Whether the influences of such receptor mediated assemblies are confined to transcriptional events in the nuclear chromatin scene or can also exert effects elsewhere in the cell (eg cell membranes, ribosomal processes) remains to be assessed. In principle, it seems reasonable to expect oestrogen receptor effects at other subcellular sites as well.

This view of receptor operations raises a number of new potentials in the field of receptor biology. Such predictions emanate from the properties of receptors themselves and the properties of their interactive components. The most basic and guiding feature is the ability of the ligand activated receptors to interact topologically with limited regions of other macromolecules (chiefly

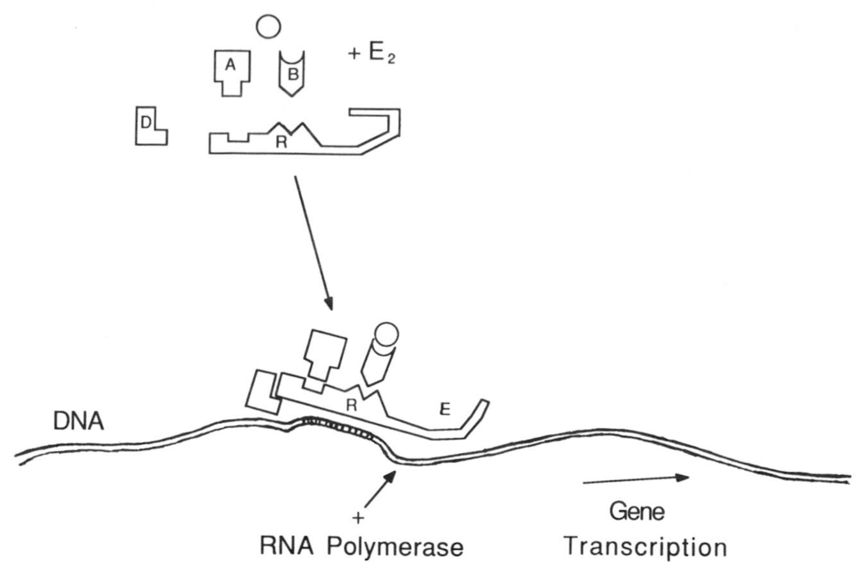


**Fig. 1.** Receptors in action. Receptors (R) are visualized as changing shape and interacting with oestradiol ($E_2$), dimerizing in most instances, exposing sites for the assembly of helper proteins (ie transcription factors such as c-*fos*, c-*jun*, A, B, D, O) and addressing the oestrogen response element of an oestrogen responsive gene. A multimolecular assembly is facilitated at or around such regulatory sites, which then act positively or negatively in the function of RNA polymerase complexes in the transcription of the downstream gene. Such assemblies are termed receptorsomes, since the components include the receptor, interact intimately through small domains and exhibit a significant stability amid the chromatin structuralist

proteins and restricted DNA segments) both to mediate and to respond to the conformational forces involved. In the best studied examples of protein-protein interactions (eg proteolytic enzyme inhibitors and antibodies), only 15–30 aminoacids are involved in addressing a target site. The fact that such interactions involve rather small sequences of aminoacids and nucleotides opens the road to intervention through manipulation of the presence and concentration of peptides or oligonucleotides that might compete in such receptor target interactions to change the usual response.

A current research effort in our laboratory is directed at exploring the relevance of this hypothesis using the transfection of plasmids to inform for the overproduction of peptide segments corresponding to such sites of interaction. These studies use information for specific segments of the receptor structure and for specific regions of the proteins that we have recently shown to have the ability to interact selectively with the oestrogen receptor.

Looking to the future, I predict that the use of molecular genetic or pharmacological techniques to produce receptor interactive components (or mimetic segments) and/or the manipulation of their physiological states (eg phosphorylated, dephosphorylated or other post-translationally modified state) will one day enable receptor mediated hormonal responses to be guided selectively to specific cells and genes. At that point, the potential for such cell and gene specific regulation could have great relevance to our need to control specific hormonal responses in such settings as breast, uterine and pituitary cancer or in such phenomena as hormonal imprinting in the control of important behavioural processes. The rate of progress and success of these approaches will depend greatly on the expansion of our basic knowledge of how the receptor interacts with the other macromolecular components of the response scene. We need to know, in particular, the topological features of such interaction sites at the level of Å dimensions. Modern nuclear magnetic resonance, X-ray crystallographic and electron microscopic techniques will provide the data to define such regions. Then the clever use of molecular genetic technology and organic chemical routes to synthesis should provide the new molecules or peptide mimetic principles that will address such interaction sites and open the door to the pharmacology of the future.

In sum, I think that the directions are clear and that exciting discoveries are on the horizon. I only regret that after 40 years of close involvement in this challenging arena of cell biology, my own research time is running out. But to young scientists I say, "Wonderful opportunities are ahead—enjoy, enjoy".

## References

Auricchio F (1989) General review: phosphorylation of steroid receptors. *Journal of Steroid Biochemistry* **32** 613–622

Bhattacharyya K, Olsen MR and Mueller GC (1987) Reactivation of affinity-purified estrogen receptors by peptides derived from histone H2B. *Biochemical and Biophysical Research Communications* **149** 823–829

Blanco J, Kimura H and Mueller GC (1987) A method for detecting protein-DNA interactions at sites of chromatin replication. *Analytical Biochemistry* **163** 537–545

Green S, Walter P, Kumar V *et al* (1986) Human estrogen receptor cDNA: sequence, expression, and homology to v-erb-A. *Nature* **320** 134–139

Jensen EV and Jacobson (1960) Fate of steroid estrogens in target tissues, In: Pincus G and Vollmer EP (eds). *Biological Activities of Steroids in Relation to Cancer,* pp 161–178, Academic Press, New York

Kim UH, Van Oosbree TR and Mueller GC (1982) Influence of tetracaine on the structure and function of estrogen receptors. *Endocrinology* **111** 260–268

Kumar V, Green S, Stack G, Berry M, Jin JR and Chambon P (1987) Functional domaines of the human estrogen receptor. *Cell* **51** 941–951

Mueller GC, Herranen AM and Jervell KF (1958) Studies on the mechanism of action of estrogens, In: Pincus G (ed). *Recent Progress in Hormone Research,* vol 14, Academic Press, New York

Mueller GC, Vonderhaar B, Kim UH and LeMahieu M (1972) Estrogen action: inroad to cell biology, In: Pincus G (ed). *Recent Progress in Hormone Research,* vol 28, Academic Press, New York

Mueller GC, Olsen MR, Bhattacharyya K and Schuh TJ (1987) Physical and functional parameters of isolated estrogen receptors, In: Roy AK and Clark JH (eds). *Gene Regulation by Steroid Hormones,* III, Springer-Verlag, New York

Riegel IL and Mueller GC (1954) Formation of a protein-bound metabolite of estradiol-16-C14 by rat liver homogenates. *Journal of Biological Chemistry* **210** 249–257

Schuh TJ (1990), PhD Dissertation, University of Wisconsin-Madison

Van Oosbree TR, Kim UH and Mueller GC (1984) Affinity chromatography of estrogen receptors on diethylstilbestrol-agarose. *Analytical Biochemistry* **136** 321–327
Wahli W and Martinez E (1991) Superfamily of steroid nuclear receptors: positive and negative regulators of gene expression *FASEB Journal* **5** 2243–2249

The author is responsible for the accuracy of the references.

# Androgen Receptor Mutants That Affect Normal Growth and Development

**A O BRINKMANN • J TRAPMAN**

*Department of Endocrinology & Reproduction and Department of Pathology, Erasmus University Rotterdam, PO Box 1738, 3000 DR Rotterdam, The Netherlands*



## INTRODUCTION

Sexual differentiation and development in the male proceed under the direct control of androgens (eg testosterone and 5α-dihydrotestosterone). Differentiation of the Wolffian duct is under testosterone control, whereas differentiation of urogenital sinus and tubercle are regulated by 5α-dihydrotestosterone (Griffin and Wilson, 1989). The actions of both testosterone and 5α-dihydrotestosterone are mediated by the same intracellular androgen receptor protein, which belongs to a superfamily of ligand dependent transcription factors (Evans, 1988; O'Malley, 1990).

The androgen receptor complex transduces the extracellular androgenic signal via binding to specific enhancer sequences referred to as androgen responsive elements present in the 5′ flanking region of androgen target genes and is supposed to control transcription via protein-DNA interactions and by protein-protein interactions with other transcription factors (Beato, 1989; Wahli and Martinez, 1991). This could result in formation of a stable preinitiation complex near the transcription start site of the target gene, which allows efficient transcription initiation by RNA polymerase II. The androgen

*Cancer Surveys* Volume 14: *Growth Regulation by Nuclear Hormone Receptors*

receptor might achieve this in the presence of androgens by stimulating the assembly of the preinitiation complex or by stabilizing the complex. This interaction can be direct or indirect, involving co-activators that mediate synergism between different transcription factors. In addition, different ligands could, as is the case for the androgen receptor, affect differently the structural constraint of the complex and therefore possibly permit interaction with different transcription factors. Tissue specific differential regulation of target genes by the same androgen receptor protein but different ligands (eg testosterone and 5α-dihydrotestosterone) could be explained in this way. This might be particularly relevant during male sexual differentiation, the process in which both testosterone and 5α-dihydrotestosterone have their own specific functions (Griffin and Wilson, 1989). An attractive alternative explanation is the tissue specific expression of a repertoire of synergistic transcription factors, which interact in a specific way with the androgen receptor irrespective of the nature of the complexed ligand.

One of the organs that develops from the urogenital sinus under the influence of 5α-dihydrotestosterone is the prostate. 5α-Dihydrotestosterone is also involved in later stages of androgen regulated growth of the prostate and regulates the expression of several prostate specific proteins (Cunha *et al*, 1987). Moreover, the growth of the majority (about 80%) of prostate tumours is initially also androgen dependent. However, after a short period of regression during endocrine therapy, almost all prostate tumours progress in an androgen independent fashion (Andriole and Catalona, 1991). The molecular cause for the failure of endocrine therapy in a later stage of prostate cancer is not known.

The cloning and characterization of complementary DNA (cDNA) encoding the human androgen receptor and the elucidation of the organization of the human androgen receptor gene have greatly increased our knowledge of the functional structure and expression of the human androgen receptor (Chang *et al*, 1988; Lubahn *et al*, 1988; Trapman *et al*, 1988; Faber *et al*, 1989; Tilley *et al*, 1989). In human prostate tissue, two androgen receptor messenger RNAs (mRNAs) have been demonstrated with different sizes (11 kb and 8.5 kb, respectively). The two mRNAs are generated by alternative splicing in the 3′ untranslated region of the mRNA (Faber *et al*, 1991). The cDNA sequence of the human androgen receptor reveals an open reading frame of 2730 nucleotides encoding a protein of 910 aminoacid residues with a calculated molecular mass of 98.5 kDa. The gene is located on the X chromosome in the Xq11-12 region and has a size of >90 kb (Brinkmann *et al*, 1989; Brown *et al*, 1989; Kuiper *et al*, 1989). The information for the protein coding region is separated over eight exons (see Fig. 1). The sequence encoding the aminoterminal domain is present in one large exon (exon 1) (Faber *et al*, 1989), whereas the DNA binding domain is encoded by exons 2 and 3. The information for the steroid binding domain is distributed over five exons (exons 4–8) ( Kuiper *et al*, 1989; Lubahn *et al*, 1989; Marcelli *et al*, 1990a).

It is generally accepted that in the X chromosome linked androgen in-

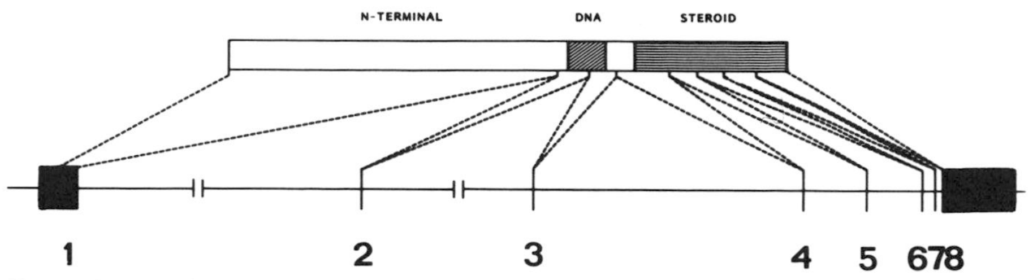


**Fig. 1.** Structural organization of the human androgen receptor gene. The exons (numbered 1–8) corresponding to the different functional domains (eg aminoterminal, DNA binding and steroid binding) are indicated

sensitivity syndrome, defects in the androgen receptor gene have prevented the normal development of both internal and external male structures in 46,XY individuals (Pinsky and Kaufman, 1987; Griffin and Wilson, 1989). The information on the molecular structure of the human androgen receptor has facilitated the study of molecular defects associated with androgen insensitivity. Naturally occurring mutations in the androgen receptor gene are an interesting source for the investigation of receptor structure-function relationships. In addition, the variation in clinical syndromes provides the opportunity to correlate a defect in the androgen receptor structure with impairment of a specific physiological function of the androgen receptor.

In this overview, we will focus on the functional domain structure of the wild type human androgen receptor and on the molecular structure of the androgen receptor from subjects with the androgen insensitivity syndrome and from a human prostate cancer cell line (LNCaP). Throughout the text, the numbering of the various codons is based on a total number of 910 aminoacid residues in the human androgen receptor. This number differs from the aminoacid content published by others: 917 residues (Tilley *et al,* 1989); 918 residues (Chang *et al,* 1988) and 919 residues (Lubahn *et al,* 1988). These differences are caused by the variation in length of the polyglutamine and polyglycine stretches in the aminoterminal domain of the receptor reported by the different research groups.

## PARTIAL AND COMPLETE DELETIONS OF THE ANDROGEN RECEPTOR GENE

Only three cases with partial or complete deletion of the androgen receptor gene have been reported, indicating the low frequency of this type of androgen receptor defect (Brown *et al,* 1988; French *et al,* 1990). All individuals in the reported cases are completely resistant to androgen. In one person, a complete deletion of the gene was found, and two other individuals had deletions of exons 5–8 and exons 3–8, respectively (French *et al,* 1990). There has been no report on possible mutations in the androgen receptor promoter region or in the 5′ and 3′ untranslated regions of the gene. In the X linked androgen insensitivity syndrome, defects in these regions cannot be excluded and might directly affect androgen receptor mRNA levels either by defective or altered transcription or by a modification of the stability of the mRNA.

## MUTATIONS IN THE AMINOTERMINAL DOMAIN

The aminoterminal domain of the human androgen receptor is characterized by the presence of several homopolymeric aminoacid stretches, for example three polyglutamine stretches with different lengths, a long polyglycine stretch, a polyproline stretch of eight residues and a polyalanine stretch of five residues (Faber *et al,* 1989). The exact function of these aminoacid repeats is still unknown, although glutamine stretches, designated opa-repeats, have been found in proteins that are involved in developmental control and/or regulation of gene expression (Wharton *et al,* 1985; Duboule *et al,* 1987). Glutamine rich regions and a polyglutamine repeat have been found in the transcription factors Sp1 and TFIID, respectively (Mitchell and Tjian, 1989; Kao *et al,* 1990). Two separate regions in the aminoterminal domain encoded by aminoacid residues 51–211 and 244–360, respectively, and containing all three glutamine repeats were identified as being essential for transcription activation in model systems (Jenster *et al,* 1991). The length of the longest polyglutamine repeat varies in different normal individuals from 17 to 26 residues, with a high frequency of 20–22 residues (see also Fig. 2) (La Spada *et al,* 1991; McPhaul *et al,* 1991). This type of polymorphism offers the possibility of linkage studies in families with supposed defective androgen receptors.

In a family with androgen resistance associated with profound hypospadias, the length of the polyglutamine repeat has been reported to be extremely shortened (only 12 glutamine residues [Fig. 2]) (McPhaul *et al,* 1991). The deletion of 8–10 glutamine residues as such is not the cause for the androgen resistance but amplifies dramatically the thermolability of the androgen receptor in combination with a point mutation found in exon 5 (Tyr→Cys; codon 754; ligand binding domain) (Table 1). This point mutation in exon 5 causes rapid ligand dissociation but no thermolability. These recently published data indicate that two mutations in different functional domains affect each other and suggest an interaction of the two separated regions in the androgen receptor. Another interesting aspect is that the defect becomes critical in only part of the androgen target tissues because androgen resistance is only partial in this family (McPhaul *et al,* 1991).

Another unique example of tissue specific effects of a defective androgen receptor is the Kennedy syndrome. This rare X linked motor neuron disease becomes manifest in men between ages 30 and 40 and is characterized by a progressive spinal and bulbar muscular atrophy associated with signs of androgen insensitivity and infertility (Harding *et al,* 1982; Arbizu *et al,* 1983; Warner *et al,* 1991). The length of the polyglutamine stretch in the androgen receptor in all investigated Kennedy syndrome patients ranges between 40 and 52 (Fig. 2) (La Spada *et al,* 1991). Whether a doubling of the length of the polyglutamine stretch directly affects motor neuron function is still unclear.

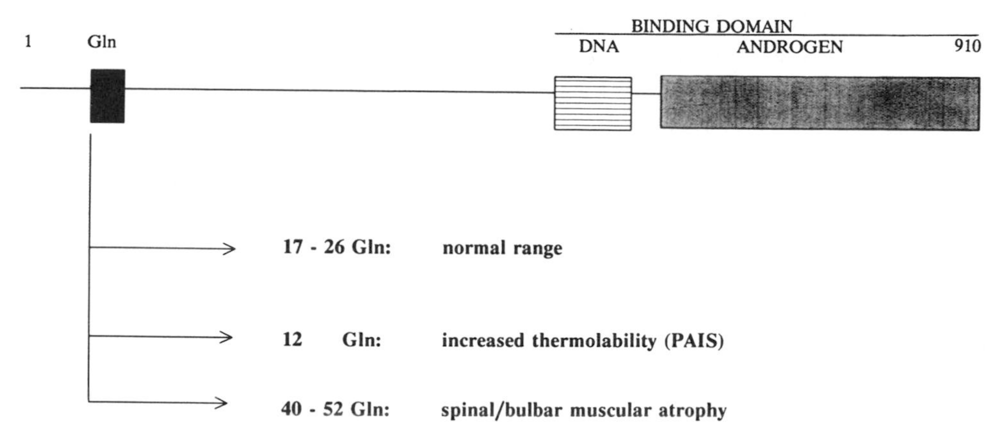


**Fig. 2.** Position of the glutamine (Gln) repeat in the aminoterminal domain of the androgen receptor, which is susceptible to variation in normal individuals, in persons with partial androgen insensitivity and in individuals with spinal/bulbar muscular atrophy respectively

Possibly a change in the structural constraint of the aminoterminal domain could result in a modified interaction with motor neuron specific transcription factors and consequently in an altered assembly of the transcription initiation complex. Other explanations, however, cannot be excluded.

**TABLE 1. Locations of different aminoacid substitutions in the androgen binding domain due to point mutations in exons 4-8 in the androgen receptor gene of individuals with the complete (CAIS) or partial (PAIS) androgen insensitivity syndrome**

| | | Aminoacid | | Nucleotide | | |
|---|---|---|---|---|---|---|
| | Position | wild type | mutant | change | Phenotype | Reference[a] |
| Exon 4 | 686 | Asp | Asn | (G→A) | CAIS | 1 |
| | 686 | Asp | His | (G→C) | CAIS | 1 |
| | 709 | Trp | Stop | (G→A) | CAIS | 2 |
| Exon 5 | 750 | Ser | Phe | (C→T) | CAIS | 3 |
| | 754 | Tyr | Cys | (A→G) | PAIS | 4 |
| Exon 6 | 765 | Arg | Cys | (C→T) | CAIS | 5,6 |
| | 787 | Trp | Stop | (G→A) | CAIS | 7 |
| Exon 7 | 822 | Arg | Stop | (C→T) | CAIS | 3 |
| | 822 | Arg | Gln | (G→A) | CAIS | 5 |
| | 831 | Arg | His | (G→A) | PAIS | 3 |
| | 846 | Arg | Cys | (C→T) | CAIS | 3 |
| | 846 | Arg | His | (G→A) | PAIS | 8 |
| | 855 | Asp | Gly | (A→G) | CAIS | 3 |
| | 857 | Val | Met | (G→A) | CAIS | 5 |
| Exon 8 | 874 | Lys | Stop | (A→T) | CAIS | 9 |

[a]1, Ris-Stalpers *et al* (1991); 2, Sai *et al* (1990); 3, De Bellis *et al* (1991); 4, McPhaul *et al* (1991); 5, Brown *et al* (1990); 6, Marcelli *et al* (1991b); 7, Marcelli *et al* (1990b); 8, Charest *et al* (1991); 9, Trifiro *et al* (1991)

Complete androgen resistance in the mouse has been known for more than two decades (Lyon and Hawkes, 1970), but only very recently has the molecular defect in the androgen receptor been established. Sequence analysis of genomic DNA from these so-called testicular feminized (Tfm/Y) mice has revealed a single base deletion in exon 1 (Charest *et al,* 1991; He *et al,* 1991). The consequence of this deletion is a frameshift, resulting in a premature termination codon and predicting the synthesis of an androgen receptor protein that completely lacks the DNA and ligand binding domains. A translation initiation sequence downstream from the stop codon could function as a potential new starting point for a shorter receptor molecule that exhibits high affinity binding to androgens and DNA. The latter could explain the observation of low levels of a smaller androgen and DNA binding protein in Tfm/Y mice (Young *et al,* 1989). The aminoterminal truncated androgen receptor, however, is unable to activate transcription, which is the direct explanation for the observed complete androgen resistance. With mouse specific RNA probes, it was assessed that androgen receptor mRNA levels in liver and kidney in the Tfm/Y mouse were 10–20-fold lower than in the corresponding tissues in wild type mice, suggesting that the *Tfm/Y* mutation results not only in the production of an aberrant protein but also in an unstable androgen receptor mRNA (Gaspar *et al,* 1990).

## MUTATIONS IN THE DNA BINDING DOMAIN

The DNA binding domain of the human androgen receptor revealed a large homology (about 80%) with those of the human glucocorticoid and progesterone receptors (Trapman *et al,* 1988). It is characterized by a high content of basic aminoacid residues and by nine conserved cysteine residues. Detailed structural information has been published on the crystal structure of the DNA binding domain of the glucocorticoid receptor complexed with DNA (Luisi *et al,* 1991). This structural information might be representative also for the other members of the steroid/thyroid hormone receptor family, including the human androgen receptor. Briefly, the DNA binding domain has a compact, globular structure in which two substructures can be distinguished. Both substructures contain centrally one zinc atom, which interacts via coordination bonds with four cysteine residues. The two zinc coordination centres are both carboxyterminally flanked by an α helix. The two zinc clusters are structurally and functionally different and are encoded by two different exons (exon 2 and exon 3). The α helix of the most aminoterminal located zinc cluster (exon 2) interacts directly with nucleotides of the hormone responsive element in the major groove of the DNA. It is known already from previous studies that three aminoacid residues at the aminoterminus of this α helix are responsible for the specific recognition of the DNA sequence of the responsive element (Umesono and Evans, 1989). These three aminoacid residues, Gly (568), Ser (569) and Val (572) (Fig. 3), are identical in the androgen, progesterone and

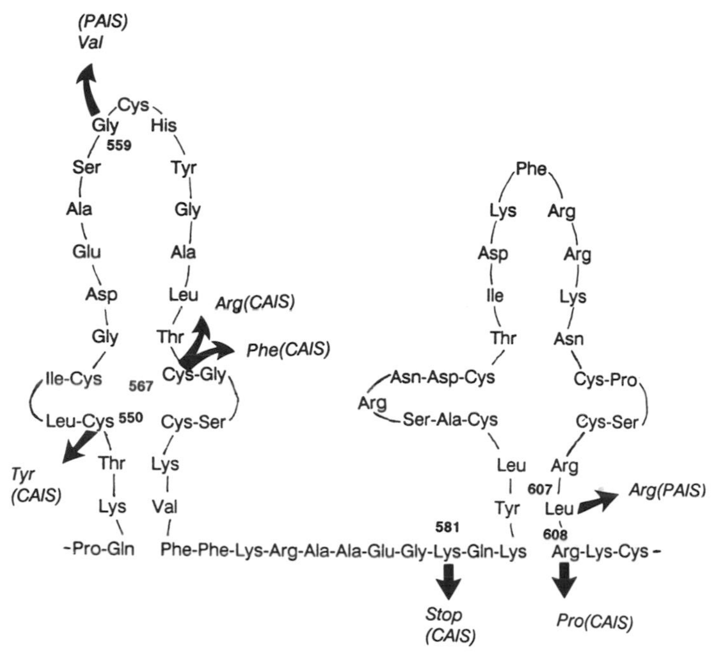


**Fig. 3.** Locations of the different aminoacid substitutions in the DNA-binding domain in exons 2 and 3 due to point mutations in the androgen receptor gene of individuals with the complete (CAIS) or partial (PAIS) androgen insensitivity syndrome

glucocorticoid receptor but are different in the oestradiol receptor. It is not surprising therefore that the androgen, progesterone and glucocorticoid receptors recognize the same responsive element. For the hormone and tissue specific responses of the different receptors, additional determinants are needed. DNA sequences flanking the hormone responsive element, receptor interactions with other proteins and the concentration of the receptor might all be important in this respect.

## Mutations in Exon 2

Recently, two different point mutations have been reported in the same cysteine codon (567) in exon 2 of the androgen receptor gene of two unrelated individuals with the complete androgen insensitivity syndrome (Fig. 3) (Chang *et al,* 1991; Zoppi *et al,* 1991). This residue is one of the four conserved cysteines involved in coordination bonding with a zinc atom. Both receptor proteins display normal ligand binding characteristics but have impaired biological activity. Another point mutation in this exon in codon 550 (Fig. 3), which changes a cysteine residue into a tyrosine residue, also results in complete androgen resistance (Zoppi *et al,* 1991). A fourth mutation in exon 2 (substitution of a glycine residue by a valine residue) is reported at codon 559 in an individual with partial androgen resistance (Fig. 3) (Chang *et al,* 1991). The consequences of these mutations in vivo are obvious and clinically well defined, but the molecular and structural interplay of a complete or partial defective androgen receptor with other proteins and DNA has still to be unravelled. Crystallographic analysis of the DNA binding domain mutated at the sites of interest and complexed with an androgen responsive element, as well as in vitro transcription studies, might be very informative in this respect.

### Mutations in Exon 3

The carboxyterminal zinc cluster (exon 3) is involved in the interaction with a second receptor molecule at the DNA (Härd *et al,* 1990; Luisi *et al,* 1991). Deletion of exon 3 results in the human in complete androgen insensitivity, as has been recently demonstrated in two 46,XY children with a female phenotype (French *et al,* 1990). The in frame, complete deletion of exon 3 still allows the synthesis of an androgen receptor protein with an intact ligand binding domain. In genital skin fibroblasts of both affected siblings, androgen receptor levels were twice normal. In the mutant receptor, the SV40 large T antigen like nuclear import signal (-Arg-Lys-Leu-Lys-Lys-) (Kalderon *et al,* 1984; Jenster *et al,* 1991), located at positions 620–624 in exon 4, is present. Nuclear localization of the mutated androgen receptor was observed after radioactive nuclear ligand binding and immunocytochemistry of cultured genital skin fibroblasts (French *et al,* 1990). This naturally occurring mutation in the androgen receptor gene indicates that exon 3 is less important for nuclear import, but indispensable for proper transcription activation. A premature termination due to a point mutation at codon 581 (Lys) in exon 3 of the androgen receptor protein has been reported in a 46,XY individual with complete androgen resistance (Fig. 3) (Marcelli *et al,* 1990a). The receptor protein is truncated in the carboxyterminal region and lacks the ligand binding domain and almost all of the second zinc cluster structure. The predicted protein is non-functional, as was established after transfection studies with a reporter gene construct (Marcelli *et al,* 1990a). The mutation also leads to a reduction in the amount of androgen receptor mRNA in genital skin fibroblasts of the affected person, suggesting decreased stability of the mRNA.

In two children with partial androgen resistance and ambiguous genitalia, a substitution of a leucine residue (codon 607) with an arginine residue has been reported (De Bellis *et al,* 1991). The consequence of this mutation is three consecutive arginine residues near the carboxyterminal of the second zinc cluster (Fig. 3). The clustering of three basic aminoacid residues might be deleterious for the structure of the predicted $\alpha$ helix in this region (Luisi *et al,* 1991) and might induce instability in higher order structures as well.

Finally, in a person with receptor positive (normal ligand binding) complete androgen insensitivity, a point mutation has been reported in codon 608, converting an arginine residue into a proline residue (Fig. 3) (Marcelli *et al,* 1991a). The position of this arginine is at the carboxyterminus of exon 3 and can be considered part of the bipartite nuclear targeting signal that is conserved in all steroid hormone receptors (Robbins *et al,* 1991). Whether this arginine residue is indispensable for nuclear targeting in the human androgen receptor is still not known. Extensive mutational analysis carried out for a similar signal in the nuclear protein nucleoplasmin revealed that a single mutation of the basic aminoacid residue at the arginine position partly inhibits nuclear import (Robbins *et al,* 1991). Another possibility is that the proline substitution disrupts the structure of the α helix, which is located at the carboxyterminal side of the second zinc cluster (Luisi *et al,* 1991) and negatively influences intramolecular interactions and receptor dimerization. In all cases, the net result is a functionally defective androgen receptor.

## MUTATIONS IN THE ANDROGEN BINDING DOMAIN

The hormone binding domain is encoded by approximately 250 aminoacid residues at the carboxyterminal end. Deletions in this domain as well as truncation by deletion of the last 12 carboxyterminal aminoacid residues completely abolished hormone binding. Deletions in the aminoterminal and DNA binding domain did not affect hormone binding. These results are in agreement with mutational analysis studies of the other members of the steroid hormone receptor family. The human androgen receptor is functionally active without the hormone binding domain (Jenster *et al,* 1991; Simental *et al,* 1991). Deletion of the androgen binding domain leads to a constitutively active androgen receptor protein, which has wild type *trans*-activation capacity also in the absence of hormone. Thus, it appears that the hormone binding domain acts as a repressor of the *trans*-activation function in the absence of hormone. The published mutations in the androgen binding domain are summarized in Table 1 and will be discussed below.

### Mutations in Exon 4

Several functional and structural features have been ascribed to exon 4, which harbours a highly conserved nucleoplasmin like nuclear localization signal, the hinge region and part of the steroid binding domain. Mutations in exon 4 could therefore affect different functions of the androgen receptor. All reported mutations to date in exon 4 have been in individuals with the complete androgen insensitivity syndrome. A mutation in codon 709 (Trp) described in three siblings changes this codon in a translational stop codon (Sai *et al,* 1990). Theoretically, the predicted receptor protein should have full constitutive transcription regulating activity. However, the phenotypic expression of this carboxyterminal truncated androgen receptor in vivo does not support the presence of a constitutively active molecule. Whether a very short living protein is synthesized in this particular case has not been investigated.

G to T mutation was found on position 1 in the splice donor site of intron 4 in the androgen receptor gene from a person with the complete form of androgen insensitivity (Ris-Stalpers *et al,* 1990). The consequence of this point mutation for RNA splicing was investigated after amplification of cDNA from

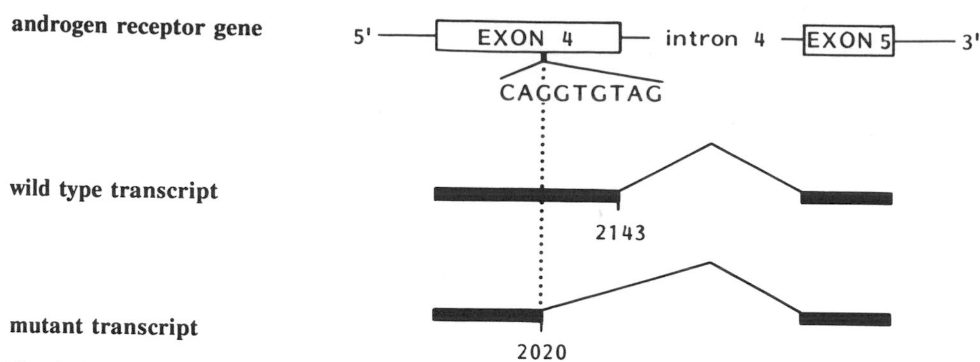


**Fig. 4.** Location of the cryptic splice donor site in exon 4 of the androgen receptor gene of a 46,XY individual with the complete form of androgen insensitivity. The wild type transcript is compared with the mutant transcript. The alternative splicing resulted in an in frame deletion of 123 nucleotides from the mRNA and consequently a deletion of 41 aminoacid residues in the receptor protein in the steroid binding domain

the affected person, using specific primers for exons 4 and 5. This resulted in the identification of a mutant fragment, which was shorter than the corresponding fragment from the wild type androgen receptor. This finding indicated an abnormal androgen receptor mRNA splicing. Sequence analysis of the mutant fragment revealed a cryptic splice donor site CAG/GTGTAG at position 2020/2021 in exon 4 of the human androgen receptor gene that is normally inactive. This cryptic splice site results in the deletion of 123 nucleotides from the mRNA (Fig. 4). Translation will consequently result in an in frame deletion of 41 aminoacid residues in the androgen receptor protein from this person. Transient expression of the mutant androgen receptor resulted in the synthesis of a protein that was approximately 5 kDa smaller than the wild type receptor. The mutant receptor was unable to bind androgens and did not activate transcription of an androgen regulated reporter gene construct (mouse mammary tumour virus promoter linked to the chloramphenicol acetyltransferase gene, *MMTV-CAT*) (Ris-Stalpers *et al,* 1990).

In two unrelated individuals with the complete form of androgen insensitivity, two different point mutations were found at the same nucleotide in codon 686 (aspartic acid) of exon 4 (Table 1) (Ris-Stalpers *et al,* 1991). One mutation (G→C) results in a substitution of histidine for aspartic acid. In this person, the androgen receptor displays low to negligible androgen binding. The other mutation (G→A) leads to a substitution of asparagine for aspartic acid. The androgen receptor in the person with the latter mutation shows normal androgen binding capacity, but the half life of the hormone receptor complex is only one fifth of that of the wild type hormone receptor complex. After transient expression, both mutant androgen receptors displayed the altered binding characteristics shown by genital skin fibroblasts from the affected subjects. In co-transfection studies, in which a glucocorticoid responsive element was linked to the thymidine kinase promoter and to the chloramphenicol acetyltransferase gene (*GRE-tk-CAT*) as a reporter gene, both mutant receptors were devoid of any *trans*-activation activity in the presence of physiological hormone concentrations (Ris-Stalpers *et al*, 1991).

### Mutations in Exon 5

Exon 5 forms an essential part of the steroid binding domain of the androgen receptor. In this exon, two point mutations have been reported. One mutation at codon 754 (Tyr) has been discussed already in combination with a shortened glutamine stretch in exon 1 (see above: Mutations in Aminoterminal Domain). The consequence of the mutation at codon 754 alone is a rapidly dissociating androgen receptor-ligand complex (McPhaul *et al*, 1991). A second more severe mutation in exon 5 at codon 750 (Ser→Phe) has also been reported (De Bellis *et al*, 1991): ligand binding by this receptor protein in the affected individual is lost completely and has resulted in complete androgen resistance (Table 1).

The molecular defect in the androgen insensitive Tfm rat has also been established. It was found that one single base alteration in the rat gene at codon 734 leads to replacement of an arginine residue by a glutamine residue (Yarbrough *et al*, 1990). Re-creation and expression of the mutant androgen receptor confirmed the reduced androgen binding capacity as observed for the androgen receptor in tissue extracts of the Tfm rat.

### Mutations in Exon 6

The two reported mutations in exon 6 have been found in androgen receptors from individuals with the complete form of androgen insensitivity (Table 1). One single nucleotide substitution identified in codon 787 (Trp) changes its sense into a stop codon (Marcelli *et al*, 1990b). The truncated protein expressed after transfection of the mutant cDNA was unable to activate transcription. The second mutation described in exon 6 refers to codon 765 (Arg), which is converted to a cysteine in the androgen receptor of a person with the receptor negative form (no androgen binding detectable) of complete androgen resistance (Marcelli *et al*, 1991b). The mRNA levels of this mutant receptor in genital skin fibroblasts of the affected persons were found to be seven times less than in those of normal persons. Whether this mutation in exon 6 directly affects the mRNA levels in the affected persons is highly unlikely and has not been investigated.

### Mutations in Exon 7

Sequence analysis of exon 7 in several persons with the complete androgen insensitivity syndrome has resulted in the detection of five different point mutations in four different codons (codon 822, Arg→Stop; codon 822, Arg→Gln; codon 846, Arg→Cys; codon 855, Asp→Gly and codon 857, Val→Met)

(Brown *et al*, 1990; De Bellis *et al*, 1991). For all reported mutant receptors, no specific androgen binding could be detected. In two other people with partial androgen resistance, arginine residues (codon 831 in one and codon 846 in the other) were substituted in both cases by a histidine residue, resulting in a considerable reduction in binding capacity of both mutant receptor proteins compared with the wild type receptor (Table 1) (Chang *et al*, 1991; De Bellis *et al*, 1991). Obviously, a relatively large number of CGX codons (arginine) are susceptible to mutagenesis in exon 7. In the whole steroid binding domain, 80% of the reported point mutations affect a G or C nucleotide (Table 1).

## Mutations in Exon 8

One mutation in exon 8 has been reported in a family with complete androgen insensitivity (Trifiro *et al*, 1991). The mutation is an adenine to thymine transversion, which changes the sense of codon 874 from a lysine to a termination codon. The mutation predicts the synthesis of a truncated receptor protein that lacks 36 aminoacid residues at the carboxyterminal end of the androgen binding domain. From mutational analysis studies, it is known that this kind of receptor truncation completely abolishes hormone binding and transcriptional activation (Rundlett *et al*, 1990; Jenster *et al*, 1991). Previous studies have established that the human prostate carcinoma cell line LNCaP can be stimulated with respect to growth not only by androgens but surprisingly also by progesterone, R5020 (a synthetic progestagen) and oestradiol (Schuurmans *et al*, 1988). Characterization of the androgen receptor in these cells revealed an altered steroid binding specificity with an increased preference for progestagens and oestradiol compared with the steroid binding specificity of the androgen receptor in normal cells (Veldscholte *et al*, 1990a). These data

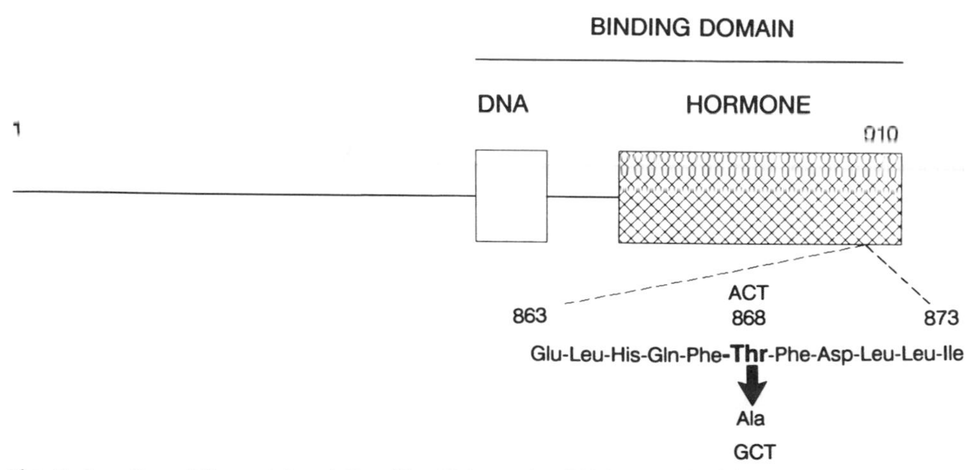


**Fig. 5.** Location of the point mutation (A→G) in codon 868 in exon 8 of the androgen receptor gene in human LNCaP cells. The mutation results in the substitution of threonine by alanine in the carboxyterminal part of the androgen binding domain

strongly suggested a modification of the androgen receptor and particularly in the steroid binding domain. Sequence analysis revealed one point mutation in codon 868 in exon 8 located at the carboxyterminal end of the steroid binding domain (Fig. 5) (Veldscholte *et al*, 1990b). The mutation (A→G) resulted in a threonine→alanine substitution. To determine whether the substitution affects the functional properties of the LNCaP androgen receptor, the mutant cDNA was cloned in an expression vector and transiently expressed in COS-1 and HeLa cells. In the transfection studies, the mutant receptor displayed increased binding affinity for progestagens and oestradiol. In addition, these ligands activate transcription at concentrations that are inactive with the wild type androgen receptor. These results confirm that the observed point mutation in the LNCaP androgen receptor is the cause of the broad steroid binding specificity. Whether such a point mutation that changes the steroid binding specificity can play a part in progressive prostate tumour growth remains to be established.

## CONCLUSIONS AND FUTURE DIRECTIONS

The information on the molecular structure of the human androgen receptor has facilitated the study of molecular defects associated with androgen insensitivity. Reports to date on androgen receptor gene structure in individuals with the complete and incomplete forms of androgen insensitivity indicate that gross deletions within the androgen receptor gene are uncommon. The locations of the different mutations reported cannot be assigned to a single site but are spread throughout the ligand binding and DNA binding domains. Further investigations on mutations in different androgen receptor linked forms of androgen resistance as well as in prostate cancer will certainly yield important new insights into the nature of the androgen receptor and will help to unravel the molecular basis of androgen action in normal and abnormal growth and development. In this respect, in vitro transcription systems for studying androgen regulated gene expression should help to define the different determinants and structural requirements. However, the application of new technologies such as the introduction of specific mutations in the androgen receptor by gene targeting in mouse embryonic stem cells is also needed for the phenotypic expression of those subtle mutations for which the present in vitro methodology is insufficient and not informative.

## SUMMARY

The elucidation of the molecular structure of the human androgen receptor has facilitated the study of molecular defects associated with androgen insensitivity. In this overview, data are presented on the functional domain structure of the wild type human androgen receptor and on the molecular structure of the androgen receptor from different subjects with the complete form of androgen insensitivity. Mutational domain analysis of the human androgen receptor has revealed that a large carboxyterminal region constitutes the hormone binding domain and that DNA binding is associated with a central basic domain. In addition, separate domains that control *trans*-activation and nuclear translocation have been identified. Reports on androgen receptor gene structure in individuals with the complete and incomplete forms of androgen insensitivity indicate that gross deletions within the androgen receptor gene are uncommon. The locations of the different point mutations reported cannot be assigned to a single site but are spread throughout the ligand binding and DNA binding domains. A point mutation found in the ligand binding domain of the human androgen receptor in a prostate tumour cell line is the cause of the altered steroid binding specificity observed for the androgen receptor in these prostate tumour cells. A considerable variation in the length of one of the polyglutamine repeats has been reported in the aminoterminal transcription regulating domain of the wild type androgen receptor. Doubling of the length of this particular polyglutamine stretch is correlated with a progressive spinal/bulbar muscular atrophy in a small group of middle aged men.

## References


Andriole GL and Catalona WJ (1991) The diagnosis and treatment of prostate cancer. *Annual Review of Medicine* **42** 9–15

Arbizu T, Santamaria J, Gomez JM, Quilez A and Serra JP (1983) A family with adult spinal and bulbar muscular atrophy, X-linked inheritance and associated testicular failure. *Journal of the Neurological Sciences* **59** 371–382

Beato M (1989) Gene regulation by steroid hormones. *Cell* **56** 335–344

Brinkmann AO, Faber PW, van Rooij HCJ *et al* (1989) The human androgen receptor: domain structure, genomic organization and regulation of expression. *Journal of Steroid Biochemistry* **34** 307–310

Brown TR, Lubahn DB, Wilson EM, Joseph DR, French FS and Migeon CJ (1988) Deletion of the steroid-binding domain of the human androgen receptor gene in one family with complete androgen insensitivity syndrome: evidence for further genetic heterogeneity in this syndrome. *Proceedings of the National Academy of Sciences of the USA* **85** 8151–8155

Brown CJ, Goss SJ, Lubahn DB *et al* (1989) Androgen receptor locus on the human X-chromosome: regional localization to Xq11-12 and description of a DNA polymorphism *American Journal of Human Genetics* **44** 264–269

Brown TR, Lubahn DB, Wilson EM, French FS, Migeon CJ and Corden JL (1990) Functional characterization of naturally occurring mutant androgen receptors from subjects with complete androgen insensitivity. *Molecular Endocrinology* **4** 1759–1772

Chang C, Kokontis J and Liao S (1988) Structural analysis of complementary DNA and amino acid sequences of human and rat androgen receptors. *Proceedings of the National Academy of Sciences of the USA* **85** 7211–7215

Chang YT, Migeon CJ and Brown TR (1991) Human androgen insensitivity syndrome due to androgen receptor gene point mutations in subjects with normal androgen receptor levels but impaired biological activity. *Program of the 73rd Annual Meeting of the Endocrine Society*, Washington DC, 37 [Abstract 28]

Charest NJ, Zhou Z, Lubahn DB, Olsen KL, Wilson EM and French FS (1991) A frameshift mutation destabilizes androgen receptor messenger RNA in the Tfm mouse. *Molecular*

*Endocrinology* **5** 573–581
Cunha GR, Donjacour AA, Cooke PS *et al* (1987) The endocrinology and developmental biology of the prostate. *Endocrine Reviews* **8** 338–362
De Bellis A, Quigley CA, Cariello NF, Ho KC and French FS (1991) DNA and steroid binding domain mutations of the androgen receptor cause partial or complete androgen insensitivity. *Program of the 73rd Annual Meeting of the Endocrine Society*, Washington DC, 316 [Abstract 1141]
Duboule D, Haenlin M, Galliot B and Mohier E (1987) DNA sequences homologous to the Drosophila opa repeat are present in murine mRNAs that are differentially expressed in fetuses and adult tissues. *Molecular and Cellular Biology* **7** 2003–2006
Evans RM (1988) The steroid and thyroid hormone receptor superfamily. *Science* **240** 889–895
Faber PW, Kuiper GGJM, van Rooij HCJ, van der Korput JAGM, Brinkmann AO and Trapman J (1989) The N-terminal domain of the human androgen receptor is encoded by one, large exon. *Molecular and Cellular Endocrinology* **61** 257–262
Faber PW, van Rooij HCJ, van der Korput JAGM *et al* (1991) Characterization of the human androgen receptor transcription unit. *Journal of Biological Chemistry* **266** 10743–10749
French FS, Lubahn DB, Brown TR *et al* (1990) Molecular basis of androgen insensitivity. *Recent Progress in Hormone Research* **46** 1–42
Gaspar ML, Meo T and Tosi M (1990) Structure and size distribution of the androgen receptor mRNA in wild-type and Tfm/y mutant mice. *Molecular Endocrinology* **4** 1600–1610
Griffin JE and Wilson JD (1989) The androgen resistance syndromes: 5α-reductase deficiency, testicular feminization and related disorders, In: CR Scriver, AL Baudet, WS Sly and D Valle (eds). *The Metabolic Basis of Inherited Disease*, pp 1919–1944, McGraw-Hill, New York
Härd T, Kellenbach E, Boelens R *et al* (1990) Solution structure of the glucocorticoid receptor DNA-binding domain. *Science* **249** 157–160
Harding AE, Thomas PK, Baraitser M, Bradbury PC, Morgan Hughes JA and Ponsford JR (1982) X-linked recessive bulbospinal neuronopathy: a report of ten cases. *Journal of Neurology, Neurosurgery and Psychiatry* **45** 1012–1019
He WW, Kumar MV and Tindall DJ (1991) A frame-shift mutation in the androgen receptor gene causes complete androgen insensitivity in the testicular-feminized mouse. *Nucleic Acids Research* **19** 2373–2378
Jenster G, van der Korput JAGM, van Vroonhoven C, van der Kwast TH, Trapman J and Brinkmann AO (1991) Domains of the human androgen receptor involved in steroid-binding, transcriptional activation and subcellular localization. *Molecular Endocrinology* **5** 1396–1404
Kalderon D, Roberts BL, Richardson WD and Smith AE (1984) A short amino acid sequence able to specify nuclear location. *Cell* **39** 499–509
Kao CC, Lieberman PM, Schmidt MC, Zhou Q, Pei R and Berk AJ (1990) Cloning of a transcriptionally active human TATA binding factor. *Science* **248** 1646–1649
Kuiper GGJM, Faber PW, van Rooij HCJ *et al* (1989) Structural organization of the human androgen receptor gene. *Journal of Molecular Endocrinology* **2** R1–R4
La Spada AR, Wilson EM, Lubahn DB, Harding AE and Fischbeck KH (1991) Androgen receptor gene mutations in X-linked spinal and bulbar muscular atrophy. *Nature* **352** 77–79
Lubahn DB, Joseph DR, Sar M *et al* (1988) The human androgen receptor: complementary deoxyribonucleic acid cloning, sequence analysis and gene expression in prostate. *Molecular Endocrinology* **2** 1265–1275
Lubahn DB, Brown TR, Simental JA *et al* (1989) Sequence of the intron/exon junctions of the coding region of the human androgen receptor gene and identification of a point mutation in a family with complete androgen insensitivity. *Proceedings of the National Academy of Sciences of the USA* **86** 9534–9538
Luisi BF, Xu WX, Otwinowski Z, Freedman LP, Yamamoto KR and Sigler PB (1991) Crystallographic analysis of the interaction of the glucocorticoid receptor with DNA. *Nature* **352**

497–505

Lyon MF and Hawkes SG (1970) X-linked gene for testicular feminization in the mouse. *Nature* **227** 1217–1219

McPhaul MJ, Marcelli M, Tilley WD, Griffin JE, Isidro-Gutierrez RF and Wilson JD (1991) Molecular basis of androgen resistance in a family with a qualitative abnormality of the androgen receptor and responsive to high-dose androgen therapy. *Journal of Clinical Investigation* **87** 1413–1421

Marcelli M, Tilley WD, Wilson CM, Griffin JE, Wilson JD and McPhaul MJ (1990a) Definition of the human androgen receptor gene structure permits the identification of mutations that cause androgen resistance: premature termination of the receptor protein at amino acid residue 588 causes complete androgen resistance. *Molecular Endocrinology* **4** 1105–1116

Marcelli M, Tilley WD, Wilson CM, Griffin JE, Wilson JD and McPhaul MJ (1990b) A single nucleotide substitution introduces a premature termination codon into the androgen receptor gene of a patient with receptor-negative androgen resistance. *Journal of Clinical Investigation* **85** 1522–1528

Marcelli M, Zopi S, Grino PB, Griffin JE, Wilson JD and McPhaul MJ (1991a) A mutation in the DNA-binding domain of the androgen receptor gene causes complete testicular feminization in a patient with receptor-positive androgen resistance. *Journal of Clinical Investigation* **87** 1123–1126

Marcelli M, Tilley WD, Zopi S, Griffin JE, Wilson JD and McPhaul MJ (1991b) Androgen resistance associated with a mutation of the androgen receptor at amino acid 772 (Arg→Cys) results from a combination of decreased messenger ribonucleic acid levels and impairment of receptor function. *Journal of Clinical Endocrinology and Metabolism* **73** 318–325

Mitchell PJ and Tjian R (1989) Transcriptional regulation in mammalian cells by sequence-specific DNA binding proteins. *Science* **245** 371–378

O'Malley BW (1990) The steroid receptor superfamily: More excitement predicted for the future. *Molecular Endocrinology* **4** 363–369

Pinsky L and Kaufman M (1987) Genetics of steroid receptors and their disorders. *Advances in Human Genetics* **16** 299–472

Ris-Stalpers C, Kuiper GGJM, Faber PW *et al* (1990) Aberrant splicing of androgen receptor mRNA results in synthesis of a nonfunctional receptor protein in a patient with androgen insensitivity. *Proceedings of the National Academy of Sciences of the USA* **87** 7866–7870

Ris-Stalpers C, Trifiro MA, Kuiper GGJM *et al* (1991) Substitution of aspartic acid 686 by histidine or asparagine in the human androgen receptor leads to a functionally inactive protein with altered hormone-binding characteristics. *Molecular Endocrinology* **5** 1562–1569

Robbins J, Dilworth SM, Laskey RA and Dingwall C (1991) Two interdependent basic domains in nucleoplasmin nuclear targeting sequence: identification of a class of bipartite nuclear targeting sequence. *Cell* **64** 615–623

Rundlett SE, Wu X-P and Miesfeld RL (1990) Functional characterizations of the androgen receptor confirm that the molecular basis for androgen action is transcription regulation. *Molecular Endocrinology* **4** 708–714

Sai T, Seino S, Chang C *et al* (1990) An exonic point mutation of the androgen receptor gene in a family with complete androgen insensitivity. *American Journal of Human Genetics* **46** 1095–1100

Schuurmans ALG, Bolt J, Voorhorst M, Blankenstein MA and Mulder E (1988) Regulation of growth and epidermal growth factor receptor levels of LNCaP prostate tumor cells by different steroids. *International Journal of Cancer* **42** 917–922

Simental JA, Sar M, Lane MV, French FS and Wilson EM (1991) Transcriptional activation and nuclear targeting signals of the human androgen receptor. *Journal of Biological Chemistry* **266** 510–518

Tilley WD, Marcelli M, Wilson JD and McPhaul MJ (1989) Characterization and expression of a cDNA encoding the human androgen receptor. *Proceedings of the National Academy of*

*Sciences of the USA* **86** 327–331

Trapman J, Klaassen P, Kuiper GGJM *et al* (1988) Cloning, structure and expression of a cDNA encoding the human androgen receptor. *Biochemical and Biophysical Research Communications* **153** 241–248

Trifiro M, Prior RL, Sabbaghian N *et al* (1991) Amber mutation creates a diagnostic MaeI site in the androgen receptor gene of a family with complete androgen insensitivity. *American Journal of Medical Genetics* **40** 493–499

Umesono K and Evans RM (1989) Determinants of target gene specificity for steroid/thyroid hormone receptors. *Cell* **57** 1139–1146

Veldscholte J, Voorhorst-Ogink MM, Bolt-de Vries J, van Rooij HCJ, Trapman J and Mulder E (1990a) Unusual specificity of the androgen receptor in the human prostate tumor cell line LNCaP: high affinity for progestagenic and estrogenic steroids. *Biochimica et Biophysica Acta* **1052** 187–194

Veldscholte J, Ris-Stalpers C, Kuiper GGJM *et al* (1990b) A mutation in the ligand binding domain of the androgen receptor of human LNCaP cells affects steroid binding characteristics and response to anti-androgens. *Biochemical and Biophysical Research Communications* **173** 534–540

Wahli W and Martinez E (1991) Superfamily of steroid nuclear receptors: positive and negative regulators of gene expression. *FASEB Journal* **5** 2243–2249

Warner CL, Griffin JE, Wilson JD *et al* (1991) X-linked spinomuscular atrophy: a kindred with associated abnormal androgen receptor binding. *Neurology* **41** (**Supplement 3130**)

Wharton KA, Yedvobnick B, Finnerty VG and Artavanis-Tsakonas S (1985) opa: a novel family of transcribed repeats shared by the notch locus and other developmentally regulated loci in D melanogaster. *Cell* **40** 55–62

Yarbrough WG, Quarmby VE, Simental JA *et al* (1990) A single base mutation in the androgen receptor gene causes androgen insensitivity in the testicular feminized rat. *Journal of Biological Chemistry* **265** 8893–8900

Young YF, Johnson MP, Prescott JL and Tindall DJ (1989) The androgen receptor of the testicular-feminized (Tfm) mutant mouse is smaller than the wild-type receptor. *Endocrinology* **124** 771–775

Zoppi S, Marcelli M, Griffin JE and Wilson JD (1991) Point mutations in the DNA-binding domain are a frequent cause of receptor positive androgen resistance. *Program of the 73rd Annual Meeting of the Endocrine Society*, Washington DC, **94** [Abstract 254]

The authors are responsible for the accuracy of the references.

# The Importance of Steroid Hormones in Prostate Cancer

**G WILDING**
*University of Wisconsin Comprehensive Cancer Center, 600 Highland Avenue, Madison, Wisconsin 53792*



## INTRODUCTION

More than 122 000 new cases of prostate cancer are diagnosed annually in the US. In fact, prostate cancer is the most common malignancy in US men in terms of both incidence and prevalence (Catalona and Scott, 1986; Scardino, 1989; Cancer Statistics, 1991; Dhom, 1991), and US black men have the highest incidence of prostate cancer in the world (Mettlin and Natarajan, 1983). Prostate cancer accounts for more than 32 000 deaths each year in the US and is second only to lung cancer as a cause of cancer deaths in US men (Feldman *et al*, 1986).

A number of perplexing characteristics emerge from epidemiological observations of prostate cancer in various populations. Firstly, the incidence of microscopic, latent prostate cancer shows a very close correlation with increasing age regardless of the population examined. This suggests a fundamental change with ageing in all men that permits the transformation of prostate epithelial cells in a large proportion of those men. Secondly, prostate cancer does not occur in eunuchs: an intact hormonal milieu seems to be important for the development of prostate cancer. Thirdly, there is tremendous variation in the incidence of clinical prostate cancer among populations around the world despite very close similarity in the incidence of latent cancer. For example, Oriental men in Japan and Taiwan have one of the lowest incidences of clinical prostate cancer in the world, whereas Oriental men living in the continental US show an incidence of clinical prostate cancer approaching the much higher rate seen in US caucasian men. The incidence of clinical prostate cancer in Oriental men residing in Hawaii is intermediate between the rates

*Cancer Surveys* Volume 14: *Growth Regulation by Nuclear Hormone Receptors*
 0-87969-371-1/92. $3.00 + .00

for the US and Asian countries. The proportion of latent cancers in Japan showing invasive characteristics is rising, although the overall incidence of latent cancer has not changed. This might presage a rise in clinical prostate cancer in future years in this population (Yatani *et al,* 1988). Similarly, although US and Nigerian black men show a similar incidence of latent cancer, US black men have the highest rate of clinical prostate cancer in the world, significantly higher than their counterparts in Nigeria.

A critical dilemma facing clinicians treating patients with prostate cancer is what to do with patients with androgen independent cancers. This dilemma results from our lack of understanding of and inability to control the growth of hormone independent prostate cancer cells. When first diagnosed, more than 75% of men with metastatic disease have tumours that are, at least in part, dependent on and responsive to androgens. However, hormone independent disease inevitably develops in virtually all such patients. Because androgen withdrawal constitutes the only effective form of systemic therapy for metastatic prostate cancer at this time, the appearance of androgen independent clones of tumour cells heralds the end of our ability to control the course of a patient's disease effectively (Labrie *et al,* 1986; Isaacs and Kyprianou, 1987).

The mechanisms by which, with increasing age and exposure to androgens, prostate epithelial cells are commonly transformed to microscopic cancer, then to clinical cancer and ultimately to hormone independent cancer are not understood. Neither are the underlying reasons for the wide variations in clinical cancer incidence with geographic changes understood. The evidence clearly points to a multistep process of carcinogenesis. One important element in the development of prostate cancer is the presence of androgens, generally over a long period of time. Surprisingly, little is known about the variation in androgen receptor (AR) expression with ageing, in men of various races, at different geographic locations or with the development of latent, clinical or hormone independent prostate cancer. In addition to a lack of information about AR messenger RNA and protein levels in these circumstances, the frequency of AR mutations has not been determined in these populations of men.

Although it is apparent that the transforming growth factors (TGF) α and β and other growth factors have autocrine and paracrine roles in the biology of hormone responsive tissues such as the prostate and breast (Dickson and Lippman, 1987; Wilding *et al,* 1989b,c; Thompson, 1990), little is known of their expression in prostate tissue under the conditions listed above. In transgenic mice, the constitutive expression of TGF-α has been shown to lead to hyperplasia and tumours in the mammary tissue (Sandgren *et al,* 1990) and the expression of TGF-β has been found to rise in prostate cancers induced by *ras/myc* transfection of prostate tissue (Thompson 1990; Thompson *et al,* 1989). The expression of these transforming factors has been shown to be regulated by oestrogens and androgens both in vivo and in vitro (Kyprianou and Isaacs, 1988, 1989; Wilding *et al,* 1989).

Prostate cancer is now first in solid tumour incidence and prevalence and second as a cause of cancer related deaths in US men. US black men have the highest rate of prostate cancer in the world (Cancer Statistics, 1991). In addition to the 122 000 US men in whom prostate cancer is diagnosed each year, benign prostatic hyperplasia affects more than two thirds of all men over age 50, approximately 20–25% of whom require surgery for the relief of urinary obstruction (Carter and Coffey, 1990). This affliction accounts for more than 400 000 transurethral resections of the prostate each year in the US. Clearly, diseases of the prostate have become a major health issue in terms of the number of men affected, the cost to the medical system and the morbidity inflicted.

## ANDROGENS IN PROSTATE BIOLOGY

What are the roles of androgens and the AR in prostate carcinogenesis? The answer has two dimensions: (a) the strong evidence that androgens and the AR play a seminal part in prostate carcinogenesis, suggesting a permissive role for androgens and (b) the actions of the AR as a transcription factor regulating the expression of proto-oncogenes, TGF and growth factor receptors in prostate epithelial cells.

Evidence exists in experimental models that androgens might act as initiators or permissive agents in prostate carcinogenesis. For example, Bosland and co-workers have demonstrated that androgens increase the incidence of N-methylnitrosourea (NMU) induced prostate cancer in rats (Bosland, 1988; Bosland *et al*, 1990). In addition, Noble (1977) has shown an increase in the incidence of prostate cancer in Nb rats through the long term administration of testosterone and oestrogen. Similarly, there is a vast literature delineating oestrogen induced carcinogenesis. Although the potential for steroid hormones, especially the oestrogens, to induce DNA damage has been addressed by Henderson *et al* (1988), the mitogenic effects of steroid hormones might place the cells in a proliferative state, thus leading to increased genetic instability or to the expression of mutated genes, which then could alter the cell's phenotype. Members of the steroid hormone receptor superfamily of genes act as transcription factors when complexed with their ligand regulating the expression of a large subset of genes such as oncogenes (eg *fos, jun* and *myc* [Wilding, *et al,* 1988; Weisz *et al,* 1990]), growth factors (eg TGF-α and TGF-β [Wilding 1989c; Jeng and Jordan, 1991]), receptors (eg epidermal growth factor receptor [Schuurmans *et al,* 1988]) and other genes coding for proteins, such as proteases (eg prostate specific antigen [Clements *et al,* 1988]) and tissue plasminogen activator (tPA), which could be involved in invasion and metastases (Dickson and Lippman, 1987). Mutation of the AR has been shown to inactivate the receptor (Brown *et al,* 1988) or broaden its ability to be activated by a wide range of ligands, including anti-androgens (Wilding *et al,* 1989a; Harris *et al,* 1991). Although the AR cannot be viewed as a proto-oncogene, the importance of its presence and its transcriptional regulation of a host of gene products proven to be important in cellular transformation cannot be overlooked.

As a male accessory sex organ, both the developing and the adult prostates are extremely sensitive and responsive to the presence of androgenic hormones and the AR (Coffey, 1988). Embryologically, the AR mediates androgenic regulation of prostate epithelial differentiation via stromal-epithelial interactions. Using tissue recombinant studies of the urogenital sinus and its epithelial and stromal components, Cunha *et al.* (1983) have shown that androgens regulate the growth, differentiation and morphogenesis of the prostate epithelium indirectly via the surrounding stroma. Analysis of the AR in these tissue components demonstrates that a functional AR in the stroma, and not the epithelium, during embryogenesis is the crucial factor for normal prostate development. The manner in which the androgen stimulated stroma mediates its influence on the developing epithelium is unclear, although extracellular matrix components and soluble growth factors are potential candidates. Chung and others have documented prostate stromal-epithelial interactions in vitro that lend credence to the possible indirect effects of androgens on the prostate epithelium (Chang and Chung 1989; Kabalin *et al,* 1989).

The regulation of the adult prostate epithelium appears to be more complex, with both direct and indirect androgenic actions taking place. Again, tissue recombination studies have shed some light on the potential influence of stroma on prostate cancer growth. Studies using fetal and neonatal urogenital mesenchymes in combination with the transplantable Dunning rat prostate adenocarcinoma have shown that the cancer cells can be induced to undergo co-ordinated growth and differentiation in the presence of specific mesenchymes (Hayashi *et al,* 1990). Thompson, using *ras* and *myc* infected components in a mouse recombinant model, has shown that *ras/myc* infected epithelium combined with uninfected stroma produces benign hyperplasia. However, *ras/myc* infected stroma combined with *ras/myc* infected epithelium results in poorly differentiated adenocarcinoma (Thompson *et al,* 1990). Thompson's studies demonstrate that normal mesenchyme is non-permissive for high frequency conversion to malignancy and suggest that *ras/myc* infected urogenital sinus mesenchyme either lacks the potential to inhibit malignant conversion of epithelium and/or promotes the conversion process (Thompson, 1990). Combinations of *ras/myc* infected mesenchyme and uninfected epithelium showed elevated levels of TGF-β. In castrated animals with *ras/myc* induced tumours, TGF-β1, TGF-β3 and tPA were all elevated two- to fourfold compared with non-castrated animals.

These results indicate that the *ras/myc* induced carcinomas responded to androgen deprivation by upregulating genes that are involved in cell growth and differentiation. In addition, mesenchyme derived TGF-β might play an important role in the conversion of premalignant epithelium to malignant carcinomas (Egawa *et al,* 1991; Ripoll *et al,* 1991). In other words, mesenchymal mediators that normally control discrete steps of development might, under the right circumstances, promote a malignant phenotype in transformed epithelial cells. Similar paracrine activities may also underlie malignant progression of prostate cancer in man.

## ANDROGENS, AGEING AND PROSTATE PATHOLOGY

If androgens and mediation of their action through the AR are important in maintaining a normal, functioning prostate gland through adulthood, what are the data on variation of androgen or AR levels with age and are there correlative changes with the development of pathological states in the gland? Firstly, it is generally agreed that there is a progressive impairment of Leydig cell function with advancing age (Nahoui and Roger, 1990). This impairment consists of a diminished capacity of Leydig cells to synthesize and secrete androgens (Serio *et al,* 1979; Takahashi *et al,* 1983; Hammar, 1985) as well as a decline in their responsiveness to chorionic gonadotropin (hCG) (Longcope, 1973). Studies of the relationship between plasma testosterone (T) and age, using a variety of assays, have been inconsistent. However, the sum of free plus albumin bound T, called non-sex hormone binding globulin (non-SHBG) or bioavailable T, has been demonstrated to be the best index for evaluating androgen activity (Cumming and Wall, 1985). Non-SHBG has been shown to be lower in elderly men than in young men (Bartsch, 1980; Nankin and Calkins, 1986; Tenover *et al,* 1987).

In general, although some data suggest a higher risk of prostate cancer in men with elevated T levels (Hsing and Comstock, 1989), other studies, using matched controls, have not confirmed this finding (Hill *et al,* 1980). Higher T levels have been associated with a western diet than with a vegetarian diet (Hill *et al,* 1979). Finally, although higher prostate cancer tissue dihydrotestosterone (DHT) levels have correlated with a higher rate of hormone response (Geller *et al,* 1984), no clear correlation has been demonstrated between DHT levels and the development of prostate cancer. For example, a prospective epidemiological study linked high plasma androgen levels with the development of prostate cancer when an increased androstenedione level was associated with a greater risk of development of prostate cancer in a cohort followed for 14 years (Barrett-Connor *et al,* 1990). On the other hand, Nomura *et al* (1988) prospectively obtained serum samples from 6860 men in Hawaii. After a surveillance period of 14 years, sera from 98 incident cases of prostate cancer and 98 matched controls were assayed for T, DHT, oestrone, oestradiol and sex hormone binding globulin. Although there was a suggestion that DHT levels were lower in prostate cancer patients, no strongly significant association could be made between hormone levels and prostate cancer. In summary, although there is evidence that alterations in androgen levels are associated with the development of prostate cancer, the data are not consistent and no conclusions can be drawn. The question thus arises as to whether variation in AR expression occurs during prostate carcinogenesis.

A broad range of T levels is observed in men with prostate cancer. Retrospective analysis from one study showed that the higher the pretreatment plasma T level, the greater the survival rate in patients with metastatic prostate cancer receiving primary hormonal therapy (Hickey *et al,* 1988). Conversely, low T levels at the time of diagnosis carry a poor prognosis (Ishikawa *et al,* 1989). Chodak and co-workers (1991) assessed the independent prognostic factors affecting survival in 240 men undergoing treatment for metastatic prostate cancer as part of a randomized clinical trial comparing a gonadotropin releasing hormone analogue with castration. In a multivariate analysis, one of the most highly significant predictors was the serum testosterone level, in addition to bone pain, performance status and serum alkaline phosphatase levels. Patients with all four factors favourable for survival had a 2 year survival rate of 84% compared with only 8% for patients with none of the four factors favourable for survival. A separate analysis of serum testosterone levels revealed that the higher the pretreatment serum testosterone level, the greater the survival rate. Compared with patients with serum testosterone levels less than 6.9 nmol/l, significant differences in survival were observed for patients with serum testosterone levels of 10.4 to 13.9, 13.9 to 17.3 and over 17.3 nmol/l.

The mainstay of hormone therapy for metastatic prostate cancer is bilateral orchidectomy. Proven by Huggins in the 1940s to be an effective palliative therapy, this therapy remains the "gold standard" by which other hormone therapies are compared. Orchidectomy removes the primary source of testosterone in men, the testicles. There exists a viewpoint that even after hormone failure, a patient should be maintained in an orchidectomized state. The rationale for this approach maintains that although a patient's tumour may no longer be dependent for growth on testosterone, it still may be responsive to testosterone stimulation.

Additive steroid therapy has also been used in the treatment of metastatic prostate cancer. Diethylstilboestrol (DES) and other oestrogens have been extensively studied. One of the primary conclusions of the Veterans Administration Cooperative Urological Research Group (VACURG) (Byar, 1973) was that 1–5 mg/day of DES was as effective as orchidectomy at palliating the symptoms of metastatic prostate cancer. Although 1 mg/day was as effective as 5 mg/day and less toxic, 1 mg/day did not achieve as great a drop in testosterone levels as was achieved with orchidectomy. Because of the availability of less toxic, although much more expensive, non-surgical hormone therapies, such as gonadotropin releasing hormone (GnRH) agonists, the use of this form of additive steroid therapy has declined in recent years. Another additive therapy that has been used less often than oestrogens is megestrol acetate, a progestin. Both of these additive therapies reduce serum testosterone levels by disrupting the hypothalamic-pituitary-gonadal axis, thus decreasing gonadotropin release from the pituitary.

Although the testes produce more than 90% of the circulating androgens in men, the adrenal glands make and secrete weak androgens that can be converted in one or two enzymatic steps to testosterone. Therefore, the concept of combining a GnRH agonist or orchidectomy with an anti-androgen was developed in an attempt to achieve maximal androgen blockade. One of the larger randomized studies addressing this issue compared the use of a GnRH analogue with or without the anti-androgen flutamide in men with metastatic prostate cancer (Crawford *et al*, 1989). The patients receiving the combined therapy had a longer progression free survival (16.5 v. 13.9 months) and an increase in the median length of survival (35.6 v. 28.3 months) compared with the patients receiving GnRH analogue alone. Most striking was the improved survival of the 41 patients with minimal metastatic disease who were treated with the combination therapy. If confirmed in subsequent studies, this would support earlier intervention with hormone therapy in patients with metastatic prostate cancer. The patients who received the combination therapy showed the greatest improvement in symptoms during the first 12 weeks of therapy, suggesting amelioration of possible GnRH induced flare effects during this period.

Primary hormone therapy ultimately fails in all patients as hormone independent disease emerges. Secondary attempts at hormone manipulation yield palliative results, generally subjective in nature, in approximately 20% of the patients and are of short duration. Secondary hormone therapies are usually aimed at reducing or blocking adrenal androgens.

## ANDROGEN RECEPTOR EXPRESSION

What is known about AR expression and protein levels in the ageing process and in prostate carcinogenesis? Until recently, the assessment of AR levels and localization in prostate tissues and cancers has been hampered by technical difficulties. Assays on homogenized tissues do not address heterogeneity issues. The condition of the tissue for example fresh, frozen, cauterized—produces variable results. Attempts to separate cellular compartments can lead to mechanical damage of cells and a shift of the receptor between the cytoplasm and nucleus of cells. With the cloning of the human AR gene and the development of monoclonal antibodies against the receptor, the issues of heterogeneity, intracellular distribution, transcriptional v. translational control, mutational frequency and structure-activity relationships can be addressed. Summarized below are the preliminary studies performed during the brief period since AR complementary DNA (cDNA) and monoclonal antibody reagents have become available.

The human AR gene is a nuclear hormone receptor. The receptor contains highly conserved regions analogous to the hormone binding and DNA binding domains of other steroid receptors such as the progesterone, mineralocorticoid, glucocorticoid, oestrogen, vitamin D and other receptors. These receptors comprise a superfamily of ligand modulated transcription factors that regulate homoeostasis, reproduction, development and differentiation. The cDNA sequence of the AR was initially reported in 1988 by several groups (Chang *et al*, 1988a,b; Lubahn *et al*, 1988a; Tilley *et al*, 1989; Faber *et al*, 1989). The coding region of the human AR encompasses approximately 3580 nucleotides from which the deduced 919 aminoacids predict a protein of 99 kDa (Lubahn *et al*, 1988b; Tan *et al*, 1988). Northern analysis surprisingly has shown hybridization of cDNA clones to messenger RNA (mRNA) species of

9.6 and 7 kb (Chang *et al*, 1989a; Lubahn *et al*, 1988b; Takeda *et al*, 1991). Immunohistochemical analysis of human prostate tissue sections with polyclonal or monoclonal antibodies has localized the AR protein predominantly in the nuclei of glandular epithelial cells (Lubahn *et al*, 1988b; Chang *et al*, 1989a,b; Takeda *et al*, 1991). The AR gene has been localized to the X chromosome between the centromere and q13.

With the development of reagents such as cDNAs and antibodies against the human AR, studies have begun to emerge that characterize the expression of the AR in human prostate tissue. Initial reports have confirmed localization of the AR predominantly to the nucleus of prostate epithelial cells in normal, benign hyperplastic and cancerous prostate and other tissues (Lubahn *et al*, 1988b; Takeda *et al*, 1989; Takeda and Chang, 1991). As noted above, a major limitation of previous studies using homogenized tissue was that the AR content was an average of all the cells in the homogenized specimen. Sadi and coworkers (1991) tried to overcome this difficulty by performing an immunohistochemical analysis of the AR in 17 patients with prostate cancer. Needle biopsy specimens were used to determine the extent of AR expression in prostate cancer specimens. The authors found that prostate cancer contains AR positive and AR negative malignant cells before androgen withdrawal therapy, but the percentage of AR positive cells did not predict the time to tumour progression after androgen withdrawal therapy.

A larger study provided additional insight into this problem (Chodak *et al*, 1992). To facilitate an understanding of how androgens participate in the genesis of benign hyperplasia and carcinoma in humans, AR was assayed in the stroma and epithelium of human prostate tissue from 57 patients. Immunohistochemical staining of human AR was performed on 106 sections of normal, hyperplastic and cancerous prostate tissue. To determine the variability of AR staining, sections taken from different portions of the gland were studied. Frozen tissue sections were incubated with monoclonal AR antibodies, and staining was completed with indirect avidin-biotin peroxidase methods. Antibody staining was found mainly in the nucleus of prostate epithelial cells, although some stromal cells also showed positive staining. Unlike normal prostate, there was a heterogeneous distribution of AR in benign prostate hyperplasia and prostate cancer. The AR content in well differentiated adenocarcinoma epithelium was significantly higher than in moderately ($p<0.05$) and poorly ($p=0.05$) differentiated carcinomas. Regardless of the origin of stromal tissue, some staining was observed. In each specimen studied, AR staining was consistent both qualitatively and quantitatively for each pathological component throughout the specimen. These findings confirm that AR is a nuclear receptor protein. Furthermore, they show the ability of monoclonal antibodies to reveal cellular/subcellular distribution of AR and demonstrate a correlation between the degree of tumour differentiation and AR content in epithelial but not stromal cells. These studies might shed some light on the variable tumour response to hormonal therapy.

The mechanism by which androgen binding activates the AR and results in the regulation of expression of a specific subset of genes has not been determined. Models have been proposed, involving the carboxyterminal region, that attempt to link the events of ligand binding, dimerization and hormone relieved transcriptional inactivation (Forman and Samuel, 1990). Since the nuclear matrix is known to bind androgen receptors in a tissue specific and steroid specific manner, Getzenberg and Coffey (1990) have proposed that the tissue specificity of the nuclear matrix arranges the DNA in a unique conformation, which may be involved in the specific interaction of transcription factors with DNA sequences, resulting in tissue specific patterns of protein expression.

Detection of abnormal AR genes (reviewed in Brinkmann and Trapman, this issue) in two patients with testicular feminization (TFM) was evaluated with a polymerase chain reaction method. Direct sequencing of the abnormal AR gene from patients with TFM was performed. Total RNA from genital skin fibroblast lines of patients was isolated with the guanidinium thiocyanate method. The first and second strand cDNAs were synthesized, and the hormone binding domains of these double stranded cDNAs were then amplified for 30 cycles. The amplified DNAs were sequenced directly. One patient was found to have a deletion of the AR gene from exon 1 to exon 7 (Trifio *et al,* 1989), and the other patient was found to have a mutation (Trp717 to stop codon) in exon 7 that abolished most of the AR binding domain (Sai *et al,* 1990).

Wilding *et al* (1989a) conducted studies on the androgen responsive human prostate cancer cell line LNCaP to determine the direct effects of a variety of androgens, anti-androgens and other steroid hormones on hormone responsive prostate cancer cells in vitro. These cells contain approximately 30 000 high affinity androgen binding sites per cell. In the absence of any androgenic stimulation, the cells were shown to be stimulated by three different anti-androgens. These cells were subsequently shown to have a point mutation in the hormone binding region of their AR by Harris *et al* (1991). The data noted above verify and confirm that androgen sensitivity can be lost or altered both through subtle changes such as point mutations and through deletions of significant portions of the AR gene.

## PEPTIDE GROWTH FACTORS

The androgen/AR complex, acting as a transcription factor, mediates its effects on a cell through transcriptional control of a specific subset of genes. The influence of the activated AR on the growth of human prostate cancer and its relationship to the surrounding stroma are controlled by complex mechanisms that are incompletely understood. Clearly, a wide range of peptide growth factors appear to play crucial parts in these processes (Thompson, 1990). I have chosen to focus on one of these factors, TGF-β and its related family members (Keski-Oja *et al,* 1988), because of its wide spectrum of biological effects, its influence on the transcription of a variety of oncogene products such as myc, jun and fos (Keski-Oja *et al,* 1988; Li *et al,* 1990; Moses *et al,* 1990), its potential interaction with suppressor genes such as *RB* (Pietenpol *et al,* 1990a,b) and its link with androgens not only through androgen mediated control of its transcription (Kyprianou and Isaacs; 1988) but also through its role in the androgen controlled process of apoptosis (Kyprianou and Isaacs, 1989).

In terms of growth, TGF-β inhibits the growth of prostate cancer cells in a cytostatic manner while stimulating the growth of critical stromal cells, such as fibroblasts (Wilding, 1991). Since the inhibitory effects of TGF-β on prostate cancer cells appear to diminish as the process of transformation progresses towards less differentiated states, the net effect on prostate tumour growth may be positive. Recent evidence suggests that the inhibitory effects, at least of TGF-β, on growth might be mediated through the RB tumour suppressor gene product and the proto-oncogene c-*myc* (Pietenpol *et al,* 1990a,b). Beyond its direct growth effects, TGF-β also alters the response of prostate cancer cells to positive mitogenic factors, such as members of the epidermal growth factor and fibroblast growth factor families (McKeehan and Adams, 1988), suggesting that growth control is, in fact, a delicate balance between positive and negative influences. Non-mitogenic responses to TGF-β by prostate cancer cells, the immune system, the stroma and the vascular system provide evidence that TGF-β might also be important in the processes of carcinogenesis, tumour establishment and metastasis (Keski-Oja *et al,* 1988). In addition, TGF-β appears to influence metabolic pathways important to drug metabolism and steroidogenesis (Chapekar *et al,* 1990; Rainey *et al,* 1990). In vivo, limited evidence suggests that TGF-β can alter the growth and differentiation of some tumour types but is very toxic when administered in high doses. A better understanding of the response of prostate cancer cells to members of the TGF-β family may open new avenues for treating and controlling this disease.

Members of the retinoic acid family might represent an opportunity to utilize steroid hormones to modulate growth factor expression in prostate cancer (Roberts and Sporn, this issue; Sporn and Roberts 1991). For example, retinoic acid has been shown to be a potent inducer of TGF-β2 in mouse keratinocytes both in vitro and in vivo but has minimal effects on TGF-β1 mRNA levels (Glick *et al,* 1989). An association between treatment with retinoic acid and the state of phosphorylation of RB protein during induction of differentiation of leukaemia cells has been reported by Mihara *et al* (1989). Additionally, as noted above, TGF-β might exert its growth inhibitory effects through a pathway involving RB. These observations suggest that retinoids might have a role in disrupting prostate carcinogenesis in humans.

Using the Lobund-Wistar rat, Pollard and co-workers (1991) demonstrated the prevention of primary prostate cancer using N-(4-hydroxyphenyl) retinamide (4-HPR). Prostate carcinomas developed in 88% of rats inoculated with NMU at 3 months of age and then chronically treated with testosterone propionate (TP), and metastases developed in 61%. Prostate carcinomas develop spontaneously in 26% of untreated animals. When 4-HPR was added to the diet of NMU/TP treated rats, the incidence of primary prostate carcinomas and metastases dropped to 21% and 16%, respectively. In addition to being a promising preventive of prostate cancer, 4-HPR appears to be well tolerated. In more than 1000 women at high risk for the development of breast cancer, 4-HPR has been tolerated very well, with a minimum of side effects during long term administration (200 mg/day) (Rotmensz *et al,* in press).

Finally, note must be made of the presence of receptors for vitamin $D_3$ in human prostate cancer cells (Miller *et al,* 1991). Not only did Miller *et al* document the presence of vitamin $D_3$ receptors in LNCaP cells, but these cells were also stimulated by vitamin $D_3$ to grow in the absence of androgens. Petkovich *et al* (1987) had previously shown that vitamin $D_3$ increases epidermal growth factor receptors and TGF-β activity in bone derived cells. These two observations raise the possibility that manipulation of vitamin D activity in prostate cancer cells might not only alter their growth but also modulate growth factor response in these cells.

## ONCOGENES IN PROSTATE CARCINOGENESIS

The factors important in prostate carcinogenesis, except for the presence of androgens, are not known. Investigation of a number of oncogenes and suppressor genes in prostate tissue for the presence of mutations, gene amplification or overexpression has failed to document a consistent genetic change in prostate cancer. One prostate cancer cell line has been shown to contain an amplified *myc* gene (Nag and Smith, 1989); no other evidence exists to implicate amplification or mutation of this gene in prostate cancer, although a number of studies have documented changes in the expression of this gene during apoptosis and with TGF-β treatment (Buttyan *et al,* 1988; Pietenpol *et al,* 1990a). In addition, Thompson has demonstrated the development of prostate tumours in mice whose urogenital sinus epithelium has been infected with activated *myc* and *ras* genes (Thompson *et al,* 1989; Thompson 1990).

Members of the *ras* oncogene family have been examined in prostate cancer. For example, we have previously shown that introduction of activated v-Ha-*ras* into the LNCaP human prostate cancer cell line results in hormone independent growth (Voeller *et al,* 1991). In this study, LNCaP cells were transfected with expression vectors that contained either the v-Ha-*ras* or the c-Ha-*ras* gene under the control of the cadmium inducible human metallothionein IIA promoter. Numerous derivative cell lines were isolated which manifested inducible expression of Ha-*ras* p21 protein when the cells were treated with $Cd^{2+}$. Several of the cell lines expressing inducible v-Ha-*ras* manifested hormone independent growth in culture when treated with $Cd^{2+}$. Induction of v-Ha-*ras* p21 by $Cd^{2+}$ was also shown to increase anchorage independent colony formation of the v-Ha-*ras* expressing cell lines tested, thus showing that the expression of a dominant mutated oncogene can change the hormone dependent growth phenotype of prostate cancer cells. Viola has shown increased expression of *ras* p21 protein in poorly differentiated prostate tumours compared with well differentiated tumours or normal epithelium (Viola *et al*, 1986). However, his antibody could not distinguish wild type from mutated *ras* p21 protein, and the increased levels of expression of this protein might merely reflect the increased growth state of the poorly differentiated cells. In the search for activated *ras* in clinical specimens, Peehl and co-workers (1987) described Ki-*ras* activation in one of eight prostate specimens when samples were screened with the 3T3 focus formation assay. Carter *et al* (1990a) screened 24 prostate adenocarcinomas and found one specimen with an activated Ha-*ras* gene. Gumerlock *et al* (1991) examined 19 prostate cancers and found a point mutation in the 61st codon of Ha-*ras* in one specimen.

Slamon (personal communication) examined 160 tissue blocks from our tissue bank for *erb-B2* amplification and failed to find a single case of *erb-B2* amplification in prostate cancer. Barker E (personal communication) similarly re-examined 50 of these cases with a monoclonal antibody and failed to document any evidence of *erb-B2* amplification.

Cancer suppressor genes have been the focus of much attention in various cancers (Hollingsworth and Lee, 1991). One human prostate cancer cell line, DU-145 has been found to have a mutated *RB* gene. Replacement with an unmutated copy of the *RB* gene reduced the tumorigenicity within these cells (Bookstein, 1990b). Subsequent analysis of seven tissue specimens has shown three to contain mutated *RB* genes (Bookstein, 1990a). Another suppressor gene, *p53*, has been reported to be mutated in prostate cancer cell lines (Gelmann, 1991; Isaacs *et al*, 1991; Robin *et al*, 1991) but may represent an artefact of in vitro growth. Thus, it appears that the *RB* gene and the growth regulatory genes it affects, such as *myc* and TGF-β, might have vital roles in prostate cancer.

A limited number of cytogenetic studies have been performed on prostate cancer tissue. Carter *et al* (1990b) examined 28 prostate cancer specimens for loss of heterozygosity at 11 different chromosomal arms including 3p, 7p, 9q, 10p, 10q, 11p, 13q, 16p, 16q, 17p and 18q; 54% (13/24) of clinically localized tumours and 4 of 4 metastatic tumours showed loss of heterozygosity on at least one chromosome, and 30% of tumours showed loss of heterozygosity at 16q and 10q. Atkin and Baker (1985) had previously shown 4 of 4 patients with late stage prostate cancer to have deletions of 10q. In addition, deletions of 10q have been observed in several prostate cancer cell lines (Gibas *et al*, 1984; Iizumi *et al*, 1987; Konig *et al*, 1988; Brothman *et al*, 1990). Brothman and coworkers (1990) examined 30 cultured primary prostate cancer specimens. The majority of specimens examined showed a normal male karyotype. The 9 specimens with abnormal karyotypes did not show consistent changes.

## SUMMARY

Steroid hormones have an important role in prostate biology. Androgens are crucial for the normal development of the prostate gland and in maintaining its functional state in the adult. It seems that the prolonged presence of androgens might also be an important factor in the development of prostate cancer. In addition, androgens and oestrogens appear to play some part in the development of benign prostatic hyperplasia, although the exact nature of their role has not been clearly defined. Stimulation of prostate cancer growth by androgens is well established, with androgen withdrawal therapy constituting the most effective therapy in men with prostate cancer. Additive steroid therapy of metastatic prostate cancer with oestrogens or progestins has also proven effective. The effects of androgens on prostate cancer cell growth might be mediated through modulation of growth factor expression and alteration of growth factor receptor levels. Androgen response can be modulated by the expression of mutated oncogenes such as *ras*. Androgen independence can occur through a loss of AR expression or mutation of the AR; however, the pat terns of AR expression in normal prostatic tissue from development to adulthood and in cancer are now just beginning to be described. Other steroids, such as the retinoids, show promise as preventive agents, possibly through the modulation of growth factors.

## References


Atkin N and Baker M (1985) Chromosome studies of five cancers of the prostate. *Human Genetics* **70** 359–364

Barrett-Connor E, Garland C, McPhillips JB, Khaw KT and Wingard DL (1990) A prospective, population based of androstenedione, estrogens and prostate cancer. *Cancer Research* **50** 169–173

Bartsch W (1980) Interrelationships between sex hormone binding globulin and testosterone, 5 α dihydrotestosterone and estradiol 17 β in blood of normal men. *Maturitas* **2** 109–118

Bookstein R, Rao P, Madreperia SA *et al* (1990a) Promoter deletion and loss of retinoblastoma gene expression in human prostate carcinoma. *Proceedings of the National Academy of Sciences of the USA* **87** 7762–7766

Bookstein R, Shew JY, Chen PL, Scully P and Lee WH (1990b) Suppression of tumorigenicity of human prostate carcinoma cells by replacing a mutated RB gene. *Science* **247** 712–715

Bosland MC (1988) The etiopathogenesis of prostate cancer with special reference to environmental factors. *Advances in Cancer Research* **51** 1–106

Bosland MC, Prinsen MK, Dirksen TJM and Spit BJ (1990) Characterization of adenocarcinomas of the dorsolateral prostate induced in Wistar rats by N-methyl-N-nitrosourea 7, 12-dimethyl benzanthracene and 3, 2′ dimethyl-4-amino biphenyl following sequential treatment with cyproterone acetate and testosterone propionate. *Cancer Research* **50** 700–709

Brothman AR, Peehl DM, Patel AM and McNeal JE (1990) Frequency and pattern of karyotypic abnormalities in human prostate cancer. *Cancer Research* **50** 3795–3803

Brown TR, Lubahn DB, Wilson EM, Joseph DR, French FW and Midgeon CJ (1988) Deletion of the steroid binding domain of the human androgen receptor gene in one family with complete androgen insensitivity syndrome: evidence for further genetic heterogeneity in this syndrome. *Proceedings of the National Academy of Sciences of the USA* **85** 8151–8155

Buttyan R, Zakeri Z, Lockshin R and Wolgemuth D (1988) Cascade induction of *c-fos, c-myc,* and heat shock 70K transcripts dur- ing regression of the rat ventral prostate. *Molecular Endocrinology* **2** 650–657

Byar DP (1973) Proceedings: the Veterans Administration Cooperative Urologic Research Group's studies of cancer of the prostate. *Cancer* **32** 1126–1130

Cancer Statistics (1991) *Ca-A Journal for Clinicians* **41** 19–36

Carter HB and Coffey DS (1990) The prostate: an increasing medical problem. *Prostate* **16** 39–48

Carter BS, Epstein JI and Isaacs WB (1990a) *ras* gene mutation in human prostate cancer. *Cancer Research* **50** 6830–6832

Carter BS, Ewing CM, Ward WS *et al* (1990b) Allelic loss of chromosomes 16q and 10q in human prostate cancer. *Proceedings of the National Academy of Sciences of the USA* **87** 8751–8755

Catalona WJ and Scott WW (1986) Carcinoma of the prostate, In: Walsh PC, Gittes RF, Perlmutter AD and Stamey TA (eds). *Campbell's Urology*, 5th ed, pp 1463-1534, WB Saunders, Philadelphia

Chang C, Kokontis J and Liao S (1988a) Molecular cloning of human and rat complementary DNA encoding androgen receptors. *Science* **240** 324–326

Chang C, Kokontis J and Liao S (1988b) Structural analysis of complementary DNA and amino acid sequences of human and rat androgen receptors. *Proceedings of the National Academy of Sciences of the USA* **85** 7211–7215

Chang C, Chodak G, Sarac E, Takeda H and Liao S (1989a) Prostate androgen receptor: immunohistochemical localization and mRNA characterization. *Steroid Biochemistry* **34** 311–313

Chang C, Whelan CT, Popovich JC, Kokontis J and Liao S (1989b) Fusion proteins containing androgen receptor sequences and their use in the production of poly- and monoclonal anti-androgen receptor antibodies. *Endocrinology* **123** 1097–1099

Chang SM and Chung LWK (1989) Interactions between prostatic fibroblast and epithelial cells in culture: role of androgen. *Endocrinology* **125** 2719–2727

Chapekar MS, Huggett AC, Chang CC, Hampton LL, Lin KH and Thorgeirsson SS (1990) Isolation and characterization of a rat liver epithelial cell line resistant to the antiproliferative effects of transforming growth factor β (Type 1). *Cancer Research* **50** 3600–3604

Chodak GW, Vogelzang NJ, Caplan RJ, Soloway M and Smith JA (1991) Independent prognostic factors in patients with metastatic (stage D2) prostate cancer. *Journal of the American Medical Association* **265** 618–621

Chodak GW, Kranc DM, Puy LA, Takeda H, Johnson K and Chang C (1992) Nuclear localization of androgen receptor in the normal, hyperplastic and neoplastic human prostate. *Journal of Urology* **147** 798–803

Clements JA, Matheson BA, Wines DR, Brady JM, McDonald RJ and Fouder JW (1988) Androgen dependence of specific kallikrein gene family members expressed in rat prostate. *Journal of Biological Chemistry* **263** 16132–16137

Coffey DS (1988) Androgen action in the sex accessary tissues, In: Knobil J and Neill J (eds). *The Physiology of Reproduction*, pp 1081-1119, Raven Press, New York

Crawford ED, Eisenerger MA, McLeod DC *et al* (1989) A controlled trial of leuprolide with and without flutamide in prostatic carcinoma. *New England Journal of Medicine* **321** 419–424

Cumming DC and Wall SR (1985) Non-sex hormone binding globulin- bound testosterone as a marker for hyperandrogenism. *Journal of Clinical Endocrinology and Metabolism* **61** 873–876

Cunha GR, Chung LWK, Shannon JM, Taquchi O and Fujii H (1983) Hormone induced morphogenesis and growth: role of mesenchymal- epithelial interactions. *Recent Progress in Hormone Research* **39** 559–598

Dhom G (1991) Epidemiology of hormone depending tumors, In: Voight KD and Knabbe C (eds). *Endocrine Dependent Tumors*, pp 1-37, Raven Press, New York

Dickson RB and Lippman ME (1987) Estrogenic regulation of growth and polypeptide growth factor secretion in human breast carcinoma. *Endocrine Review* **8** 29–43

Egawa S, Kadmon D, Greene DR, Scardino PT and Thompson TC (1991) Adaptive changes in gene expression following androgen ablation in clonal mouse adenocarcinoma. *Journal of*

*Urology* **145** 539

Faber PW, Kuiper GGJM, vanRoij HCJ, vanderKorput JAGM, Brinkman AO and Trapman J (1989) The N terminal domain of the human androgen receptor is encoded by one large exon. *Molecular and Cellular Endocrinology* **61** 257–262

Feldman AR, Kessler L, Myers MH and Naughton MD (1986) The prevalence of cancer: estimates based on the Connecticut Tumor Registry. *New England Journal of Medicine* **315** 1394–1397

Forman BM and Samuel HH (1990) Interactions among a subfamily of nuclear hormone receptors: the regulatory zipper model. *Molecular Endocrinology* **4** 1293–1301

Geller J, Albert JD, Nachtsheim DA and Loza D (1984) Tissue dihydrotestosterone levels and clinical response to hormonal therapy in patients with advanced prostate cancer. *Journal of Clinical Endocrinology and Metabolism* **58** 36–40

Gelmann EP (1991) Oncogenes and growth factors in prostate cancer. *Journal NIH Research* **3** 62–64

Getzenberg RH and Coffey DS (1990) Tissue specificity of the hormone response in sex accessary tissues is associated with nuclear matrix protein patterns. *Molecular Endocrinology* **4** 1336–1352

Gibas Z, Becher R, Kawinski E, Horoszewicz J and Sandberg AA (1984) High resolution study of chromosome changes in a human prostatic carcinoma cell line (LNCAP). *Cancer Genetics and Cytogenetics* **11** 399–404

Glick AB, Flanders KC, Danielpour D, Yuspa H and Sporn M (1989) Retinoic acid induces transforming growth factor β2 in cultured keratinocytes and mouse epidermis. *Cell Regulation* **1** 87–97

Gumerlock PH, Poonamalee UR, Meyer FJ and deVereWhite RW (1991) Activated *ras* alleles in human carcinoma of the prostate are rare. *Cancer Research* **51** 1632–1637

Hammar M (1985) Impaired *in vitro* testicular endocrine function in elderly men. *Andrologia* **17** 444–449

Harris S, Harris MA, Rong Z *et al* (1991) Androgen regulation of HGBF I (aFGF) and characterization of the androgen receptor mRNA in the human prostate carcinoma cell line LNCaP/A-dep, In: Karr JP, Coffey DS, Smith RG, Tindall DJ (eds). *Molecular and Cellular Biology of Prostate Cancer*, pp 315-330, Plenum Press, New York

Hayashi N, Cunha GR and Wong YC (1990) Influence of male genital tract mesenchyme on differentiation of Dunning prostatic adenocarcinoma. *Cancer Research* **50** 4747–4754

Henderson BE, Ross R and Bernstein L (1988) Estrogens as a cause of human cancer. *Cancer Research* **48** 246–253

Hickey D, Todd B and Soloway MS (1986) Pretreatment testosterone levels: significance in androgen deprivation therapy. *Journal of Urology* **136** 1038–1040

Hill P, Wynder E, Garbaczewski L, Garnes H and Walker ARP (1979) Diet and urinary steroids in black and white North American men and black South African men. *Cancer Research* **39** 5101–5105

Hill P, Wynder E, Garnes H and Walker A (1980) Environmental factors, hormone status and prostate cancer. *Preventive Medicine* **9** 657–666

Hollingsworth RE and Lee WH (1991) Tumor suppressor genes: new prospects for cancer research. *Journal of the National Cancer Institute* **83** 91–96

Hsing A and Comstock G (1989) Serum hormones and risk of subsequent prostate cancer. *American Journal of Epidemiology* **130** 829

Iizumi T, Yazaki T, Kanoh S, Ikuko I and Koiso K (1987) Establishment of a new prostate carcinoma cell line (TSU-PRI). *Journal of Urology* **137** 1304–1306

Isaacs JT and Kyprianou N (1987) Development of androgen- independent tumor cells and their implication for the treatment of prostate cancer. *Urolologic Research* **15** 133–138

Isaacs WB, Carter BS and Ewing CM (1991) Wildtype p53 suppresses growth of human prostate cancer cells containing mutant p53 alleles. *Cancer Research* **51** 4716–4720

Ishikawa S, Soloway MS, Van Der Zwang R and Todd B (1989) Prognostic factors in survival free of progression after androgen deprivation therapy for treatment of prostate cancer. *Journal of Urology* **141** 1139–1142

Jeng MH and Jordan VC (1991) Growth stimulation and differential regulation of TGF β1, TGF β2 and TGF β3 messenger RNA levels by norethindrone in MCF-7 breast cancer cells. *Molecular Endocrinolology* **5** 1120–1128

Kabalin JN, Peehl DM and Stamey TA (1989) Clonal growth of human prostate epithelial cells is stimulated by fibroblasts. *Prostate* **14** 251–263

Keski-Oja J, Postlethawaite AE and Moses HL (1988) Transforming growth factors in the regulation of malignant cell growth and invasion. *Cancer Investigation* **6** 705–724

Konig JJ, Hagemeijer A, Smit B, Kamst E, Romijin J and Schroder FH (1988) Cytogenetic characterization of an established xenografted prostatic adenocarcinoma cell line (PC82). *Cancer Genetics and Cytogenetics* **34** 91–99

Kyprianou N and Isaacs JT (1988) Identification of a cellular receptor for transforming growth factor β in rat ventral prostate and its regulation by androgens. *Endocrinology* **123** 2124–2131

Kyprianou N and Isaacs JT (1989) Expression of transforming growth factor beta in the rat ventral prostate during castration induced program cell death. *Molecular Endocrinology* **3** 1515–1522

Labrie F, Dupont A and Giguere M (1986) Advantages of the combination therapy in previously untreated and treated patients with advanced prostate cancer. *Journal of Steroid Biochemistry* **25** 877–883

Li L, Hu JS and Olsen EN (1990) Different members of the *jun* proto- oncogene family exhibit distinct patterns of expression in response to type β transforming growth factor. *Journal of Biological Chemistry* **265** 1556–1562

Longcope C (1973) The effect of human chorionic gonadotropin on plasma steroid levels in young and old men. *Steroids* **21** 583–592

Lubahn DB, Joseph DR, Sullivan PM, Willard HF, French FS and Wilson EM (1988a) Cloning of human androgen receptor complementary DNA and localization to the X chromosome. *Science* **240** 327–330

Lubahn DB, Joseph DR, Sar M *et al* (1988b) The human androgen receptor: complementary deoxyribonucleic acid cloning, sequence analysis and gene expression in prostate. *Molecular Endocrinology* **2** 1265–1275

Mettlin C and Natarajan N (1983) Epidemiologic observations from the American College of Surgeons survey on prostate cancer. *Prostate* **4** 323–331

Mihara K, Cao X-R, Yen A *et al* (1989) Cell cycle-dependent regulation of phosphorylation of the human retinoblastoma gene product. *Science* **246** 1300–1303

Miller GJ, Stapleton GE, Houmiel KL and Ferrara JA (1991) Specific receptors for vitamin D3 in human prostatic carcinoma cells. In: Karr JP, Coffey DS, Smith RG, Tindall DJ (eds). *Molecular and cellular Biology of Prostate Cancer*, pp 253-260, Plenum Press, New York

Moses HL, Yang EY and Pietenpol JA (1990) TGF β stimulation and inhibition of cell proliferation: new mechanistic insights. *Cell* **63** 245–247

Nag A and Smith RG (1989) Amplification, rearrangement and elevated expression of *c-myc* in the human prostatic carcinoma cell line LNCAP. *Prostate* **15** 115–122

Nahoui K and Roger M (1990) Age-related decline of plasma bioavailable testosterone in adult men. *Journal of Steroid Biochemistry* **35** 293–299

Nankin HR and Calkins JH (1986) Decreased bioavailable testosterone in aging normal and impotent men. *Journal of Clinical Endocrinology and Metabolism* **63** 1418–1420

Noble K (1977) The development of prostatic adenocarcinoma in Nb rats, following prolonged sex hormone administration. *Cancer Research* **37** 1929–1933

Nomura A, Heilbrun LK, Stemmermann GN and Judd HL (1988) Prediagnostic serum hormones and the risk of prostate cancer. *Cancer Research* **48** 3515–3517

Peehl DM, Wehner N and Stamey TA (1987) Activated *Ki-ras* oncogene in human prostatic adenocarcinoma. *Prostate* **10** 281–289

Petkovich PM, Wrang JL, Grigoriadis AE, Heersche JNM and Sodek J (1987) 1, 25 dihydroxy vitamin $D_3$ increases epidermal growth factor receptors and transforming growth factor beta-like activity in a bone-derived cell line. *Journal of Biological Chemistry* **262** 13424–13428

Pietenpol JA, Holt JT, Stein RW and Moses HL (1990a) Transforming growth factor β, suppression of *c-myc* gene transcription: role in inhibition of keratinocyte proliferation. *Proceedings of the National Academy of Sciences of the USA* **87** 3758–3762

Pietenpol JA, Stein RW, Moran E *et al* (1990b) TGF β1 inhibition of *c-myc* transcription and growth in keratinocytes is abrogated by viral transforming proteins with pRB binding domains. *Cell* **61** 777–785

Pollard M, Luckert PH and Sporn MB (1991) Prevention of prostate cancer in Lobund-Wistar rats by 4-hydroxyphenylretinamide. *Cancer Research* **51** 3610–3611

Rainey WE, Naville D, Saez JM *et al* (1990) Transforming growth factor β inhibits steroid 17αhydroxylase cytochrome p-450 expression in ovine adrenocortical cells. *Endocrinology* **127** 1910–1915

Ripoll E, Morz VW, Kadmon D *et al* (1991) Mesenchyme-derived factors are responsible for the conversion of oncogene-transformed epithelial cells to malignant adenocarcinomas in mouse prostate. *Journal of Urology* **145** 540

Robin SJ, Hallahan DE, Ashman CR *et al* (1991) Two prostate carcinoma cell lines demonstrate abnormalities in tumor suppressor genes. *Journal of Surgical Oncology* **46** 31–37

Rotmensz N, DePalo G, Formell F *et al* Longterm tolerability of fenretinide (4HPR) in breast cancer patients. *European Journal of Cancer* (in press)

Sadi MV, Walsh PC and Barrack ER (1991) Immunohistochemical study of androgen receptors in metastatic prostate cancer. *Cancer* **67** 3057–3064

Sai T, Seino S, Chang C *et al* (1990) An exonic point mutation of the androgen receptor gene in a family with complete androgen insensitivity. *American Journal of Human Genetics* **46** 1095–1100

Sandgren EP, Luettke NC, Pelmeter RD, Brinster RL and Lee DC (1990) Overexpression of TGF α in transgenic mice: induction of epithelial hyperplasia, pancreatic metaplasia and carcinoma of the prostate. *Cell* **61** 1121–1135

Scardino PT (1989) Early detection of prostate cancer. *Urology Clinics of North America* **16** 635–655

Schuurmans ALG, Bolt J and Mulder E (1988) Androgens stimulate both growth rate and epidermal growth factor receptor activity of the human prostate tumor cell line LNCaP. *Prostate* **12** 55–64

Serio M, Gonnell P and Borelli D (1979) Human testicular secretion with increasing age. *Journal of Steroid Biochemistry* **11** 893–897

Sporn MB and Roberts AB (1991) Interactions of retinoids and transforming growth factor β in regulation of cell differentiation and proliferation. *Molecular Endocrinology* **5** 3–7

Takahashi J, Higash Y, LaNasa JA *et al* (1983) Stimulation measurement of nine testicular steroids: evidence for reduced mitochondrial function in testis of elderly men. *Journal of Clinical Endocrinology and Metabolism* **56** 1178–1187

Takeda H and Chang C (1991) Immunocytochemical and in situ hybridization analysis of androgen receptor expression during the development of the mouse prostate gland. *Journal of Endocrinology* **129** 83–89

Takeda H, Chodak G, Mutchnik S, Nakamoto T and Chang C (1990) Immunohistochemical localization of androgen receptors with mono- and polyclonal antibodies to the androgen receptor. *Journal of Endocrinology* **126** 17–25

Takeda H, Nakamoto T, Kokontis J, Chodak GW and Chang C (1991) Autoregulatiuon of androgen receptor expression in rodent prostate: immunohistochemical and in situ hybridization analysis. *Biochemical and Biophysical Research Communications* **177** 488–496

Tan J, Joseph DR, Quarmby VE *et al* (1988) The rat androgen receptor: primary structure, autoregulation of its messenger ribonucleic acid, and immunocytochemical localization of the receptor protein. *Molecular Endocrinology* **2** 1276–1285

Tenover JS, Matsumoto AM, Plymate SR and Bremmer WJ (1987) The effects of aging in normal men on bioavailable testosterone and leutinizing hormone secretion: response to clomiphene citrate. *Journal of Clinical Endocrinology and Metabolism* **65** 1118–1126

Thompson TC (1990) growth factors and oncogenes in prostate cancer. *Cancer Cells* **2** 345–354

Thompson TC, Southgate J, Kitchner G and Land H (1989) Multistage carcinogenesis induced by *ras* and *myc* oncogenes in a reconstituted organ. *Cell* **56** 917–930

Tilley WK, Marcelli M, Wilson JD and McPhaul MJ (1989) Characterization and expression of a cDNA encoding the human androgen receptor. *Proceedings of the National Academy of Sciences of the USA* **86** 327–331

Trifio M, Prior L, Pinsky L *et al* (1989) A single transition at an exonic CpG site apparently abolishes androgen receptor binding activity in a family with complete androgen insensitivity. *American Journal of Human Genetics* **45 (Supplement)** A225

Viola MV, Fronowitz F, Oravez S *et al* (1986) Expression of *ras* oncogene p24 in human prostate cancer. *New England Journal of Medicine* **314** 133–137

Voeller HJ, Wilding G and Gelmann EP (1991) v-H-ras expression confers hormone independent *in vitro* growth to LNCAP prostate carcinoma cells. *Molecular Endocrinology* **5** 209–216

Weisz A, Cicatiello L, Persico E, Scalona M and Bresciani F (1990) Estrogen stimulates expression of *c-jun* proto-oncogene. *Molecular Endocrinology* **4** 1041–1050

Wilding G (1991) Response of prostate cancer cells to peptide growth factors: transforming growth factor-β. *Cancer Surveys* **11** 147–163

Wilding G, Lippman ME and Gelmann EP (1988) The effects of steroid hormones and peptide growth factors on *c-fos* oncogene expression in human breast cancer cells. *Cancer Research* **48** 802–805

Wilding G, Chen M and Gelmann EP (1989a) Abertent responsein *vitro* of hormone responsive prostate cancer cells to antiandrogens. *Prostate* **14** 103–115

Wilding G, Knabbe C, Zugmaier G, Flanders K and Gelmann EP (1989b) Differential effects of TGF β on human prostate cancer cells *in vitro*. *Molecular and Cellular Endocrinology* **62** 79–87

Wilding G, Valvarius E, Knabbe C and Gelmann EP (1989c) The role of transforming growth factor alpha in human prostate cancer cell growth. *Prostate* **15** 1–12

Yatani R, Shiraishi T, Nakakuki K *et al* (1988) Trends in the frequency of latent prostate carcinoma in Japan from 1965-1979 to 1982-1986. *Journal of the National Cancer Institute* **80** 683–687

The author is responsible for the accuracy of the references

# Progression from Steroid Sensitive to Insensitive State in Breast Tumours

**R J B KING**

*Breast Biology Group, Imperial Cancer Research Fund, School of Biological Sciences, University of Surrey, Guildford, Surrey GU2 5XH*



## INTRODUCTION

The pioneering work of Foulds (1969) on mammary tumours clearly showed that changes accompanying the transition from normal to neoplastic epithelium did not stop with the generation of the initial cancer cells (Fig. 1). Additional changes took place that affected the subsequent behaviour of those cells, one aspect of which was loss of hormone sensitivity. Foulds demonstrated that multiple pathways led to this insensitive state, a biological observation that has been further highlighted by more recent molecular studies. This article will review current data on events involved in progression to the insensitive state with special emphasis on the involvement of steroid receptors.

Foulds generated his models of progression by analysing growth behaviour of spontaneous mouse mammary tumours that were initially pregnancy dependent, regressed at parturition and regrew at the same site in the next pregnancy. Some tumours retained this behaviour through several pregnancies, whereas others only partly regressed or continued growing at parturition. Subsequent work involving serial transplantation of mammary tumours into endocrinologically manipulated animals established two features (Kim and Depowski, 1975). Firstly, there were sequential changes in growth sensitivity,

*Cancer Surveys* Volume 14: *Growth Regulation by Nuclear Hormone Receptors*

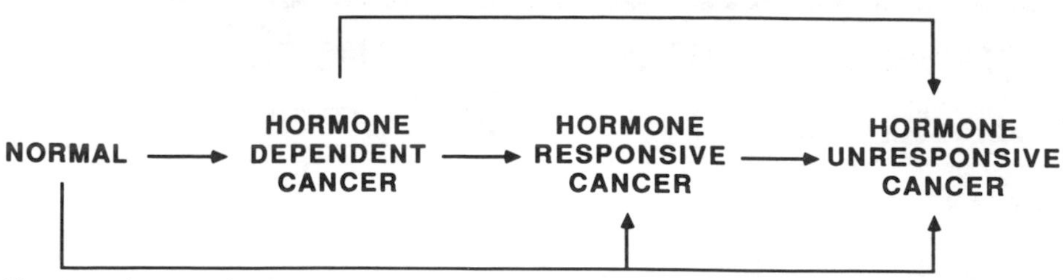


**Fig. 1.** Pathways of progression of normal hormone sensitive cells to unresponsive cancer cells

which in the case of the methylcholanthrene induced rat mammary tumour, manifested itself as a progression from dependence on prolactin→oestrogen→independence→androgen→complete hormone independence. Secondly, multiple transplants from the same tumour gave rise to a spectrum of behaviours. Such data fitted with clinical observations on the varied behaviour of multiple metastases in one patient with a good endocrine response being followed by regrowth despite continuation of treatment.

Characterization and quantitation of steroid receptors in tumours provided clues as to how changes in sensitivity might be generated. In breast and other endocrine related tumours, absence of relevant receptor was associated with an insensitive phenotype (Jensen *et al,* 1975), whereas various mutant glucocorticoid receptors were identified in unresponsive mouse lymphoma cells (Yamamoto *et al,* 1976). However, the notion of loss of response being due to loss of functional receptor proved too simple to explain the clinical data, and, with the expansion of available model systems and increasing sophistication of techniques for dissecting events, several pathways to independence have been identified.

## BIOLOGICAL CONSIDERATIONS

With two possible exceptions (Minesita and Yamaguchi, 1965; Kim and Depowski, 1975), progression always takes place from the sensitive to insensitive phenotype, which implies a selective growth advantage for the later state. It is certainly true that both in vivo and in cell culture, unresponsive tumours grow faster than their responsive counterparts (Darbre and King, 1988; King and Darbre, 1989). This important change in biological behaviour has to be incorporated into any model that attempts to explain mechanisms of response loss. Clinically, hormone manipulation aimed at decreasing growth of the sensitive cells also provides a selective advantage to the insensitive cells (Fig. 2). Figure 2 also illustrates the terminology used to describe the growth behaviour of such cells. Dependency means an absolute requirement for hormone, whereas responsive indicates altered proliferation above a basal growth rate. Sensitive implies an effect without qualifying its type. It should be emphasized that these are operational terms that apply only to the particular property being analysed. There are examples from both cell culture and clini-

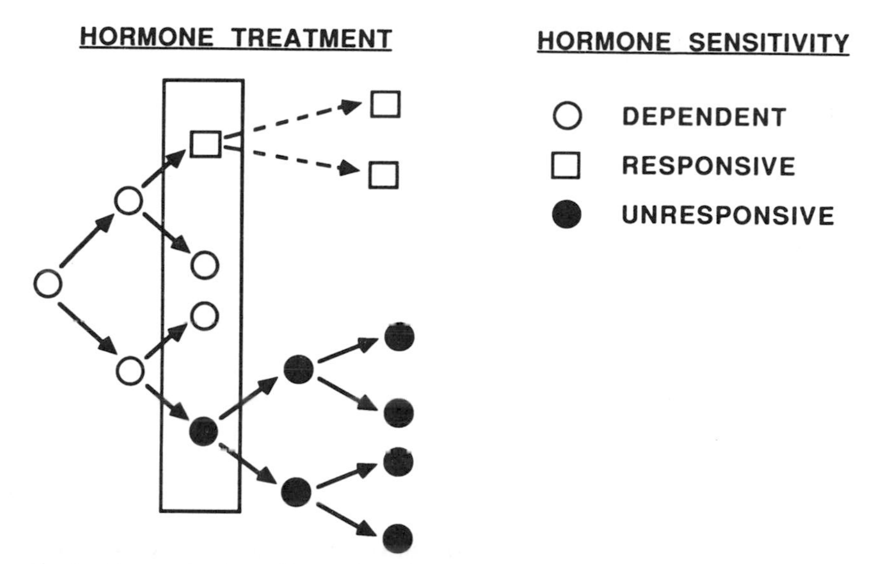


**Fig. 2.** Progression by cell selection. Hormone treatment stops growth of dependent cells and slows growth of responsive cells but has no effect on unresponsive cells. Adapted from Isaacs (1982)

cal studies in which the cells are unresponsive for growth but responsive for induction of certain proteins. Not all steroid sensitive pathways are uniformly affected by changes accompanying tumour progression.

Potential ways in which hormone sensitivity can be changed are listed in Fig. 3. These have been discussed elsewhere (King, 1990), and this article will be devoted to inheritable events related to steroid receptors.

## RELATION OF RECEPTOR CHANGES TO LOSS OF RESPONSE

The three general models covering the possible relationships are (a) loss of receptor is the causal event, (b) abnormal receptors are generated that are active in the absence of ligand and (c) loss of receptor is not required and changes observed may be a consequence of other events (Fig. 4). Each of these models will be discussed.

### Receptor Loss Associated with Hormone Independence

Two lines of evidence pointed to the relevance of such an association. Analysis of receptors in normal tissues indicated their absence in insensitive cells, and several basic and clinical studies indicated that low detectability of appropriate receptor correlated with insensitivity. The general interpretation of such results was that receptor loss accompanied response loss and that the former event might be causally related to the latter. Neither explanation distinguished between cause and effect: was receptor loss a cause or consequence of response loss? A third experimental system based on murine lymphoma pro-

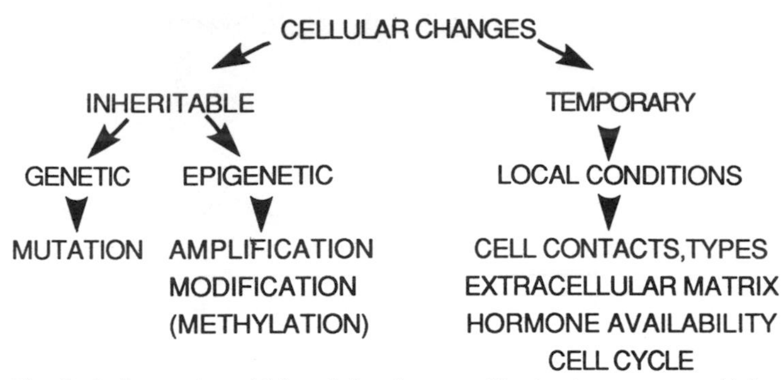


**Fig. 3.** Pathways by which cellular changes affecting hormone sensitivity can be generated. Reproduced from King (1990)

vided the link suggesting causality. These lymphoma cells are killed by glucocorticoids in cell culture. Experimental mutagenesis generated a panel of resistant clones, the majority of which either had lost glucocorticoid receptor activity or had receptor with aberrant nuclear binding properties (Yamamoto *et al*, 1976).

There was not, however, an unqualified acceptance of the model that loss of receptor resulted in hormone independent growth. Breast cancer cells are growth stimulated by oestrogen, whereas lymphomas are killed by glucocorticoids, so the relevance of the two systems to each other is questionable. Loss of glucocorticoid receptor (GR) would naturally result in absence of killing effect, but why should diminished oestradiol receptor (ER) accelerate growth of breast cancer? Models can be suggested based on relief of inhibition by the liganded complex, but to date, no evidence exists for such a mechanism with

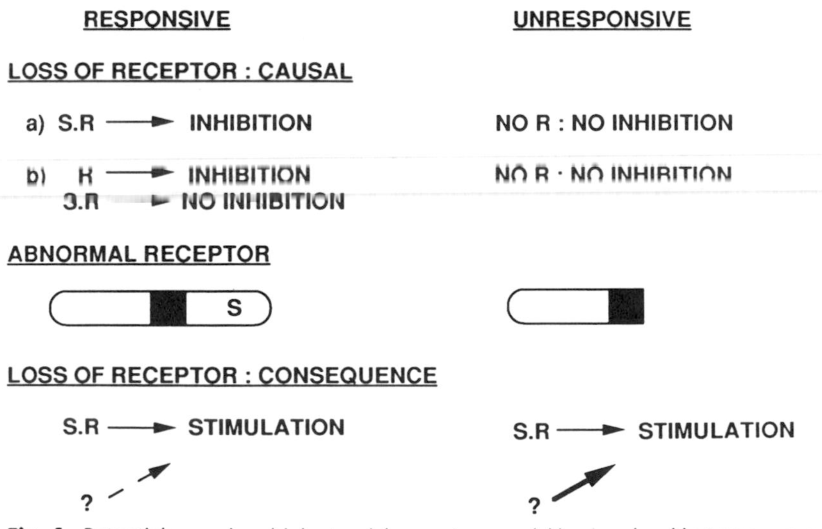


**Fig. 4.** Potential ways in which steroid receptors could be involved in tumour progression. S = steroid; R = receptor; solid box in the domain structure of the receptor is the DNA binding domain

steroids. However, thyroid hormone can act by such a mechanism (Chatterjee and Tata, this issue), so the possibility must be noted (Fig. 4).

Generation of drug resistance in cultured human breast cancer cells by increasing concentrations of drugs such as adriamycin results in concomitant loss of ER and an increase in both proliferation and expression of epidermal growth factor (EGF) receptor (Vickers *et al,* 1988). This observation satisfies the biological requirement for increased growth accompanying loss of steroid receptor, but the faster growth may be a consequence of increased EGF receptor rather than absent ER. Another method of provoking change from responsive to unresponsive tumours in culture is to deprive the cells of steroid. This is usually associated with receptor retention but can also result in loss of ER (Murphy *et al,* 1990).

Several lines of evidence thus indicate a link between loss of both growth response and receptor; it is therefore important to define events resulting in loss of receptor. Currently, the relevant data are sparse, the only conclusion being that loss of ER is not due to gross deletion of the whole ER gene. Loss of ER could take place in several ways, and thus far, three possibilities have been analysed in a preliminary manner: methylation, mutation of the gene and defective transcription. Examples of each possibility have been shown, but their biological relevance remains to be established (Murphy, 1990; King, 1992; McGuire *et al,* this issue).

All original work on receptor quantitation used radiolabelled ligands; this prompted the question of whether lack of binding indicated absence of receptor or receptor that had lost its binding function. This question has largely been answered by comparison of results obtained with ligand binding assays and immunoassays. In most cases, the two assays agree (Blankenstein, 1990), although discrepancies exist, which may indicate abnormal receptor but could also be due to technical errors in the assays. This is not a trivial point given the observation that receptors genetically engineered without a ligand binding domain can be constitutively active (King, 1992). The lymphoma cells previously mentioned contain a variant GR that does not bind ligand, but this is biologically inactive (Meisfeld, 1989).

A second question about lack of receptor is whether it is truly absent or below the detection limit of the assay. Both situations probably exist, but it is not an easy question to answer. Again, its resolution might have implications for future drug therapies. Potential ways of upregulating receptors to increase hormonal sensitivity would be more likely to work with a partially active gene than with one that has been inactivated. It is worth noting that a higher concentration of receptors per cell might be required for effects on cell proliferation than for other functions (Rabindran *et al,* 1987).

## Abnormal Receptor as a Cause of Hormone Independence

The biological basis for such a possibility was firmly established with the GR/lymphoma experiments (Meisfeld, 1989) and was reaffirmed with molecu-

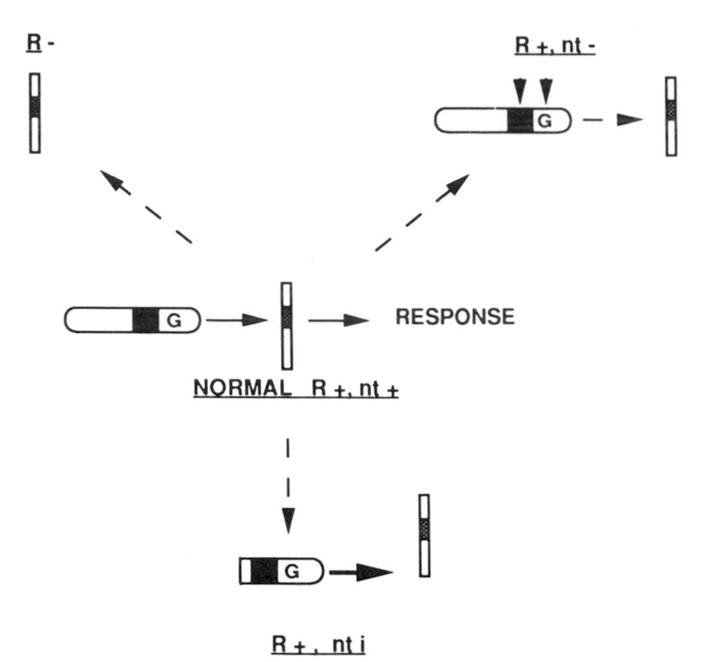


**Fig. 5.** Glucocorticoid receptor variants in mutagenized S49 mouse lymphoma cells. The wild type, responsive phenotype is receptor positive (R+) and nuclear transfer positive (nt+). Unresponsive clones can be either receptor negative (R–) or receptor positive (R+). The latter can be nuclear transfer defective (nt–) with point mutations (small arrows) in both DNA binding (solid box) and glucocorticoid (G) binding domains or they can exhibit increased nuclear binding (nti) due to a truncated receptor. DNA with specific glucocorticoid response element is represented by the stippled box

lar studies on domain structures of receptors and of receptors in genetically determined disease states such as familial vitamin D resistant rickets type II, testicular feminization and generalized resistance to thyroid hormone (King, 1992). These diseases are relevant in the sense that they establish the natural existence of receptor mutations resulting in insensitivity to ligand, whereas the experiments on receptor domains established one way in which loss of ligand binding, due to deletion of the whole domain, resulted in an active receptor. However, it should be emphasized that truncated receptors of this type do not stimulate gene function to the same extent as the wild type receptor.

Since the lymphoma experiments are more directly relevant to the present topic and have had a major influence on thinking about receptor function and sensitivity of tumours, they will be discussed in more detail. Two glucocorticoid resistant phenotypes associated with receptor changes were characterized at the molecular level after mutagenesis of cultured lymphoma cells (Fig. 5) (Meisfeld, 1989). The nuclear transfer increased (nti) variant binds more avidly to DNA than the wild type receptor and lacks the aminoterminal portion of the receptor, because of splicing errors. The receptor is functionally inactive, probably because the increased binding is to non-specific DNA rather than to specific genes. The nuclear transfer defective receptor (nt–) has a

**TABLE 1. Mammary tumour progression in culture due to long-term steroid deprivation**

| Cultural system | Functional receptor | Growth changes | | Reference |
|---|---|---|---|---|
| | | −steroid | +steroid | |
| S115 androgen | Yes | ↑ | No change | Darbre and King (1988) |
| MCF7 oestrogen | Yes | ↑ | No change | Katzenellenbogen *et al* (1987) |
| ZR75 oestrogen | Yes | ↑ | No change | Darbre and Daly (1989) |
| T47D oestrogen | Yes | ↑ | No change | Darbre and Daly (1989)<br>Daly *et al* (1990) |

point mutation in the DNA binding domain that decreases its affinity for DNA and another point mutation in the glucocorticoid binding domain that diminishes ligand affinity.

As more sophisticated molecular techniques are developed, examples of abnormal ERs in breast tumours are being documented and are reviewed by McGuire *et al* and by Horwitz in this issue and by Murphy (1990).

## Steroid Resistance Unrelated to Receptor Changes

In culture, ER positive but oestrogen resistant cells can be generated either by spontaneous changes in rapidly growing populations or by steroid deprivation resulting in slower proliferation. In the former case, clones of cells are generated with a growth advantage over the responsive cells (Reddel *et al,* 1988; Horwitz, this issue), some, but not all, of which have abnormal ER.

Steroid deprivation of several breast cancer cell lines generates variants that proliferate rapidly, are not growth stimulated by steroid agonists and, by all the standard tests, retain functional receptors. Examples of this type (Table 1), suggest that this behaviour is a general property of cancer cells but has not yet been reported for normal cells. This difference may be real, but it could simply reflect the poor availability of steroid sensitive model culture systems for normal cells.

The transition from a responsive to an unresponsive state in culture involves an ordered sequence of changes in many functions, which in the initial stages can be reversed after readdition of steroid (King and Darbre, 1989). Prolonged steroid deprivation results in permanently insensitive cells. Such unresponsive cells could be generated by spontaneous mutations, unrelated to hormone withdrawal, that yield clones of cells with a growth advantage over the responsive cells (Reddel *et al,* 1988; Graham *et al,* 1990), but two features of the experimental system listed in Table 1 suggest otherwise. Oestrogen deprivation of ZR75 human breast cancer cells results in growth cessation after about two cell generations. The cells remain static, but after some months in the absence of steroid, colonies appear and visual inspection suggests a frequency of generation of unresponsive clones of about 1 per $10^3$ cells plated

(Darbre and Daly, 1989). In the S115 murine mammary tumour model, virtually all cells change within 3 weeks (Darbre and King, 1988). Comparison of such figures with the classical frequencies of about one spontaneous mutation per $10^7$ cell generations (Baker, 1979) indicates the much higher rate of change in the steroid deprived cultures. The mechanism(s) underlying these events are ill defined, but an epigenetic explanation has been argued (King and Darbre, 1989).

In all of the examples listed in Table 1, the relevant steroid receptor remains functional as judged from transfection of sensitive genes or induction of endogenous genes that seem to be unrelated to the growth process (Darbre and King, 1988; King and Darbre, 1989). The number of receptors per cell also remains relatively constant. However, formal proof of molecular identity of the relevant receptors in parent and progeny has not been provided.

A second feature common to each test system is that the basal growth rate in the absence of steroid increases until it approaches that exhibited by the parent cells in the presence of steroid (Table 1 ). This accelerated basal growth is accompanied by changes in growth factor production (Daly *et al,* 1990; Dickson, 1990). It has been argued that this upregulation, either by the switching off of negative signals or the switching on of positive ones, is the driving force towards the insensitive state (King and Darbre, 1989). With this hypothesis, the receptor positive insensitive cells grow optimally in the absence of hormone and are unable to exceed that growth rate when the receptor pathway is activated; steroid sensitive pathways unrelated to proliferation would remain sensitive, and sensitivity to inhibitory compounds such as antioestrogens would be unchanged. It is not clear whether the accelerated growth is driven by new pathways or by constitutive expression of previously steroid sensitive genes.

### Conclusions

So many examples exist of loss of growth response—associated with loss of receptor, abnormal receptor or presence of a functional receptor—that attempts to formulate a common mechanism may be counterproductive. The biology of tumour progression clearly indicates that multiple pathways exist, and the same may be true for changes in receptor phenotype.

## ALTERED GENE SENSITIVITY AS A CAUSE OF UNRESPONSIVENESS

### Hyposensitivity

The best documented example of hyposensitivity is in the murine S115 mammary tumour cell line (Darbre and King, 1988), which has functional androgen receptor (AR) and GR and in which loss of androgen response is accompanied by loss of glucocorticoid sensitivity; addition of either type of steroid will protect against loss of response to both classes of compounds. Furthermore, a

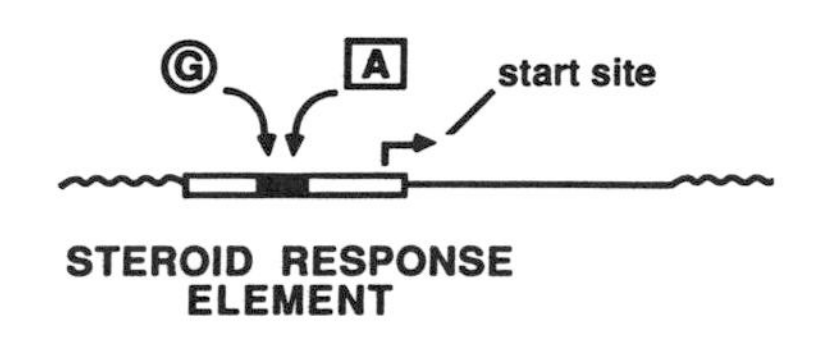


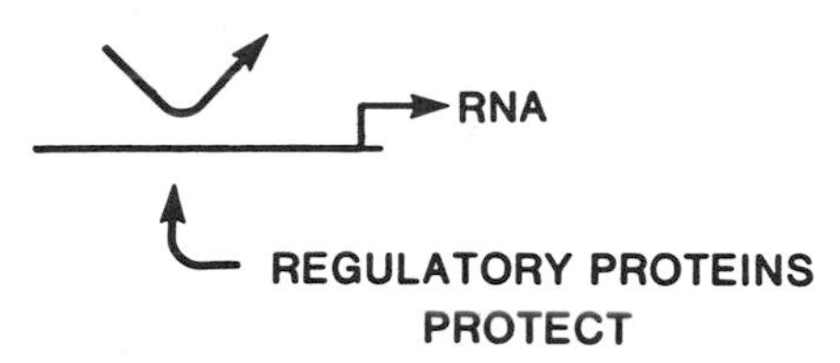


**Fig. 6.** Loss of steroid response. The DNA steroid response element (solid box) upstream from the start site of RNA transcription can bind either glucocorticoid (G) or androgen (A) receptor. Regulatory proteins such as steroid receptor complexes, when bound to such elements, protect against inactivation. In the absence of ligand or receptor, protection is lost

transgene driven by an androgen response element (ARE) is inactivated in a time dependent manner unless cells are maintained in androgen containing medium (Darbre and King, 1987). The model that best explains these data is illustrated in Fig. 6. Response elements for AR also bind GR (King, 1992) so agonist driven occupancy of those elements by either AR or GR protects against inactivation. (It is interesting, in this context, that the pure antagonist ICI 176,334 is not protective [Darbre and King, 1990], whereas another anti-androgen, hydroxyflutamide, is [Luthy and Labrie, 1987].) Changed nuclear binding of AR in some autonomous S115 variants transplanted in vivo is compatible with the hyposensitivity model, although one variant in which AR is nuclear in the absence of androgen could suggest constitutive gene expression rather than hyposensitivity (Bruchovsky and Rennie, 1978).

The androgen sensitive transgene experiments suggest that inactivation is a continual process, but its precise nature requires more study. Hypermethylation of the response elements accompanies desensitization, but a cause and effect relationship has not been established (King and Darbre, 1989).

There is no a priori reason to link increased growth rate with loss of steroid sensitivity, so what mechanisms might be involved? One view that reconciles these two features is the well known inverse correlation between differentiation and growth. When cells dedifferentiate, they usually proliferate more rapidly, and there is no doubt that the ability to respond to steroids is a differentiated function. Dedifferentiation is, by definition, the switching off of gene functions, and it is plausible that the prevention of gene inactivation by receptor steroid response element interaction should be seen in this context. What remains to be established is whether the accompanying increase in basal growth is passively related to an increased availability of the cell's economic resources or whether specific inhibitory genes are switched off. This question cannot be answered until we know more about the postreceptor events involved in hormonal regulation of growth. As growth factors and their receptors are clearly involved in this type of proliferation, an understanding of the regulatory interrelations between genes involved in steroid sensitive processes and growth factors and their receptors would be informative (see Roberts and Sporn, this issue). There is abundant evidence that steroids influence growth factor production by mammary tumours (Daly *et al,* 1990; Dickson, 1990; Nonomura *et al,* 1990), but to date, there are no indications that this involves direct interaction of steroid receptors with genes for the growth factor/receptor. A well studied indirect link exists between ER and EGF receptor. Clinically, there is an inverse relationship between the levels of these two receptors; higher numbers of the EGF receptor are associated with more aggressive tumours (Nicholson *et al,* 1990). The same association occurs in model culture systems (Davidson *et al,* 1987), the most clearcut example being the generation of drug resistant human breast cancer cells by selection in adriamycin-containing medium. This cell line is accompanied by loss of ER and upregulation of EGF receptor (Vickers *et al,* 1988). The processes involved are not known.

There are both similarities and differences between the androgen/glucocorticoid sensitive S115 cells just described and the human models. Although they all show resistance to growth stimulation after steroid deprivation, the generalized steroid insensitivity noted in murine cells is not found in human cells. Oestrogen receptor mediated stimulation of genes such as *PR* and *pS2* still takes place in the growth insensitive human cells (Darbre and Daly, 1989). Growth inhibition of the MCF7 human cell line by anti-oestrogens such as tamoxifen is retained by variants that are resistant to growth stimulation by oestrogen (Katzenellenbogen *et al,* 1987). An important explanatory point is that so called anti-oestrogens can influence gene function in the absence of oestrogen. Thus, tamoxifen increases transforming growth factor-β (TGF-β) production, which might explain the growth inhibitory effects of antagonists in the absence of agonist (Knabbe *et al,* 1987; Rochefort, 1987). Clearly, the term "anti-oestrogen" inadequately describes their properties.

These differences between human and murine models may reflect fundamental differences, but an alternative explanation might lie in the different kinetics of change. S115 cells rapidly lose sensitivity, whereas longer periods of steroid withdrawal are required for the human cells (King and Darbre, 1989). It may be that more complete steroid insensitivity could be generated in the human cells by even longer periods of withdrawal. As judged from the growth rate of parent cells in the short term absence of steroid, one could further suggest that the parent murine cells are farther along the progression pathway than their human counterparts. Additional support for the idea of progressive changes in sensitivity comes from work on MCF7 cells maintained in tamoxifen treated nude mice (Gottardis and Jordan, 1988). Oestrogen receptor positive variants arise that can be either tamoxifen sensitive or resistant, and these authors suggest that the former may be a precursor of the resistant phenotype. Resolution of these difficulties requires more data. Gene inactiva-

**TABLE 2. Main features of tumour progression model**

1. Increased steroid independent growth is accompanied by minimal change in steroid stimulated proliferation
2. Downregulation of genes having negative influence on growth (dedifferentiation) may be as important as upregulation of genes mediating positive pathways
3. Not all steroid sensitive genes need be changed
4. One class of steroid can protect against desensitization to another class of steroid provided the appropriate receptors are present. Conversely, loss of response to one steroid is paralleled by loss of response to other steroids; this is not true for growth antagonists
5. Progression is driven by absence of steroid receptor complex in regulatory regions (steroid response elements) of specific genes; this could be achieved either by steroid deprivation without loss of receptor or independently by loss of functional receptor

tion can result from steroid deprivation, but there is selectivity as to which genes are affected.

### Hypersensitivity

A possible reason for apparent autonomy from the influence of exogenous hormones could be that the cells have become hypersensitive to low endogenously produced hormone. Such a model is difficult to reconcile with clinical data that indicate refractoriness to one hormone accompanied by growth insensitivity to all types of endocrine therapies including anti-oestrogens.

## MODEL OF TUMOUR PROGRESSION DERIVED FROM CULTURED CELLS

Table 2 and Fig. 7 illustrate the principal features of a model that explains many of the points described so far. Points 1-4 in Table 2 have been discussed in preceding sections and elsewhere (Darbre and King, 1988; King and Darbre, 1989), whereas point 5 requires additional comment. Most culture experiments indicate that absence of steroid enhances progression with no major loss in functional receptor. We have previously hypothesized (King and Darbre, 1989) that absence of receptor may be a late event arising as a consequence, rather than a cause, of loss of sensitivity. There would have to be selectivity as to which steroid sensitive genes were inactivated, a process that may vary according to species and possibly the cell type involved.

An alternative view is that loss of receptor, for whatever reason, would have the same end result as steroid withdrawal, namely exposure of regulatory elements of the sensitive genes to inactivation. This hypothesis would be compatible with the clinical observation that metastases have a higher proportion of ER negative tumours than do primary samples and that a limited degree of progression from receptor positive to negative state takes place in individual patients (King and Darbre, 1989). Receptor loss in cultured mammary tumour cell lines has also been shown.

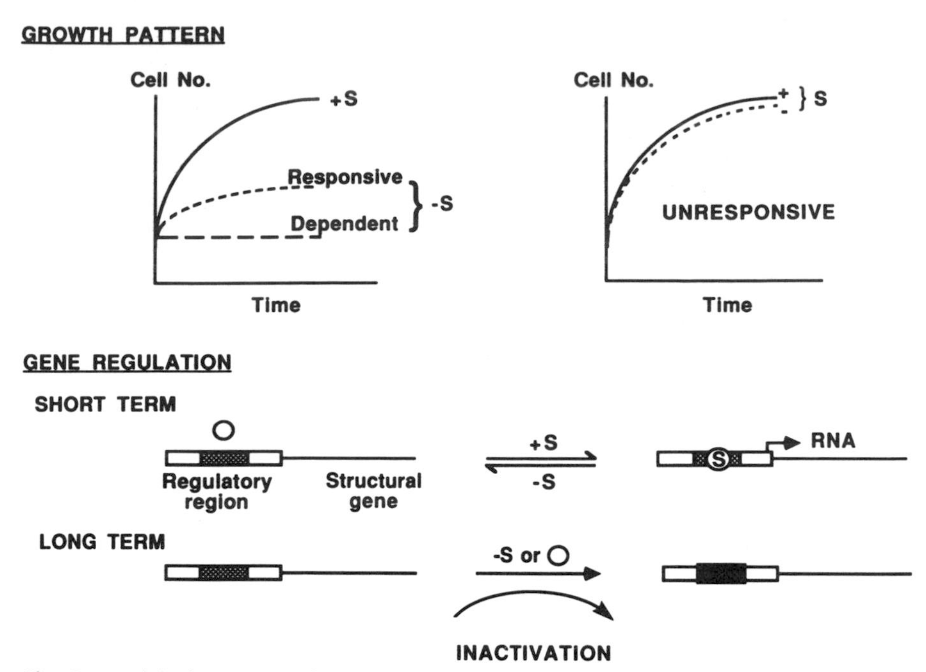


**Fig. 7.** Model of events involved in progression from responsive to unresponsive cells. The growth patterns of sensitive cells indicate either an absolute (dependent) or partial (responsive) requirement for steroid (S). Long term loss of either steroid (S) or its receptor (O) leads to increased growth. At the level of gene regulation, short term manipulation of steroid levels with sensitive cells changes transcription via interaction of the steroid receptor complex with specific regulatory regions (steroid response element, stippled box) of the gene. Long term loss of steroid or receptor exposes this region(s) to inactivation (solid box). Reproduced from King and Darbre (1989)

That genes transfected into unresponsive but receptor positive cells can lose their initial steroid sensitivity hints at a continual process of desensitization. If confirmed by more substantial data, this notion is an important one that raises the question of whether or not the process is confined to cancer cells. The only pertinent data relate to endometrium from postmenopausal women, which can be made to mimic all the features of premenopausal endometrium by re-exposure to oestrogens and progestins (King and Whitehead, 1983). This finding suggests differences between normal and neoplastic cells, but differences between endometrium and breast caution against overemphasis of this point (King, 1992).

## CLINICAL ASPECTS

The speculative model outlined in the preceding section has a number of clinical implications that are worth identifying further. The most useful biochemical markers of breast cancer behaviour are oestrogen and progestin receptors, together with indices of cell proliferation such as percentage of cells in S phase, [$^3$H]thymidine labelling index or mitotic activity (King and Darbre, 1989). The model described in the preceding section should therefore be put in the context of these markers.

The idea of postreceptor defects in the oestrogen response pathway led to the identification of ER+PR− tumours with a poorer response rate to hormone therapy than ER+PR+ tumours. However, neither category provides ideal predictions, since about one third of ER+PR+ tumours do not respond, whereas about one fourth of ER+PR− tumours exhibit good endocrine responses (Horwitz *et al,* 1985). There are undoubtedly multiple explanations for these deficiencies, such as cellular heterogeneity, strict criteria for defining a response and suboptimum analytical methods, but our model may help to define additional biological reasons that can be exploited. If there is an increasing rate of steroid insensitive proliferation during progression, then, at suboptimum steroid levels, such as exist in postmenopausal women, one would predict that slower growing ER+ tumours would have a higher response rate than faster proliferating tumours of similar ER phenotype. Inclusion of indices of proliferation is not a novel suggestion and supportive data exist for that prediction (McGuire, 1987; Paradiso *et al,* 1988).

Preferential selection of unresponsive clones undoubtedly takes place, but it is difficult to reconcile such a mechanism with the clinical observation that a good response to a first round of endocrine therapy followed by relapse is frequently indicative of response to a second endocrine treatment. Such behaviour might be anticipated if epigenetic mechanisms were involved and the initial stage of progression were reversible.

The age adjusted incidence of breast cancer rises through the premenopausal years, then temporarily plateaus or even falls at about the time of the menopause (Clemeson's hook), and then resumes its rise postmenopausally. This behaviour can be explained by two-disease models in which premenstrual and postmenopausal breast cancers exist as different entities (Farewell, 1979). The incidence changes over the menopausal period might reflect changes in growth rate of occult cancer cells and thereby changes in cancer detectability. The steroid replete premenopausal environment would promote growth of responsive cells. At the climacteric, steroid decline would have two effects: slow the growth rate and promote the generation of unresponsive cells. Tumours first detected in the perimenopausal period respond poorly to hormones and have an aggressive natural history (Hayward, 1970). This discussion assumes a fall in biologically available levels of oestrogens at the menopause. However, endogenous oestrogens are surprisingly high, in the nanomolar range, in both breast tumours and uteri from postmenopausal women. Despite these high overall concentrations, they must be biologically inactive because in the case of endometrium, the cells are atrophic (King, 1990). The basis of this accumulation of oestradiol in a biologically inert state has not been established, but the data on endogenous steroid levels in postmenopausal tissues do not invalidate these ideas.

Histochemical methods have highlighted the remarkable heterogeneity within individual human breast tumours, observations that extend to proportions of hormone sensitive and insensitive cells within a tumour mass (King, 1990). The pattern of heterogeneity does not always fit with that expected from outgrowth of clonally derived unresponsive cells. Frequently, one sees numerous individual cells or small foci of cells that stain differently to the bulk of the tumour for oestradiol receptor, progesterone receptor or oestradiol receptor related proteins. Such a pattern is compatible both with epigenetic and reversible changes in individual cells with simultaneous field changes in many cells and with data indicating that responsive and unresponsive cells in the same environment can communicate with each other (King, 1990).

## SUMMARY

Progression from a hormone sensitive to insensitive tumour can take place by several independent pathways, some of which involve loss of receptor, whereas, others may be due to the presence of abnormal receptors. Examples now exist of insensitivity despite seemingly normal receptors, such cells being generated either by spontaneous changes or by steroid deprivation. Breast cancer is a multifactorial set of diseases with steroids being only one component of the jigsaw; it is likely that further mechanisms of response loss will be identified when experimental systems are refined and additional parts of the jigsaw identified.

## References


Baker RM (1979) Nature and use of ouabain-resistant mutants, In: Hsie AW, O'Neill JP and McElheny VK (eds). *Banbury Report 2. Mammalian Mutagenesis: The Maturation of Test Systems*, pp 237–247, Cold Spring Harbor Laboratory Press, New York

Blankenstein MA (1990) Comparison of ligand binding assay and enzyme immunoassay of oestrogen receptor in human breast cancer cytosols: experience of the EORTC. *Breast Cancer Research and Treatment* **17** 91–98

Bruchovsky N and Rennie PS (1978) Classification of dependent and autonomous variants of Shionogi mammary carcinoma based on heterogeneous patterns of androgen binding. *Cell* **13** 273–280

Daly RJ, King RJB and Darbre DP (1990) Interaction of growth factors during progression towards steroid independence in T-47-D human breast cancer cells. *Journal of Cellular Biochemistry* **43** 199–211

Darbre PD and King RJB (1987) Progression to steroid insensitivity can occur irrespective of the presence of functional steroid receptors. *Cell* **51** 521–528

Darbre PD and King RJB (1988) Steroid hormone regulation of cultured breast cancer cells, In: ME Lippman and RB Dickson (eds). *Breast Cancer: Cellular and Molecular Biology*, pp 307–341, Kluwer Academic Publishers, Boston

Darbre PD and Daly RJ (1989) Effects of oestrogen on human breast cancer cells in culture. *Proceedings of the Royal Society in Edinburgh* **95B** 119–132

Darbre PD and King RJB (1990) Antiandrogen ICI 176 334 does not prevent development of androgen insensitivity in S115 mouse mammary tumour cells. *Journal of Steroid Biochemistry* **36** 385–389

Davidson NE, Gelman EP, Lippman ME *et al* (1987) Epidermal growth factor receptor gene expression in estrogen receptor-positive and negative human breast cancer cell lines. *Molecular Endocrinology* **1** 216–223

Dickson RB (1990) Stimulatory and inhibitory growth factors and breast cancer. *Journal of Steroid Biochemistry and Molecular Biology* **37** 795–803

Farewell VT (1979) Statistical methods and mathematical models for research in breast disease, In: RD Bulbrook and DJ Taylor (eds). *Commentaries on Research in Breast Disease,* pp 193–232, Alan R Liss, New York

Foulds L (1969). *Neoplastic Development 1,* pp 46–90, Academic Press, London

Gottardis MM and Jordan VC (1988) Development of Tamoxifen-stimulated growth of MCF-7 tumors in athymic mice after long-term antiestrogen administration. *Cancer Research* **48** 5183–5187

Graham ML, Krett NL, Miller LA *et al* (1990) T47Dco cells, genetically unstable and containing estrogen receptor mutations are a model for the progression of breast cancers to hormone resistance. *Cancer Research* **50** 6208–6217

Hayward J (ed) (1970) Hormones and human breast cancer, In: . *Recent Results in Cancer Research,* Springer Verlag, Berlin

Horwitz KB, Wei LL, Sedlacek SM and D'Arville CM (1985) Progestin action and progesterone receptor structure in human breast cancer: a review. *Recent Progress in Hormone Research* **41** 249–317

Isaacs J (1982) Mechanisms for and implications of the development of heterogeneity of androgen sensitivity in prostatic cancer, In: Owens AH, Coffey DS and Baylin SB (eds). *Tumor cell Heterogeneity: Origins and Implications,* pp 99–111, Academic Press, New York

Jensen EV, Polley TZ and Smith S (1975) Prediction of hormone dependency in human breast cancer, In: McGuire WL, Carbone PP and Vollmer EP (eds). *Estrogen Receptors in Human Breast Cancer,* pp 37–56, Raven Press, New York

Katzenellenbogen BS, Kendra KL, Norman MJ and Berthois Y (1987) Proliferation, hormonal responsiveness, and estrogen receptor content of MCF-7 human breast cancer cells grown in the short-term and long-term absence of estrogens. *Cancer Research* **47** 4355–4360

Kim U and Depowski MJ (1975) Progression from hormone dependence to autonomy in mammary tumors as an in vitro manifestation of sequential clonal selection. *Cancer Research* **35** 2068–2077

King RJB (1990) Heterogeneity of endocrine response within tissues. *Annals of the New York Academy of Sciences* **595** 242–250

King RJB (1991) A discussion of the roles of oestrogen and progestin in human mammary carcinogenesis. *Journal of Steriod Biochemistry and Molecular Biology* **39** 811–818

King RJB (1992) Effects of steroid hormones and related compounds on gene transcription. *Clinical Endocrinology* **36** 1–14

King RJB and Whitehead MI (1983) Estrogen and progestin effects on epithelium and stroma from pre- and postmenopausal endometria: application to clinical studies of the climacteric syndrome, In: Jasonni VM, Nenci I and Flamigni C (eds). *Steroids and Endometrial Cancer,* pp 109–115, Raven Press, New York

King RJB and Darbre PD (1989) Progression from steroid responsive to unresponsive state in breast cancer, In: Cavalli F (ed). *Endocrine Therapy of Breast Cancer III,* pp 1–15, Springer Verlag, Berlin

Knabbe C, Lippman ME, Wakefield LM *et al* (1087) Evidence that transforming growth factor-β is a hormonally regulated negative growth factor in human breast cancer cells. *Cell* **48** 417–428

Luthy I and Labrie F (1987) Development of androgen resistance in mouse mammary tumor cells can be prevented by the antiandrogen flutamide. *The Prostate* **10** 89–94

McGuire WL (1987) Prognostic factors for recurrence and survival in human breast cancer. *Breast Cancer Research and Treatment* **10** 5–9

Miesfeld RL (1989) The structure and function of steroid receptor proteins. *Critical Reviews in*

*Biochemistry and Molecular Biology* **24** 101–117
Minesita T and Yamaguchi K (1965) An androgen-dependent mouse mammary tumor. *Cancer Research* **25** 1168–1175
Murphy CS, Pink JJ and Jordan VC (1990) Characterisation of a receptor-negative hormone unresponsive clone derived from a T47D human breast cancer cell line kept under estrogen-free conditions. *Cancer Research* **50** 7285–7292
Murphy LC (1990) Estrogen receptor variants in human breast cancer. *Molecular and Cellular Endocrinology* **74** C83--C86
Nicholson S, Wright C, Sainsbury JRC *et al* (1990) Epidermal growth factor receptor (EGFr) as a marker for poor prognosis in node-negative breast cancer patients: neu and tamoxifen failure. *Journal of Steroid Biochemistry and Molecular Biology* **37** 811–814
Nomomura N, Lu J, Tanaka A *et al* (1990) Interaction of androgen-induced autocrine heparin-binding growth factor with fibroblast growth factor receptor on androgen-dependent shionogi carcinoma 115 cells. *Cancer Research* **50** 2316–2321
Paradiso A, Lorusso V, Tommasi S, Svhittulli F, Maiello E and De Lena M (1988) Relevance of cell kinetics to hormonal response of receptor positive advanced breast cancer. *Breast Cancer Research and Treatment* **11** 31–36
Rabindran SK, Danielson M and Stallcup MR (1987) Glucocorticoid-resistant lymphoma cell variants that contain functional glucocorticoid receptors. *Molecular and Cellular Biology* **7** 4211–4217
Reddel RR, Alexander IE, Koga M, Shine J and Sutherland RL (1988) Genetic instability and the development of steroid hormone insensitivity in culture T47D human breast cancer cells. *Cancer Research* **48** 4340–4347
Rochefort H (1987) Do antiestrogens and antiprogestins act as hormone antagonists or receptor-targeted drugs in breast cancer? *Trends in Pharmacological Science* **8** 126–128
Vickers PJ, Dickson RB, Shoemaker R and Cowan KH (1988) A multidrug-resistant MCF-7 human breast cancer cell line which exhibits cross-resistance to antiestrogens and hormone-independent tumor growth in vivo. *Molecular Endocrinology* **2** 886–892
Yamamoto KR, Gehring U, Stampfer MR and Sibley CH (1976) Genetic approaches to steroid hormone action. *Recent Progress in Hormone Research* **32** 3–32

The author is responsible for the accuracy of the references.

# Thyroid Hormone Receptors and Their Role in Development

**V K K CHATTERJEE[1] • J R TATA[2]**

*[1]Department of Medicine, University of Cambridge Clinical School, Addenbrooke's Hospital, Hills Road, Cambridge CB2 2QQ; [2]Laboratory of Developmental Biochemistry, National Institute for Medical Research, Mill Hill, London NW7 1AA*



## INTRODUCTION

A major characteristic of thyroid hormones is the multiplicity of cellular functions they regulate in virtually every type of vertebrate tissue (Pitt-Rivers and Tata, 1959; Tata, 1974, 1984, 1985; Baulieu and Kelly, 1990). The hormonal function has been retained through evolution and manifests itself in different ways according to the species and tissue studied (Tata, 1985). These diverse responses to thyroid hormone (TH) can be divided into two major categories: (a) regulation of metabolic activity, energy consumption and muscular activity in adult mammals and (b) the regulation of postembryonic or perinatal growth and development, including, in particular, the functional differentiation of the central nervous system and bone in mammals and the obligatory induction and maintenance of metamorphosis in amphibia and other poikilotherms (Tata, 1974, 1991; Beckingham Smith and Tata, 1976; Dussault and Walker, 1983; DeLong *et al*, 1989). In recent years, it has emerged quite clearly that, despite


 0-87969-371-1/92. $3.00 + .00

**TABLE 1. Major growth and developmental actions of thyroid hormones**

| Species | Action |
|---|---|
| Mammals and birds | Rate of overall growth via regulation of synthesis and secretion of growth hormone |
| Mammals and birds | Postembryonic maturation of central nervous system and bone |
| Amphibia, fish and reptiles | Obligatorily required for metamorphosis and developmental switch in most tissues |
| Mammals | Regulation of synthesis of some mitochondrial structures and respiratory enzymes |

this wide diversity, most, if not all, actions of thyroid hormones can be explained by their interaction with nuclear receptors that have been highly conserved through evolution (Glass and Holloway, 1990; Chin, 1991).

In this review, we restrict ourselves to the growth and developmental actions of thyroid hormones and focus on the role of their receptors in mediating these actions. The presentation below is divided into three parts. The first briefly describes the major growth and developmental actions of thyroid hormones; the second summarizes current knowledge of the structure and function of thyroid hormone receptors (TR) in the context of normal and abnormal responses to the hormones; and the third part covers the initiation and regulation of amphibian metamorphosis as a developmental model for studying thyroid hormone action, particularly the regulation of expression of its receptor genes.

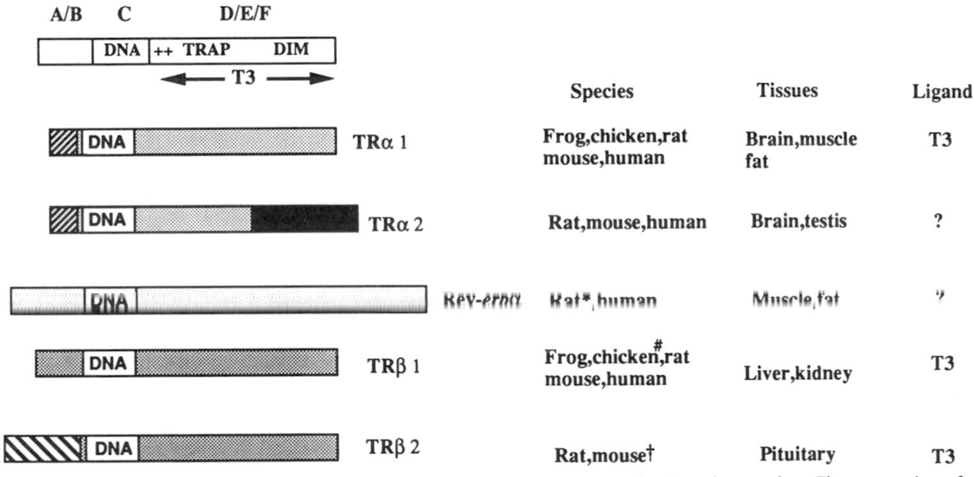


**Fig. 1.** Structural organization of thyroid hormone receptors and related proteins. They consist of conserved domains (A to F), which mediate distinct functions such as DNA binding, ligand binding, receptor dimerization (DIM), nuclear localization (++) and interaction with thyroid hormone receptor accessory factor (TRAP). The proteins are aligned via the DNA binding domains that are most homologous, and regions with divergent sequences are indicated. For each receptor isoform the putative ligand, tissues in which they are highly expressed and species from which they have been cloned are indicated. Symbols (asterisk, dagger, number symbol) indicate species in which aminoterminally truncated forms of the proteins shown are expressed

## GROWTH AND DEVELOPMENTAL ACTIONS OF THYROID HORMONES

Table 1 lists some of the major growth and developmental actions of thyroid hormones. Among the most notable actions in mammals is the fetal or late embryonic development of the central nervous system, best exemplified by cretinism in humans caused by thyroid hormone deficiency in fetal life and during the early perinatal period. A voluminous literature has accumulated over the years concerning the role of thyroid hormones in mammalian brain development (see Dussault and Walker, 1983; DeLong *et al*, 1989; Tata, 1991). This developmental action of thyroid hormones is particularly relevant to this review, since different regions of the mammalian brain display a high degree of tissue specificity in the distribution of different isoforms of TRs (Fig. 1).

In considering the role of the receptor in mediating the action of a hormonal signal, it is important to distinguish between direct and indirect pathways of hormonal action. Only a few of the developmental actions of thyroid hormones are manifestations of direct interaction between the hormone and given tissue or cell type. Most are indirect, ie exerted via other hormone or growth factors, or brought about co-operatively with other hormones or developmental factors.

A convincing demonstration of a direct hormone-target cell interaction is to reproduce the developmental or physiological action in tissue culture. For example, undifferentiated rodent Purkinje cells from thyroidectomized animals fail to grow and differentiate in culture, but the addition of thyroid hormone to the culture medium can restore this function (see Kandel and Schwartz, 1985; Siegel *et al*, 1989). We shall see below how the addition of thyroid hormone alone can induce and sustain both morphogenesis and cell death in tissue culture, characteristic of the normal processes in amphibian metamorphosis.

Some of the growth promoting effects of thyroid hormones are indirectly mediated via growth hormone (GH) released from the pituitary. This action is particularly enhanced in the action of thyroid hormone on cell lines derived from pituitary tumours, and much work has been done on the role of TR in regulating GH gene expression in this tissue culture system (Ye *et al*, 1988; Chin, 1991). An example of a positive co-operative action of thyroid hormone with another hormone is the regeneration of liver regulated by thyroid hormone and growth hormone acting in concert, whereas an example of negative co-operativity is the interaction between thyroid hormone and glucocorticoids in hepatic regulation of serum protein gene expression.

## THYROID HORMONE RECEPTORS

### Properties and Cloning of Thyroid Hormone Receptors

Some of the biochemical properties of TRs had been characterized before the cloning of the complementary DNAs (cDNA) encoding these proteins. Most mammalian cells contain approximately 2000–10 000 receptors that are associated with nuclear chromatin and capable of binding thyroid hormone with high affinity. These proteins also exhibit a hierarchy of relative affinities for other iodothyronine analogues which parallels their biological potency as follows; triiodothyroacetic acid (TRIAC)>3,5,3′-triiodo-L-thyronine (L-T3)> 3,5,3′,5′-tetraiodo-L-thyronine (L-T4)>3,3′,5′-triiodo-L-thyronine (reverse L-T3) (Chin, 1991). The observation that thyroid hormone administration increased messenger RNA (mRNA) synthesis in rat liver first led to the suggestion that these hormones might act by controlling gene expression (Tata and Widnell, 1966). Subsequent studies have confirmed that a number of mammalian genes (rat growth hormone, malic enzyme, a-myosin heavy chain) are induced, whereas others (thyrotropin α and β subunits) are repressed by thyroid hormone (Samuels *et al,* 1988). These effects occur largely at the transcriptional level and have been shown to correlate directly with occupancy of receptor by ligand (Shupnik *et al,* 1986).

Two groups simultaneously reported the cloning of cellular homologues (c-*erbA*) of the viral oncogene v-*erbA* from chick embryo and human placental tissue, respectively (Sap *et al,* 1986; Weinberger *et al,* 1986). They also noted that when expressed in vitro, the protein products of these cDNAs bound thyroid hormone and its analogues in a manner characteristic of the native receptor in vivo. The ability of c-*erbA* to mediate activation of the rat growth hormone gene promoter in the presence of thyroid hormone (Koenig *et al,* 1988) provided proof that it was indeed a TR. Inspection of the predicted aminoacid sequence of these proteins showed that they consist of a variable aminoterminal region followed by a highly homologous central cysteine rich domain and then a carboxyterminal ligand binding domain (Fig. 1). This structural organization is similar to that of other steroid hormone receptors, so that these proteins are also part of this superfamily of ligand inducible transcription factors (Evans, 1988).

## Diversity of Thyroid Hormone Receptors

The chicken and human c-erbA proteins were designated α and β, respectively, because they were structurally dissimilar, particularly in the aminoterminal region. Subsequent studies have confirmed the existence of these two receptor isoforms, which are encoded by separate genes in many species (Chin, 1991; Wood *et al,* 1991; Yaoita *et al,* 1990). However, each of these genes undergoes alternate splicing to generate distinct proteins with differing properties and tissue distributions. The structural diversity and functional properties of these splice variants are summarized in Fig. 1. The major products of the α gene are a receptor isoform designated TRα1 together with a second protein called c-erbAα2, which does not bind thyroid hormone. Remarkably, there is also evidence that the opposite strand of the α gene is also transcribed to produce yet another member of this receptor superfamily, called Rev-erbAα (Lazar *et al,* 1989), the ligand of which is unknown. The function of these other proteins remains unclear. One possibility is that they represent "orphan" receptors whose authentic ligands and target genes have yet to be identified. However, c-erbAα2 has been shown to inhibit the action of both α1 and β1 receptor isoforms when they are co-expressed, suggesting that it may act as a negative modulator of receptor action in vivo (Koenig *et al*, 1989). Similarly, it has been suggested that Rev-erbAα may regulate transcription of the sense strand of the α gene or influence alternate splicing (Lazar *et al*, 1990). The β receptor gene also generates multiple proteins differing in aminoterminal composition such that the major receptor isoform TRβ1 has an approximately 100 aminoacid aminoterminal domain, whereas a second variant protein TRβ2 has a more extended aminoterminus (Hodin *et al*, 1989).

Several functional differences between these receptor proteins have been noted in mammalian tissues. Firstly, the various isoforms are expressed in a tissue specific manner. TRα1 is found predominantly in brain, cardiac and skeletal muscle and brown fat, whereas TRβ1 is most abundant in kidney and liver and TRβ2 is found exclusively in the pituitary (Chin, 1991). Furthermore, it has been shown in both the rat and chicken that levels of TRα1 in brain remain relatively constant during embryonic development, whereas TRβ1 expression is induced markedly at birth or following hatching (Forrest *et al*, 1991; Mellstrom *et al*, 1991). Secondly, TRβ1 and TRα1 exhibit differing affinities and responsiveness to triiodothyronine and TRIAC (Thompson and Evans, 1989; Scheuler *et al*, 1990). Thirdly, thyroid hormone has divergent effects on TR gene expression, causing a marked suppression of TRβ2 mRNA levels, with lesser effects on TRα1 and TRβ1 mRNAs (Hodin *et al*, 1990). Similarly, TRα and TRβ genes are also differentially regulated by thyroid hormone during amphibian metamorphosis. Thus, this diversity of receptor proteins probably contributes significantly to tissue specific variations in thyroid hormone responsiveness.

## Structural Determinants of Receptor Action

Although TR isoforms have aminoterminal A/B domains of varying length and composition (Fig. 1), the role of this region is poorly understood. Thyroid hormone receptor α1 is known to be phosphorylated at two serine residues within this domain (Goldberg *et al*, 1988), but the functional significance of this post-translational modification remains unclear. Furthermore, the other receptor isoforms (TRβ1, TRβ2) do not contain homologous sequences. Transfection studies using both positively and negatively regulated target genes show that the receptor isoforms are functionally equivalent (Hodin *et al*, 1989; Rentoumis *et al*, 1990) and that removal of the aminoterminus does not impair receptor dependent *trans*-activation (Thompson and Evans, 1989). Nevertheless, the possibility remains that as more thyroid hormone responsive genes are tested, promoter or cell type specific differences in the action of these receptor isoforms will become apparent.

The central C domain of the receptors consists of 68 aminoacids, including

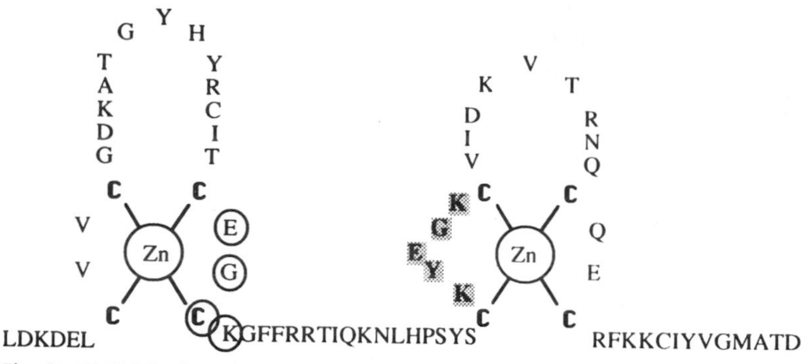


**Fig. 2.** DNA binding domain of the thyroid hormone receptor. Single letter aminoacid codes are used, and a central zinc ion is shown co-ordinated tetrahedrally by conserved cysteine residues to form a "zinc finger". Aminoacids E,G,C and K (circled) constitute the "P" box and are important for DNA sequence recognition. Residues within the "D" box (shaded) are involved in discrimination of spacing and orientation of the DNA sequence motifs

nine cysteine residues that are highly conserved among all members of the steroid/thyroid hormone superfamily (Evans, 1988; Green and Chambon, 1988). By analogy with the proposed structure for *Xenopus* transcription factor IIIA, these invariant cysteines are thought to co-ordinate zinc to form "fingers" (Fig. 2) that mediate binding of TR to a specific DNA recognition sequence or thyroid response element (TRE), usually located in the promoter regions of responsive genes. Point mutagenesis of these conserved cysteine residues abolishes receptor-TRE interactions as well as function (Chatterjee *et al,* 1989, 1991). Umesono and Evans (1989) have also elegantly demonstrated that two aminoacid sequence motifs located at the base of the first finger and in the stem of the second finger, designated the P and D boxes, respectively (Fig. 2), play an important part in DNA sequence recognition. Structural studies of the C domains of the glucocorticoid and oestrogen receptors (Schwabe and Rhodes, 1991) confirm that aminoacids within the P box form part of an α helix, which interacts with DNA, and that TR is likely to adopt a similar conformation.

The carboxyterminal D/E/F domain of the receptor encodes many functions. Immunocytochemical studies show that, unlike other members of the steroid receptor superfamily, TR is constitutively located in the cell nucleus (Macchia *et al,* 1990), probably reflecting the fact that it is not associated with cytosolic heat shock proteins (Dalman *et al,* 1990). Nevertheless, the translocation of TR from the cytoplasm into the nucleus is thought to be mediated by a specific sequence of aminoacids resembling those identified in other proteins, such as the SV40 large T antigen or progesterone receptor (Guichon-Mantel *et al,* 1989). A short basic peptide sequence within the receptor D domain is indeed capable of targeting a heterologous cytoplasmic protein to the nucleus (Dang and Lee, 1989).

Experiments in which various domains of TR were coupled to a heterologous DNA binding domain indicate that the hormone binding and *trans-*

activation functions of TR are encompassed within the D/E/F domains (Thompson and Evans, 1989). However, further deletions within either the D domain (Lin *et al,* 1991) or truncation of aminoacids at the carboxyterminus (Chatterjee *et al,* 1991) lead to a dramatic loss of ligand binding capacity. Single point mutations within this region are equally deleterious (Chatterjee *et al,* 1991), suggesting that the ability to bind ligand is distributed widely across this region of the receptor. Forman *et al* (1989) observed that truncated transcriptionally inactive TRα1 receptors lacking A/B and C domains were capable of inhibiting wild type receptor function when they were co-expressed. Deletion analyses showed that a region between aminoacids 199 and 408 is required to elicit this inhibition. Inspection of aminoacid sequences within this domain revealed the presence of eight heptad repeat motifs containing hydrophobic aminoacids at the first, fifth and eighth positions. The authors suggested that if these heptads were to fold into coiled coil helical structures, a dimerization interface analogous to the "leucine zipper" motif found in other proteins would be generated. Other studies have provided biochemical confirmation of homodimer formation both between TRs as well as of heterodimers between thyroid hormone and retinoic acid receptors (Glass and Holloway, 1990) when bound to a DNA recognition sequence.

The mechanism by which TR activates gene transcription remains obscure. A helical wheel plot indicates that the nine aminoacids at the extreme carboxyterminus of TR could form an amphipathic helix made up of hydrophobic and acidic residues analogous to activation domains present in proteins such as GAL4 and VP16 (Zenke *et al,* 1990) and may therefore represent the ligand dependent *trans*-activation domain of the receptor. Other groups have observed that TR is capable of interacting with auxiliary proteins (TRAPs) in nuclear extracts and that a highly conserved sequence within the E domain mediates this interaction (Rosen *et al,* 1991). Some mutations in this region are capable of selectively abolishing interaction with TRAP and *trans*-activation, whereas ligand binding is unaffected. This has led to the suggestion that TRAP functions as an adaptor which, upon activation by ligand-occupied receptor, stimulates the transcription initiation complex so that the conserved region in the receptor E domain can be considered a *trans*-activation sequence.

## Positive and Negative Regulation of Gene Expression

The induction of rat growth hormone (rGH) gene expression in cultured pituitary cells has proved to be a useful model system to examine positive transcriptional regulation by thyroid hormone. Deletion and mutational analyses of the gene promoter by several groups has identified a broad region between –209 and –166 bp as being important for hormone responsiveness (Samuels *et al,* 1988). The cloning of TR has enabled studies to be carried out on the interaction between DNA sequences in this region and the expressed receptor protein. Glass *et al* (1988) identified a 16 bp element between –177 and –162 bp that bound receptor and also noted that it shared homology with a near-

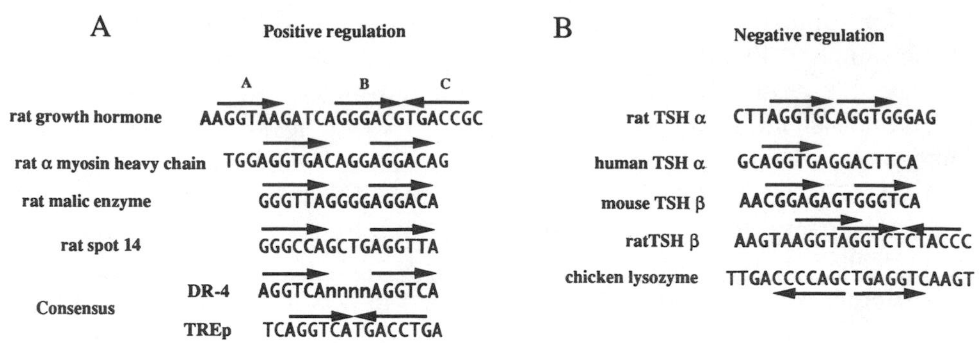


**Fig. 3.** Positive (A) and negative (B) thyroid response elements. Nucleotide sequences from the promoter regions of different target genes that are induced or repressed by thyroid hormone are shown. (Positive regulation sequence information from Brent *et al* [1989] and Umesono *et al* [1991]. Negative regulation, see references in text.) Arrows overlie sequences that interact with a receptor monomer. Consensus positive TREs that have been derived are shown

palindromic oestrogen response element in the vitellogenin gene. A systematic series of mutations enabled a synthetic palindromic sequence to be derived (TREp, Fig. 3A) that represents an optimized TRE capable of conferring marked thyroid hormone responsiveness to a heterologous gene promoter. Subsequently, TREp has also been shown to be responsive to retinoic acid (Umesono *et al,* 1988). However, several observations do not fit this model for positive regulation. Firstly, many naturally occurring TREs appear to adopt a configuration consisting of tandem direct repeats rather than palindromic sequences (Fig. 3A), and, secondly, unlike the synthetic TREp, these elements are not responsive to retinoic acid. In addition, Brent *et al* (1989) have shown that a third upstream receptor binding site between −189 and −184 bp is required for maximal responsiveness of the rGH promoter to thyroid hormone. Taking all of these observations into account, it has been suggested that both palindromic response elements consisting of the motif AGGTCA or elements containing direct repeats of this sequence separated by four nucleotides (DR-4) are capable of mediating a positive transcriptional response to thyroid hormone. According to this model, the rGH promoter contains a tripartite TRE consisting of both direct repeat (A,B) and palindromic (B,C) motifs.

Many groups have noted that in the absence of ligand, TR is capable of occupying positive TREs and suppressing the basal level of transcription from the promoter. Such a mechanism is thought to operate in the chicken lysozyme gene (Bahniahmad *et al,* 1990), where an inverted palindromic DNA sequence (Fig. 3B) acts as a "silencer" when bound to unoccupied TR. By contrast, negative regulation in the context of other gene promoters is a strictly hormone dependent process. This is best exemplified by the genes encoding the α and β subunits of thyroid stimulating hormone (TSHα and TSHβ), which are subject to negative transcriptional regulation by the end organ product—thyroid hormone—in classical feedback fashion. Several groups have used transient expression assays as well as receptor-DNA binding studies to

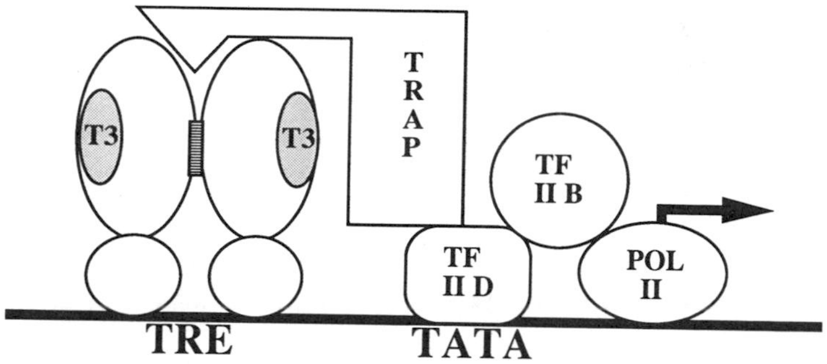


**Fig. 4.** Proposed model for thyroid hormone receptor action. The transcription initiation complex consists of RNA polymerase II (POL II) together with factors (TFIIB, TFIID) that recognize a "TATA" motif in the gene promoter. An auxiliary protein (TRAP) may transduce the signal from ligand activated TR bound to a thyroid response element (TRE)

delineate negative thyroid response elements (nTREs) that mediate inhibition of the *TSHα* and *TSHβ* genes from a variety of species (Burnside *et al,* 1989; Chatterjee *et al,* 1989; Darling *et al,* 1989; Wood *et al,* 1989), as shown in Fig. 3B. One common feature is that they appear to be tandemly repeated motifs that either overlap or exhibit minimal spacing, but these elements have not been characterized sufficiently to derive a consensus sequence. The nTREs all localize to regions adjacent to the TATA box or in the 5′ untranslated regions of these genes, suggesting that negative regulation may occur by displacement or inhibition of the transcription initiation complex.

The ability of TR to bind to a diverse array of response elements with variable spacing and orientation has led to the suggestion that there may be considerable flexibility in the interaction between receptor monomers when bound to a DNA template (Schwabe and Rhodes, 1991). Figure 4 depicts our current understanding of the way in which the receptor modulates gene transcription. According to this model, the receptor interacts with DNA as a dimer. Hormone binding induces a conformational change and alters interactions with accessory proteins, thus modulating the activity of the transcription initiation complex.

## Thyroid Hormone Resistance Syndromes

The cloning of TRs has led to the elucidation of the pathogenetic defect that causes the syndrome of generalized thyroid hormone resistance in humans. This disorder, usually inherited as an autosomal dominant, is characterized by elevated circulating levels of thyroid hormone together with central and peripheral refractoriness to hormone action. It is now known that affected individuals have single aminoacid point mutations affecting one of their two β receptor gene alleles (Usala and Weintraub, 1991). The clinical manifestations in these cases are relatively mild but can include short stature as well as mental abnormalities such as hyperactivity and attention deficit, mild dyslexia and a reduced IQ. By contrast, a single family with a recessive form of the disorder exhibits a more severe phenotype (Refetoff *et al,* 1967). Short stature is associated with radiological evidence of epiphyseal dysplasia. Mental abnormalities include sensorineural deafness as well as mutism. Abnormal somatic features such as bird like facies, an abnormally protruberant sternum and winged scapulae have also been noted. The genetic defect in this kindred consists of a deletion of most of the coding exons of TRβ in both alleles. These observations suggest that TRα1 receptors alone are capable of promoting fetal survival and provide direct evidence for the role of TRβ1 in normal human growth and development. Heterozygous members of this family with a single deleted β receptor allele were completely normal when assessed with both biochemical and clinical criteria (Takeda *et al* 1991). This argues against mere deficiency of functional β receptor as the cause of resistance in the dominantly inherited forms of the disorder described above. Indeed, there is now evidence that the mutant receptors inhibit the activity of their normal counterparts in a "dominant negative" manner to generate the abnormal phenotype in these cases (Chatterjee *et al,* 1991). Further evidence for such a mechanism is provided by another family, in which an individual inherited two alleles of a "dominant negative" mutant β receptor (Usala and Weintraub, 1991), resulting in severe growth and mental retardation together with biochemical evidence of extreme resistance to thyroid hormone action. The nature of this inhibitory interaction between wild type and mutant receptor has yet to be elucidated.

## THYROID HORMONE RECEPTORS AND AMPHIBIAN METAMORPHOSIS

Metamorphosis is a dramatic example of postembryonic development in oviparous animals (Gilbert and Frieden, 1981). The process is obligatorily controlled by hormones, with great similarities between invertebrates and vertebrates; ecdysone and thyroid hormone induce and sustain metamorphosis, whereas juvenile hormone and prolactin delay or arrest it (Granger and Bollenbacher, 1981; White and Nicoll, 1981). The major advantages of metamorphosis as a model developmental system are: (a) many features are also common to mammalian postembryonic or fetal development leading to establishment of the adult phenotype; (b) the process is easily manipulated by hormones in both directions; (c) the same hormonal signal simultaneously sets in motion different developmental programmes in different types of target cells and (d) it is an excellent model for studying developmental hormone action and the functional analysis of receptors.

### Biochemical and Cellular Responses of Amphibian Cells to Thyroid Hormones

Considerable information is available on the biochemical changes and cellular reorganization of amphibian tadpole tissues in response to thyroid hormones (Weber, 1967; Cohen, 1970; Frieden and Just, 1970; Beckingham Smith and

**TABLE 2. Some cellular and biochemical responses during thyroid hormone induced metamorphosis**

| Tissue | Response |
|---|---|
| Brain | Structural and functional remodelling |
| Liver | Induction of urea cycle enzymes, albumin; larval to adult haemoglobin |
| Skin | Induction of collagen and 63 kDa keratin |
| Eye | Porphyropsin to rhodopsin; induction of β-crystallin |
| Limb bud | Limb formation |
| Pancreas, gut | Cell death and replacement |
| Tail | Complete regression |

Tata, 1976; Gilbert and Frieden, 1981). Table 2 lists some that have been particularly intensively studied. It is clear that almost every single cell type in the tadpole responds to the same hormonal signal (via the same receptors) to activate such diverse developmental programmes as remodelling of the central nervous system, morphogenesis of limbs and cell death in tail and gills. Furthermore, the changes induced by thyroid hormone in central nervous system and visual structures, the activation of albumin and adult haemoglobin genes and programmed cell death all have their counterparts during postembryonic and perinatal development in mammals.

## Thyroid Hormone Receptors in Amphibia

Unlike the extensive information available on the structure and functional domains of mammalian thyroid hormone receptors and their mRNA (Ye *et al*, 1988; Glass and Holloway, 1990; Chin, 1991), relatively little is known about amphibian TR genes and their products. Brooks *et al* (1989) were the first to clone a full length TR cDNA from *Xenopus laevis*: from its sequence and ligand binding activity, these authors showed that it was similar in its domain structure to that of mammalian and avian TRα. Yaoita *et al* (1990) described the presence of two genes each for TRα and TRβ. The two TRα genes were shown to be very similar to their mammalian counterparts, although no alternatively spliced TRα cDNA clones were detected. In sharp contrast, *Xenopus* TRβ mRNA exhibited a highly complex pattern of alternative splicing within the 5′ untranslated region, as well as multiple transcriptional start sites. These workers thus identified eight exons that are alternatively spliced, giving rise to a minimum of two aminotermini for each of the two TRβ proteins. Both TRα and TRβ genes contain multiple AUG codons with short open reading frames. However, the functional significance of the multiplicity of *Xenopus* TRβs remains unknown.

## Developmental, Hormonal and Spatial Expression of *Xenopus* TR

Normal metamorphosis in *Xenopus* tadpoles is only initiated after developmental stage 54 (2–3 months after fertilization) when the tadpole thyroid gland begins to secrete thyroid hormones (Nieuwkoop and Faber, 1967; Leloup and Buscaglia, 1977). However, the tadpole acquires a response to exogenous thyroid hormone in the first week after fertilization (Tata, 1968; Mathisen and Miller, 1987; Moskaitis *et al*, 1989). This suggests that the receptor is expressed well before the target cells would be exposed to endogenous thyroid hormones.

Functional TR protein has not been identified in premetamorphic tadpoles, but TR like T3 binding activity has been demonstrated in Rana catesbeiana tadpoles (Moriya *et al*, 1984; Galton and St Germain, 1985). However, with the cloning of *Xenopus* TRα and TRβ cDNAs, it has been possible to establish how TR genes are expressed during development. Using an assay in which the TR mRNA is converted to cDNA, followed by northern blotting, Yaoita and Brown (1990) demonstrated the presence of TRα transcripts by stage 44 and TRβ transcripts at later stages. Similar results were obtained by RNase protection assays by Kawahara *et al* (1991), which showed that the transcripts accumulated rapidly with development, reaching their maximum levels by metamorphic climax (stages 58–62), then declining to almost undetectable levels upon completion of metamorphosis. This pattern is compatible with the differential and rapidly increasing sensitivity of tadpole tissues to thyroid hormones as metamorphosis progresses, as well as the virtual absence of response of adult amphibia to the hormone (Tata, 1968; Gilbert and Frieden, 1981).

In situ hybridization analysis confirmed the presence of TR mRNAs in early developmental stages of *Xenopus* tadpoles (Kawahara *et al*, 1991). As early as 1 week after fertilization (stage 44), TR mRNAs were found predominantly located in the tadpole brain, spinal cord, intestinal epithelium, tail and liver. At mid metamorphosis, strong hybridization signals were also recorded in the hind limb buds. Thus, TR genes are particularly extensively expressed in tissues programmed for morphological and biochemical remodelling (brain, liver), de novo morphogenesis (limb buds) and cell death (tail, intestine) during T3 regulated metamorphosis.

It is relevant to compare the results obtained by Vennstrom's group, combining biochemical and in situ analysis, on the differential expression of TRα and TRβ genes in different regions during chick embryonic brain development (Forrest *et al*, 1991). These workers noted that TRα was expressed from early embryonic stages (by day 9), well before the tissue acquires hormone sensitivity, whereas TRβ was sharply induced after day 19, which coincides with the known hormone sensitive period. Furthermore, they also observed that TRα and TRβ mRNAs were differentially expressed in different regions of the brain, particularly in the cerebellum. In *Xenopus* development, whereas the TR transcripts disappeared rapidly from most tissues upon completion of metamorphosis, in situ hybridization analysis revealed, quite surprisingly, high levels of accumulation in stage 1 and 2 oocytes of the developing froglet ovary (Kawahara *et al*, 1991). Although these transcripts did not accumulate further with growth of the oocytes, they were exceptionally stable, thus raising the

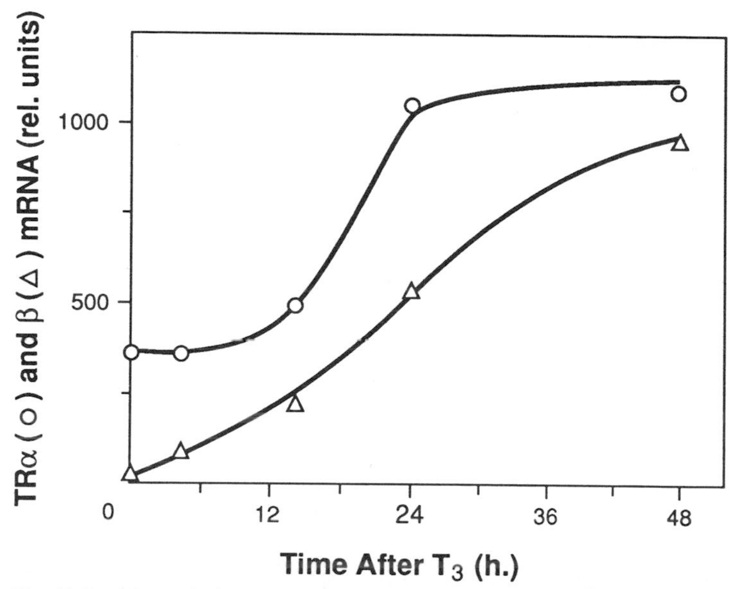


**Fig. 5.** Rapid autoinduction of TRα and TRβ mRNAs by $10^{-9}$ mol/l T3 in stage 50 (premetamorphic) *Xenopus* tadpoles. Details in Baker and Tata (in press)

likelihood that the trace levels of TR mRNAs found in early *Xenopus* embryos were of maternal origin and that the response to thyroid hormone of tadpoles at early stages (42–44) may be effected via TR synthesized on maternally derived mRNA, as will be considered below.

### Autoinduction of TR

An intriguing finding in the above studies is the phenomenon of autoinduction of TR genes by thyroid hormone (Yaoita and Brown, 1990; Kawahara *et al*, 1991). Exposure of tadpoles at premetamorphic stages (48–52) to exogenous T3 substantially enhanced the accumulation of TR mRNA, which could explain the accelerated increase in sensitivity of tadpoles to thyroid hormones at the onset of natural metamorphosis. The autoinduction of TR mRNA was more marked for TRβ than for TRα. As shown in Fig. 5, $10^{-9}$ mol/l T3 administered to stage 52 tadpoles increased the relative amount of TRα transcripts by 2–4-fold, whereas that of TRβ was augmented by 20–50-fold. Most importantly, the autoinduction of TR was found to be extremely rapid; that of TRβ mRNA could be detected by 4 hours after stage 52 tadpoles were exposed to exogenous T3 (Baker and Tata, 1992). This is the most rapid biochemical response to T3 involving gene expression yet described for amphibian larvae. In a wider context, it is worth considering that the phenomenon of autoinduction may not be restricted to TR but may occur with other developmentally important nuclear receptors. Upregulation of oestrogen receptor and its mRNA has been reported in *Xenopus* liver and oviduct (Perlman *et al*, 1984; Baker and Tata, 1990; Varriale and Tata, 1990), and a retinoic responsive element has been identified in the promoter of retinoic acid receptor β gene, which may also explain its auto induction (de Thé *et al*, 1990).

## Juvenilizing Action of Prolactin and Autoinduction of TR mRNA

An important question that arises is whether the upregulation by T3 of its own receptor would be of any physiological significance. During early tadpole development in many amphibia, substantial amounts of prolactin are detected in the pituitary and blood. Both the circulating and hypophyseal prolactin disappear very precipitously around the onset of metamorphosis, the kinetics of its disappearance exhibiting a close reciprocal relationship to the rapidly increasing appearance of thyroid hormones in blood (White and Nicoll, 1981; Duellman and Trueb, 1986).

The significance of this developmental control of prolactin synthesis and secretion lies in the property of prolactin to prevent the induction of metamorphosis by endogenous or exogenous thyroid hormone (Nicoll, 1974; Tata *et al*, 1991). The kinetics of prolactin secretion and its anti-metamorphic properties exhibit a striking parallel with the interplay between juvenile hormone and ecdysteroids in invertebrate development (Gilbert and Frieden, 1981; Granger and Bollenbacher, 1981). Prolactin can therefore be considered as a truly juvenilizing hormone. It is therefore significant that when prolactin was added to organ cultures of *Xenopus* limb buds and tails, it prevented both morphogenesis and cell death induced by T3 in these two tissues (Tata *et al*, 1991).

Since the autoinduction of TR mRNA is the most rapid biosynthetic response to T3 known in amphibian metamorphosis, the TRα and TRβ genes can be considered as "early" genes activated by the hormone. By contrast, the 63 kDa keratin gene, which is only expressed in adults and whose induction by T3 requires 2–3 days, can be considered a "late" gene activated via TR (Mathisen and Miller, 1989). Baker and Tata (1992) were able to show that exogenous prolactin administered to stage 50–54 *Xenopus* tadpoles or to tail organ cultures completely abolished the rapid T3 induced autoinduction of TRα and TRβ mRNAs. This inhibition, shown only for TRβ mRNA in Fig. 6, was followed by that of the de novo activation of the 63 kDa keratin gene by T3. This sequential inhibition of the expression of the "early" and "late" genes has led to the proposal of a testable model whereby prolactin exerts its juvenilizing action in *Xenopus* by preventing the amplification of TR by the autoinduction by T3 of its mRNA.

## A Model for the Physiological Role of Autoinduction of TR

According to the proposed model (Fig. 7) the presence of TR transcripts in oocytes, embryos and early stages of tadpoles (Kawahara *et al*, 1991) indicates that traces of functional TR of maternal origin could be present in stages beyond 44 of premetamorphic *Xenopus* tadpoles. If one makes the verifiable assumption that the receptor-thyroid hormone complex interacts with high affinity with putative TRE(s) in the TR gene promoter, then the earliest consequence of first trickle of thyroid hormones secreted by the tadpole thyroid gland at around stage 54 would be to increase rapidly the amount of TR mRNA and thereby the functional receptor. The TR gene is here considered

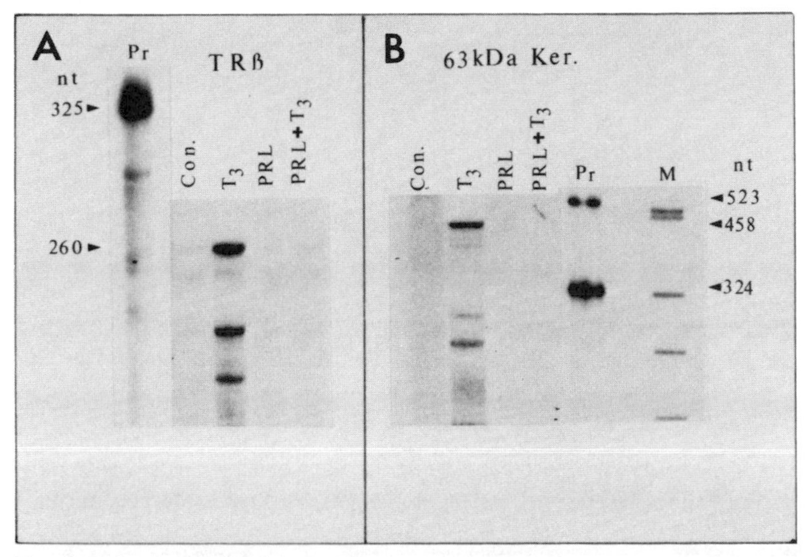


**Fig. 6.** Prolactin (PRL) inhibits the activation by T3 of (A) TRβ and (B) 63 kDa keratin genes in stage 52 *Xenopus* tadpoles. Batches of 18 tadpoles were treated with T3 and PRL alone or together. The tadpoles were treated for 4 days with 0.2 iu/ml PRL, whereas $10^{-9}$ mol/l T3 was administered for the last 3 days. RNA was extracted and TRβ (protected band 260 nt) and 63 kDa keratin mRNAs (protected band 458 nt) in the same RNA sample were assayed by means of RNase protection with the respective probes (325 and 523 nt). Note that the usual doublet protected bands for TRβ mRNA at 260 nt has not been resolved in this electrophoretic gel. In B, the position of a 324 nt 5S RNA cRNA probe used for an internal standard is also shown. M = markers. Autoradiogram was exposed for 4 days. Other details as in Baker and Tata (1992)

as an "immediate early" gene. The upregulation of TR would then lead to the relatively slow activation of the target genes whose products, such as albumin, globin and 63 kDa keratin, constituting the establishment of the adult phenotype, could be considered as "late" genes. Although the mechanism of action of prolactin is not known (Rillema, 1987), the decline in circulating prolactin and the simultaneous rise in the secretion of thyroid hormones into the blood (Nicoll, 1974; Duellman and Trueb, 1986) would cause a marked amplification of TR accumulation. The autoinduction of thyroid hormone receptor by thyroid hormone would thus escape from the anti-metamorphic or juvenilizing action of prolactin. It raises the important question of whether or not prolactin or the related placental lactogen, known to be present in fetal blood, may also exert a similar control over thyroid hormone regulated processes during mammalian postembryonic development.

## SUMMARY

That most major physiological actions of thyroid hormones could be mediated via hormonal regulation of gene expression has been known for more than 25 years. The localization of TR in the cell nucleus, first reported almost 20 years

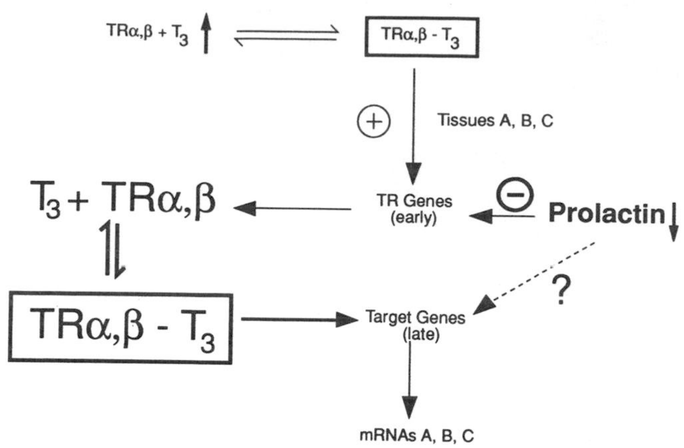


**Fig. 7.** A working model to explain the significance of the inhibition of autoinduction of TR by prolactin during amphibian metamorphosis. At the onset of metamorphosis, the first traces of thyroid hormone bind to the small amount of TR to upregulate its own receptor. The combined effect of rapidly increasing amounts of circulating thyroid hormone (T3↑) and release from the inhibition of autoinduction due to loss of prolactin in blood (Prolactin↓) would lead to high levels of TRα and TRβ during metamorphic climax. At high concentrations, TRα and TRβ will induce different sets of preprogrammed genes to produce tissue specific phenotypic changes. The different sizes of lettering for TRα,β, T3 and TRα,β - T3 represent the relative amounts of the receptor, hormone and functionally active hormone-receptor complex, respectively. See text for other details

ago, confirmed this concept. But it is only since the cloning of the TR gene and its identification as the c-*erbA* oncogene, accomplished 6 years ago, that we have begun to understand the details of the interaction between the hormone and its receptor and between the receptor and its target gene. Perhaps the most significant concept to emerge from the molecular studies is that the TR belongs to the superfamily of nuclear receptors for steroid hormones and morphogens such as retinoids. It highlights the evolutionary conservation of a major network of cellular signalling and intracellular regulatory pathways and which has helped bring us closer to a unified concept of the action of many growth and developmental hormones.

Several important questions to be solved in the future become obvious from this brief review of the role of thyroid hormone in regulating developmental processes. Among these is the explanation for the high degree of tissue specificity of hormonal regulation of gene expression. The discovery of differential expression of the two major TR genes and the generation of multiple isoforms of the receptor by alternate splicing go some way to answering this question, but this is clearly not the sole factor determining tissue specificity. It will be most important to find out more about the interaction between the receptor and other transcription factors or nuclear proteins, some of which may be tissue specific and others expressed ubiquitously. The autoinduction of TR during amphibian metamorphosis, described above, emphasizes the intriguing question of how receptor genes are regulated during development. We know little as yet about the promoters and regulatory factors involved in this process. Finally, we have also described the recent recognition of genetic defects in TRs that underlie thyroid hormone linked diseases in humans. Future studies on the molecular genetics of receptors will enhance the importance in clinical practice of receptor linked diseases.

## References


Baker BS and Tata JR (1990) Accumulation of proto-oncogene c-*erb*-A related transcripts during *Xenopus* development: association with early acquisition of response to thyroid hormone and estrogen. *EMBO Journal* **9** 879–885

Baker BS and Tata JR (1992) Prolactin prevents the auto-induction of thyroid hormone receptor mRNAs during amphibian metamorphosis. *Developmental Biology* **149** 463–467

Baniahmad A, Steiner C, Kohne AC and Renkawitz R (1990) Modular structure of a chicken lysozyme silencer: involvement of an unusual thyroid hormone receptor binding site. *Cell* **61** 505–514

Baulieu E-E and Kelly PA (eds) (1990) *Hormones: From Molecules to Disease*, Hermann, Paris

Beckingham Smith K and Tata JR (1976) Amphibian metamorphosis, In: Graham CF and Wareing PF (eds). *Developmental Biology of Plants and Animals*, pp 232–245, Blackwell, Oxford

Brent GA, Harney JW, Chen Y, Warne RL, Moore DD and Larsen PR (1989) Mutations of the rat growth hormone promoter which increase and decrease response to thyroid hormone define a consensus thyroid hormone response element. *Molecular Endocrinology* **3** 1996–2004

Brooks AR, Sweeney G and Old RW (1989) Structure and functional expression of a cloned *Xenopus* thyroid hormone receptor. *Nucleic Acids Research* **17** 9395–9405

Burnside J, Darling DS, Carr FE and Chin WW (1989) Thyroid hormone regulation of the rat glycoprotein hormone α-subunit gene promoter activity. *Journal of Biological Chemistry* **264** 6886–6991

Chatterjee VKK, Lee J-K, Rentoumis A and Jameson JL (1989) Negative regulation of the thyroid stimulating hormone a gene by thyroid hormone: receptor interaction adjacent to the TATA box. *Proccedings of the National Academy of Sciences of the USA* **86** 9114–9118

Chatterjee VKK, Nagaya T, Madison LD, Datta S, Rentoumis A and Jameson JL (1991) Thyroid hormone resistance syndrome: inhibition of normal receptor function by mutant thyroid hormone receptors. *Journal of Clinical Investigation* **87** 1977–1984

Chin WW (1991) Nuclear thyroid hormone receptors, In: Parker M (ed). *Nuclear Hormone Receptors*, pp 79–102, Academic Press, London

Cohen PP (1970) Biochemical differentiation during amphibian metamorphosis. *Science* **168** 533–543

Dalman FC, Koenig RJ, Perdew GH, Massa E and Pratt WB (1990) In contrast to the glucocorticoid receptor the thyroid hormone receptor is translated in the DNA binding state and is not associated with hsp90. *Journal of Biological Chemistry* **265** 3615–3618

Dang CV and Lee WMF (1989) Nuclear and nucleolar targeting sequences of c-erbA, c-myb, N-myc, p53, hsp70 and HIV tat proteins. *Journal of Biological Chemistry* **264** 18019–18023

Darling DS, Burnside J and Chin WW (1989) Binding of thyroid hormone receptors to the rat thyrotropin-β gene. *Molecular Endocrinology* **3** 1359–1368

DeLong GR, Robbins J and Condliffe PG (eds) (1989) *Iodine and the Brain*, Plenum Press, New York

de Thé H, Vivanco-Ruiz M, Tiollais P, Stunnenberg H and Dejean A (1990) Identification of a retinoic acid responsive element in the retinoic acid receptor beta gene. *Nature* **343** 177–180

Duellman WE and Trueb L (1986) Metamorphosis, In: *Biology of Amphibians*, pp 173–194, McGraw-Hill, New York

Dussault JH and Walker P (eds) (1983) *Congenital Hypothyroidism*, Marcel Dekker Inc, New York

Evans RM (1988) The steroid and thyroid hormone receptor superfamily. *Science* **240** 889–895

Forman BM, Yang C, Au M, Casanova J, Ghysdael J and Samuels HH (1989) A domain containing leucine-zipper-like motifs mediates novel in vivo interactions between the thyroid hormone and retinoic acid receptors. *Molecular Endocrinology* **3** 1610–1626

Forrest D, Hallbook F, Persson H and Vennstrom B (1991) Distinct functions for thyroid hormone receptors α and β in brain development indicated by differential expression of receptor genes. *EMBO Journal* **10** 269–275

Frieden E and Just JJ (1970) Hormonal responses in amphibian metamorphosis, In: Litwack G (ed). *Biochemical Actions of Hormones*, vol I, pp 1–52, Academic Press, New York

Galton VA and St Germain DL (1985) Putative nuclear triiodothyronine receptors in tadpole liver during metamorphic climax. *Endocrinology* **117** 912–916

Gilbert LI and Frieden E (eds) (1981) *Metamorphosis. A Problem in Developmental Biology*, Plenum Press, New York

Glass CK and Holloway JM (1990) Regulation of gene expression by the thyroid hormone receptor. *Biochimica et Biophysica Acta* **1032** 157–176

Glass CK, Holloway JM, Devary OV and Rosenfeld MG (1988) The thyroid hormone receptor binds with opposite transcriptional effects to a common sequence motif in thyroid hormone and estrogen response elements. *Cell* **54** 313–323

Goldberg Y, Glineur C, Gesquiere J-C *et al* (1988) Activation of protein kinase C or cAMP-dependent protein kinase increases phosphorylation of the c-erbA encoded thyroid hormone receptor and of the v-erbA encoded protein. *EMBO Journal* **7** 2425–2433

Granger NA and Bollenbacher WE (1981) Hormonal control of insect metamorphosis, In: Gilbert LI and Frieden E (eds). *Metamorphosis. A Problem in Developmental Biology*, pp 105–137, Plenum Press, New York

Green S and Chambon P (1988) Nuclear receptors enhance our understanding of transcription regulation. *Trends in Genetics* **4** 309–314

Guichon-Mantel A, Loosfelt H, Lescop P *et al* (1989) Mechanisms of nuclear localization of the progesterone receptor: evidence for interaction between monomers. *Cell* **57** 1147–1154

Hodin RA, Lazar MA, Wintman BI *et al* (1989) Identification of a thyroid hormone receptor that is pituitary specific. *Science* **244** 76–79

Hodin RA, Lazar MA and Chin WW (1990) Differential and tissue-specific regulation of multiple rat c-erbA messenger RNA species by thyroid hormone. *Journal of Clinical Investigation* **85** 101–105

Kandel ER and Schwartz JH (eds) (1985) *Principles of Neural Science*, Elsevier, New York

Kawahara A, Baker BS and Tata JR (1991) Developmental and regional expression of thyroid hormone receptor genes during *Xenopus* metamorphosis. *Development* **112** 933–943

Koenig RJ, Warne RL, Brent GA, Harney JW, Larsen PR and Moore DD (1988) Isolation of a cDNA clone encoding a biologically active thyroid hormone receptor. *Proccedings of the National Academy of Sciences of the USA* **85** 5031–5035

Koenig RJ, Lazar MA, Hodin RA *et al* (1989) Inhibiton of thyroid hormone action by a non-hormone binding c-erbA protein generated by alternative mRNA splicing. *Nature* **337** 659–661

Lazar MA, Hodin RA, Darling DS and Chin WW (1989) A novel member of the thyroid/steroid hormone receptor family is encoded by the opposite strand of the rat c-erbA alpha transcriptional unit. *Molecular and Cellular Biology* **9** 1128–1136

Lazar MA, Hodin RA, Cardona G and Chin WW (1990) Gene expression from the c-erbAα/Rev-erbAα genomic locus: potential regulation of alternate splicing by alternate strand transcription. *Journal of Biological Chemistry* **265** 12859–12863

Leloup J and Buscaglia M (1977) La triiodothyronine, hormone de la metamorphose des amphibiens. *Comptes rendus hebdomadaires des Séances de l'Académie des Sciences, Paris* **284** 2261–2263

Lin K-H, Parkison C, McPhie P and Cheng S-Y (1991) An essential role of domain D in the thyroid hormone binding activity of human β1 thyroid hormone nuclear receptor. *Molecular Endocrinology* **5** 485–492

Macchia E, Nakai A, Janiga *et al* (1990) Characterization of site-specific polyclonal antibodies to c-erbA peptides recognising the human thyroid hormone receptors α1, α2, and β and native 3,5,3′-triiodothyronine receptor and a study of tissue distribution of the antigen. *Endocrinology* **126** 3232–3239

Mathisen PM and Miller L (1987) Thyroid hormone induction of keratin genes: a two-step activation of gene expression during development. *Genes and Development* **1** 1107–1117

Mathisen PM and Miller L (1989) Thyroid hormone induces constitutive keratin gene expression during *Xenopus* laevis development. *Molecular and Cellular Biology* **9** 1823–1831

Mellstrom B, Naranjo JR, Santos A, Gonzalez AM and Bernal J (1991) Independent expression of the α and β c-erbA genes in developing rat brain. *Molecular Endocrinology* **5** 1339–1350

Moriya T, Thomas CR and Frieden E (1984) Increase in 3,5,3′-triiodothyronine (T3)-binding sites in tadpole erythrocyte nuclei during spontaneous and T3-induced metamorphosis. *Endocrinology* **114** 170–175

Moskaitis JE, Sargent TD, Smith LH Jr, Pastori RL and Schoenberg DR (1989) *Xenopus* laevis serum albumin: sequence of the complementary deoxyribonucleic acids encoding the 68- and 74-kilodalton peptides and the regulation of albumin gene expression by thyroid hormone during development. *Molecular Endocrinology* **3** 464–473

Nicoll CS (1974) Physiological actions of prolactin, In: Knobil E and Sawyer WH (eds) *Handbook of Physiology,* Section 7, vol 4, part 2, pp 253–292, American Physiological Society, Washington

Nieuwkoop PO and Faber J (1967) *Normal Table of Xenopus laevis (Daudin),* 2nd Edition, North-Holland, Amsterdam

Perlman AJ, Wolffe AP, Champion J and Tata JR (1984) Regulation by estrogen receptor of vitellogenin gene transcription in *Xenopus* hepatocyte cultures. *Molecular and Cellular Endocrinology* **38** 151–161

Pitt-Rivers R and Tata JR (1959) *The Thyroid Hormones,* Pergamon Press, London

Refetoff S, DeWind LT and DeGroot LJ (1967) Familial syndrome combining deaf-mutism, stippled epiphyses, goiter and abnormally high PBI: possible target organ refractoriness to thyroid hormone. *Journal of Clinical Endocrinology and Metabolism* **27** 279–294

Rentoumis A, Chatterjee VKK, Madison LD *et al* (1990) Negative and positive transcriptional regulation by thyroid hormone receptor isoforms. *Molecular Endocrinology* **4** 1522–1531

Rillema JA (ed) (1987) *Actions of Prolactin on Molecular Processes,* CRC Press, Boca Raton

Rosen ED, O'Donnell AL and Koenig RJ (1991) Protein-protein interactions involving erbA superfamily receptors: through the TRAP door. *Molecular and Cellular Endocrinology* **78** C83–C88

Samuels HH, Forman BM, Horowitz ZD and Ye Z-S (1988) Regulation of gene expression by thyroid hormone. *Journal of Clinical Investigation* **81** 957–967

Sap J, Munoz A, Damm K *et al* (1986) The c-erbA protein is high affinity receptor for thyroid hormone. *Nature* **324** 635–640

Scheuler PA, Schwartz HL, Strait KA, Mariash CN and Oppenheimer JH (1990) Binding of 3,5,3'-triiodothyronine (T3) and its analogs to the in vitro translational products of the c-erbA protooncogenes: differences in the affinity of the α and β forms for the acetic acid analog and failure of the human testis and kidney α-2 products to bind T3. *Molecular Endocrinology* **4** 227–234

Schwabe JWR and Rhodes D (1991) Beyond zinc fingers: steroid receptors have a novel structural motif for DNA recognition. *Trends in Biochemical Sciences* **16** 291–296

Shupnik MA, Ardisson LJ, Meskell MJ, Bornstein J and Ridgway EC (1986) Triiodothyronine regulation of thyrotropin subunit gene transcription is proportional to T3 nuclear receptor occupancy. *Endocrinology* **118** 367–371

Siegel GJ, Agranoff BW, Albers RW and Molinoff P (eds) (1989) *Basic Neurochemistry,* 4th

Edition, Raven Press, New York

Takeda K, Balzano S, Sakurai A, DeGroot LJ and Refetoff S (1991) Screening of nineteen unrelated families with generalized resistance to thyroid hormone for known point mutations in the thyroid hormone receptor β gene and the detection of a new mutation. *Journal of Clinical Investigation* **87** 496–502

Tata JR (1968) Early metamorphic competence of *Xenopus* laevis. *Developmental Biology* **18** 415–440

Tata JR (1974) Growth and developmental action of thyroid hormones at the cellular level. *Handbook of Physiology-Endocrinology* III, pp 469–478, American Physiological Society, Washington DC

Tata JR (1984) (The action of growth and developmental hormones, In: Goldberger RF and Yamamoto KR (eds). *Biological Regulation and Development*, vol 3B, pp 1–58, Plenum Publishing Corporation, New York

Tata JR (1985) Evolution of hormones and their actions. *Nobel Symposium on Molecular Evolution of Life, Chemica Scripta* 2B 179–190

Tata JR (1991) Brain development and molecular genetics. *Journal of the Pontifical Council for Pastoral Assistance to Health Care Workers* **6** 28–36

Tata JR and Widnell CC (1966) Ribonucleic acid synthesis during the early action of thyroid hormone. *Biochemical Journal* **98** 604–620

Tata JR, Kawahara A and Baker BS (1991) Prolactin inhibits both thyroid hormone-induced morphogenesis and cell death in cultured amphibian larval tissues. *Developmental Biology* **146** 72–80

Thompson CC and Evans RM (1989) Trans-activation by thyroid hormone receptors: functional parallels with steroid hormone receptors. *Proccedings of the National Academy of Sciences of the USA* **86** 3494–3498

Umesono K and Evans RM (1989) Determinants of target gene specificity for steroid/thyroid hormone receptors. *Cell* **57** 1139–1146

Umesono K, Giguere V, Glass CK, Rosenfeld MG and Evans RM (1988) Retinoic acid and thyroid hormone induce gene expression through a common responsive element. *Nature* **336** 262–265

Umesono K, Murakami KK, Thompson CC and Evans RM (1991) Direct repeats as selective response elements for the thyroid hormone, retinoic acid and vitamin D3 receptors. *Cell* **65** 1255–1266

Usala SJ and Weintraub BD (1991) Thyroid hormone resistance syndromes. *Trends in Endocrinology and Metabolism* **2** 140–144

Varriale B and Tata JR (1990) Autoinduction of estrogen receptor is associated with FOSP-1 mRNA induction by estrogen in primary cultures of *Xenopus* oviduct cells. *Molecular and Cellular Endocrinology* **71** R25–R31

Weber R (1967) Biochemistry of amphibian metamorphosis, In: Weber R (ed). *The Biochemistry of Animal Development*, pp 227–301, Academic Press, New York

Weinberger C, Thompson CC, Ong ES, Lebo R, Gruol DJ and Evans RM (1986) The c-erbA gene encodes a thyroid hormone receptor. *Nature* **324** 641–646

White BA and Nicoll CS (1981) Hormonal control of amphibian metamorphosis, In: Gilbert LI and Frieden E (eds). *Metamorphosis: A Problem in Developmental Biology*, pp 363–396, Plenum Press, New York

Wood WM, Kao MY, Gordon DF and Ridgway EC (1989) Thyroid hormone regulates the mouse thyrotropin β-subunit gene promoter in transfected primary thyrotropes. *Journal of Biological Chemistry* **264** 14840–14847

Wood WM, Ocran KW, Gordon DF and Ridgway EC (1991) Isolation and characterization of mouse complementary DNAs encoding α and β thyroid hormone receptors from thyrotrope cells: the mouse pituitary specific β2 isoform differs at the amino terminus from the coresponding species from rat pituitary tumour cells. *Molecular Endocrinology* **5** 1049–1061

Yaoita Y and Brown DD (1990) A correlation of thyroid hormone receptor gene expression with amphibian metamorphosis. *Genes and Development* **4** 1917–1924

Yaoita Y, Shi Y-B and Brown DD (1990) *Xenopus* laevis α and β thyroid hormone receptors. *Proceedings of the National Academy of Sciences of the USA* **87** 7090–7094

Ye ZS, Forman BM, Park HY, Casanova J and Samuels HH (1988) Rat growth hormone gene expression: both cell-specific and thyroid hormone response elements are required for thyroid hormone regulation. *Transactions of the Association of American Physicians* **101** 42–53

Zenke M, Munoz A, Sap J, Vennstrom B and Beug H (1990) V-erbA oncogene activation entails the loss of hormone-dependent regulator activity of c-erbA. *Cell* **61** 1035–1049

The authors are responsible for the accuracy of the references.

# The Leukaemia Oncogene v-*erbA*: A Dominant Negative Version of Ligand Dependent Transcription Factors That Regulates Red Cell Differentiation?

**JACQUES GHYSDAEL**[1] • **HARTMUT BEUG**[2]
[1]*Institut-Curie, Section de Biologie, Bât 110, Centre Universitaire F-91405 Orsay, France*
[2]*Research Institute of Molecular Pathology, Dr Bohrgasse 7, A-1030, Vienna, Austria*



## INTRODUCTION

Leukaemic cells seem to differ from their normal counterparts in at least three respects. Firstly, their growth properties may be altered. Secondly, their requirement for regulators of normal growth and/or differentiation (haematopoietic or other growth factors) could be abrogated or qualitatively changed. Finally, their ability to differentiate into functional haematopoietic cells may be partly or completely inhibited.

Although alterations in cell proliferation are thought to be a general characteristic of tumour cells (an assumption that is not always supported by experimental evidence) and abrogation and/or change in growth factor responsiveness is a frequent, well studied event during leukaemogenesis (reviewed in Beug and Graf, 1989), a clear understanding of the exact contribution of haematopoietic differentiation arrest or alteration to the leukaemic cell phenotype is still lacking. The possibly most clearcut case of an oncogene that specifically affects haematopoietic differentiation and thus contributes to leukaemic transformation is the v-*erbA* oncogene, a mutated, ligand independent version of a ligand dependent nuclear transcription factor, the avian thyroid hormone receptor α (c-erbA/TRα). Our aim is to review possible mechanisms of how the v-*erbA* oncogene specifically alters the differentiation

*Cancer Surveys* Volume 14: *Growth Regulation by Nuclear Hormone Receptors*
 0-87969-371-1/92. $3.00 + .00

programme of avian erythroid cells, how the c-erbA/TR$\alpha$ function as a ligand regulated transcription factor is altered by its activation as an oncogene and how v-erbA protein function relates to possible roles of *c-erbA/TR*$\alpha$ gene and related receptors in proliferation/differentiation control of normal erythroid progenitors.

## CELL TRANSFORMATION BY AVIAN ERYTHROBLASTOSIS VIRUS

Avian erythroblastosis virus (AEV) is an acutely transforming retrovirus that, depending on the mode of inoculation in chickens, induces erythroleukaemias, sarcomas or both. These oncogenic properties are faithfully reproduced in vitro since AEV can induce transformation of both fibroblasts and bone marrow cells in tissue culture. Leukaemic cells from chickens as well as bone marrow cells transformed in vitro represent a homogeneous population of rapidly proliferating erythroblasts that are highly arrested at an immature state of differentiation (reviewed in Graf and Beug, 1983). The genome of AEV contains two oncogenes, v-*erbA* and v-*erbB* (Vennström *et al,* 1980). The v-*erbB* oncogene encodes a transmembrane glycoprotein (Hayman *et al,* 1983) that is a truncated version of the avian epidermal growth factor/transforming growth factor-$\alpha$ (EGF/TGF-$\alpha$) receptor (Downward *et al,* 1984; Lax *et al,* 1988). Unlike its cellular homologue, which displays a ligand dependent protein tyrosine kinase activity, the activity of the v-erbB glycoprotein is constitutive (Gilmore *et al,* 1985). The product of v-*erbA* is a truncated and mutated version of the chicken thyroid hormone receptor $\alpha$ (c-erbA/TR$\alpha$) that has lost its ability to bind and respond to thyroid hormone (T3) and thus possibly functions as a constitutive repressor of c-*erbA/TR*$\alpha$ regulated genes (Sap *et al,* 1986).

Early studies on the biological properties of mutant AEV strains or recombinant retroviruses, in which either the v-*erbA* or v-*erbB* genes had been deleted, allowed identification of the respective contributions of each oncogene to the in vitro transforming and oncogenic properties of AEV. Expression of v-*erbB* was sufficient to induce erythroleukaemia and to transform fibroblasts and erythroblasts in tissue culture (Frykberg *et al,* 1983). V-*erbB* transformed fibroblasts display transformation indices and growth properties similar to those induced by other oncogenes encoding protein tyrosine kinases (Royer-Pokora *et al,* 1978). In erythroid progenitors, v-*erbB,* as well as other oncogenes encoding protein tyrosine kinases, induces extensive self renewal and an incomplete arrest in maturation probably because of partial repression of many, if not all, late erythrocyte genes (Beug et al, 1985, Knight *et al,* 1988). Unlike AEV transformed erythroblasts, which can proliferate in standard growth medium, erythroblasts transformed by v-*erbB* alone grow efficiently only under restricted pH and ionic conditions (Kahn *et al,* 1986). These specific growth conditions are in fact the hallmark of erythroblasts transformed by other protein tyrosine kinase oncogenes or activated *ras* and *mil* oncogenes (Kahn *et al,* 1984, 1986). V-*erbB* oncogene transformed cells also differ from normal erythroid progenitors in that they grow independently of erythropoietin (Frykberg *et al,* 1983; Kahn *et al,* 1986). Recently, certain normal erythroid progenitors were shown to express endogenous c-*erbB* and to self renew in response to TGF-α (Pain *et al,* 1991) as well as to ligands of other receptor tyrosine kinase genes (insulin, insulin like growth factor 1; Schroeder C, unpublished).

In contrast to v-*erbB,* the v-*erbA* oncogene seems to be non-leukaemogenic or sarcomatogenic in young chicks (Frykberg *et al,* 1983; Gandrillon *et al,* 1989). However, v-*erbA* seems to alter the growth factor requirement of chick fibroblasts in tissue culture and enhances and/or alters the phenotype of v-*erbB* transformed fibroblasts both in vivo and in vitro (Gandrillon *et al,* 1987; Jansson *et al,* 1987). Expression of v-*erbA* in v-*erbB* transformed erythroblasts affects both their growth and differentiation properties. V-*erbA* relieves the strict growth condition requirements of v-*erbB* transformed cells, blocks their spontaneous differentiation into mature erythrocytes and increases their growth rate (Kahn *et al,* 1986; Zenke *et al,* 1988). Similarly, v-*erbA* blocks the spontaneous differentiation of erythroblasts transformed by other oncogenic protein tyrosine kinase oncogenes, Ha-*ras* and the human EGF receptor (Kahn *et al,* 1986; Khazaie *et al,* 1988). V-*erbA* also specifically inhibits the normal, erythropoietin dependent differentiation of erythroblasts transformed by temperature sensitive mutants of v-*erbB* and v-*sea* oncogenes when shifted to the non-permissive temperature. Such v-*erbA* expressing cells showed a limited proliferation ability in response to erythropoietin and a phenotype that combined characteristics of immature and mature erythroid cells (Kahn *et al,* 1986, Schroeder *et al,* 1990). Finally, v-*erbA* can arrest the differentiation of both TGF-α dependent and independent normal erythroid progenitors (Gandrillon *et al,* 1989, Schroeder *et al,* 1990), inducing a phenotype that is distinct from that caused by transforming tyrosine kinase oncogenes (Schroeder *et al,* 1990). This suggests that v-*erbA* can block the differentiation of normal early erythroid progenitors responsive to endogenous hormones or growth factors.

## C-*erbA/TR*α PROTO-ONCOGENE ENCODES A THYROID HORMONE RECEPTOR

The progenitor of v-*erbA* is the chick c-*erbA*α proto-oncogene. The major translation product of c-*erbA*α is c-erbA/TRα, a nuclear protein with a molecular mass of 46 kDa, which is a high affinity receptor for the thyroid hormone 3,5,3′ triiodothyronine (T3) (Sap *et al,* 1986). The genome of vertebrates contains, in addition to c-*erbA*α, other *erbA* related genes that encode distinct T3 receptors, for example c-erbAβ/TRβ (Weinberger *et al,* 1986; Koenig *et al,* 1988; Forrest *et al,* 1990) or orphan receptors for which ligand, if any, remains to be identified, such as c-erbAα2 (Izumo and Mahdavi, 1988; Lazar *et al,* 1988; Mitsuhashi *et al,* 1988) and Rev-erbA (Lazar *et al,* 1989;

Miyajima *et al,* 1989). Thyroid hormone receptors belong to a superfamily of sequence specific, ligand dependent transcriptional regulators including steroid and retinoid receptors that act to coordinate metabolic homoeostasis, embryonic development, differentiation and reproduction (reviewed in Evans, 1988). All of these receptors have a common modular organization with a characteristic carboxyterminal ligand binding domain and a central region that folds into two zinc finger motifs that regulate specific binding to DNA.

The DNA binding specificity of hormone receptors has been studied in detail with idealized versions of natural hormone responsive elements (HREs). Receptors were found to bind with high affinity to palindromic HREs with a binding specificity influenced both by the sequence of the palindrome half site and by the spacing between half sites. Although sequence specificity between glucocorticoid receptors on the one hand and oestrogen or thyroid hormone receptors on the other is governed by three aminoacids in the stem of the first finger motif (Danielsen *et al,* 1989; Mader *et al,* 1989; Umesono and Evans, 1989), spacing specificity in the context of palindromic HREs seems to be conferred by a distinct subdomain (D box) of the second zinc finger (Umesono and Evans, 1989). Most HREs, however, are composed of both inverted and direct repeats of specific half sites. Recent evidence from idealized HREs indicates that relative orientation and spacing between half sites determines to some extent DNA binding specificity and to a major extent transcriptional responsiveness of HREs to specific receptors (Umesono *et al,* 1991; Näär *et al,* 1991). The configuration of HREs in responsive genes is therefore likely to govern both the overlapping specificity and redundancy of the hormonal response of these genes. All thyroid hormone response elements (T3REs) defined in these studies bind thyroid hormone receptors in both unliganded and liganded state. Unliganded receptors act as transcriptional repressors and addition of hormone converts them into transcriptional activators (Izumo and Mahdavi, 1988; Koenig *et al,* 1988; Damm *et al,* 1989; Sap *et al,* 1989). A remarkable exception to this model has been provided by unspaced direct repeats that confer transcriptional activation to unliganded thyroid hormone receptors, an activity that is suppressed upon hormone binding (Näär *et al,* 1991). Such negative T3REs seem to account for the well known negative regulation of the thyroid stimulating hormone β (*TSH*β) gene in pituitary cells (Wood *et al,* 1989; Näär *et al,* 1991).

No simple relation seems to exist between the selectivity/affinity of thyroid hormone receptors for a given T3RE and the ability of this element to act as a *cis*-acting transcriptional regulatory element. For example, although thyroid hormone receptors bind to inverted repeats or 3 bp gapped inverted or direct repeats of a TCAGGTCA half site, only the former can confer transcriptional activation to *cis* linked promoters (Glass *et al,* 1988; Näär *et al,* 1991). The mechanism involved in this restriction is unknown but could be linked to the differential permissiveness of different HREs to the binding of receptors as monomers, homodimers or heterodimers between thyroid hormone receptors and other transcriptional regulators (Näär *et al,* 1991; Forman *et al,* in press).

The ability of thyroid hormone receptor to form homodimers or heterodimers with other receptors of the thyroid/retinoid receptor family has been ascribed to their related carboxyterminal domain (Forman *et al*, 1989; Glass *et al*, 1989). In addition to the subdomains directly involved in ligand binding (Munoz *et al*, 1988), this domain has been proposed to include α helical structures reminiscent of the leucine zippers found in other transcriptional regulators that define a dimerization interface for receptors of the thyroid/retinoic acid family (Forman *et al*, 1989; Glass *et al*, 1989).

The extreme carboxyterminus of thyroid hormone receptors plays a major part in transcriptional activity of T3RE bound receptors. This region is important for both transcriptional repression in the absence of ligand and transcriptional activation in the presence of hormone (Damm *et al*, 1989; Holloway *et al*, 1990; Zenke *et al*, 1990). This domain has been proposed to fold into an α helical structure (Zenke *et al*, 1990) that could form a distinct dimerization interface for hormone regulated interactions with other transcriptional regulators (Glass *et al*, 1989; Burnside *et al*, 1990; O'Donnell *et al*, 1991), which, in turn, could contribute specific contacts with factors of the basal transcription machinery. The importance of the cellular context in the transcriptional response by thyroid hormone receptors is emphasized by the ability of the unliganded receptor to act as an activator of a co-transformed reporter plasmid in yeast cells (Privalsky *et al*, 1990).

The extreme aminoterminal domains of the α and β subtype thyroid hormone receptors are unrelated and do not seem to be required for transcriptional activity. However, the aminoterminal region of the chick α receptor contains several phosphorylated residues (Goldberg *et al*, 1988; Glineur *et al*, 1989), and phosphorylation at two adjacent serine residues in this domain is an important determinant of the ability of the unliganded receptor (our unpublished data) or its v-*erbA* oncogenic version (Glineur *et al*, 1991) to function as an oncogene in erythroblast transformation as well as a transcriptional repressor of at least a subset of responsive genes.

## V-*erbA* ONCOGENE ACTIVATION

The oncogenic v-erbA protein differs from c-erbA/TRα by a nine aminoacid internal deletion and nine aminoacid substitutions in the carboxyterminal ligand binding domain. In addition, two aminoacids are substituted in both the DNA binding domain and the aminoterminal region and the first 12 aminoterminal residues are deleted. Furthermore, v-*erbA* sequences are fused at their aminoterminus to about 30 kDa of retroviral *gag* gene sequences to generate a 75 kDa fusion protein (Sap *et al*, 1986).

Because of the extensive mutations in its carboxyterminal domain, the v-erbA protein fails to bind hormone (Munoz *et al*, 1988). Furthermore, the nine aminoacid carboxyterminal deletion leads to loss of a T3 dependent transcriptional activation function of erythrocyte specific genes (Zenke *et al*, 1990).

The first mutation (Gly to Ser) in the DNA binding domain of v-erbA is localized in the zinc finger region that governs sequence selectivity among receptors for glucocorticoids on the one hand and those for the oestrogen/T3/retinoids families on the other. The second mutation (Lys to Thr) in the DNA binding domain of v-erbA is localized in the D box (see previous section). Although the v-erbA protein has been noted to bind less efficiently than c-erbA/TRα to a T3RE present in the Moloney murine leukaemia virus long terminal repeat (Sap *et al,* 1989) or to a palindromic T3RE (our unpublished observations), it is unclear at present whether these differences in element recognition are linked to either or both of these mutations in the DNA binding domains of v-erbA.

These aminoacid substitutions could confer on v-erbA the ability to bind *cis*-acting DNA elements that differ slightly either in half site sequence or spacing as compared with the c-erbA/TRα thyroid hormone receptor. Such a model is indeed claimed to apply for the interference by v-erbA of the action of related receptors such as retinoic acid receptors or oestrogen receptor (Sharif and Privalski, 1991). These researchers report that v-erbA represses transcriptional activation of both receptors on idealized HRE elements and that the Gly to Ser mutation in the v-erbA DNA binding domain abolishes both erythroid transformation and repression of retinoic receptor action on these idealized HREs (Privalsky *et al,* 1988; Bonde *et al,* 1991; Sharif and Privalsky, 1991).

Despite these findings, which are based on HREs not found in genes regulated by hormone receptors in vivo, the significance of the DNA binding domain mutations in v-erbA for differentiation arrest and gene regulation in cells remains obscure. Firstly, an in vitro engineered overexpressed *gag*-c-erbAα protein arrested erythroid differentiation in the absence of hormone in a fashion similar to but not identical with bona fide v-erbA (Zenke *et al,* 1990). Secondly, a v-erbA protein (VCV6, Beug H and Bjorn Vennström, unpublished observations) lacking the two v-erbA-specific mutations in the DNA binding domain was essentially indistinguishable from v-erbA in its biological activity. Finally, overexpressed c-erbA/TRα, but not v-erbA, suppressed the biological response (differentiation induction, upregulation of erythroid genes) of erythroid progenitors to endogenous, ligand activated retinoic acid receptor (Schroeder *et al,* 1992).

The deletion of the first 12 aminoacids in v-erbA removes a serine residue that is phylogenetically conserved in erbA/TRα-type thyroid hormone receptors and that is a substrate for phosphorylation by casein kinase II (Glineur *et al,* 1989). Attempts to demonstrate a role of this modification in c-erbA/TRα protein function have failed so far (Glineur *et al,* 1989; our unpublished observations). The fact that an unliganded, *gag* fused c-erbA/TRα protein devoid of these 12 aminoacids arrested differentiation, erythroid gene expression and *trans*-activation of an HRE in the carbonic anhydrase gene promoter in a fashion similar to that of v-erbA (Zenke *et al,* 1990, Disela *et al,* 1991) suggests that both loss of serine 12 and the adjunction of retroviral *gag* sequences to v-*erbA* play a minor part, if any, in v-erbA protein function. By contrast, both serine 28 and 29 of c-erbBα are retained in the oncogenic v-erbA protein, and phosphorylation of these residues is essential for the oncogenic activity of v-erbA and its ability to repress erythroid specific gene expression (Glineur *et al*, 1991).

In conclusion, oncogenic activation of v-erbA involves the conservation of several domains of the c-erbA/TRα thyroid hormone receptor that are important for its activity as a transcriptional repressor in its unliganded state. Furthermore, v-erbA has accumulated in its carboxyterminal domain a series of mutations that impair both the relief of repressor activity on ligand binding and its ability to act as a transcriptional activator. As a consequence, v-erbA acts as a dominant repressor of T3 mediated transcriptional activation of idealized T3RE (Damm *et al*, 1989; Sap *et al*, 1989). In line with these observations, v-erbA blocks erythroblast differentiation, whereas a similarly overexpressed c-erbBα protein accelerates differentiation in the presence of T3 (Zenke *et al*, 1988; Pain *et al*, 1990; Zenke *et al*, 1990). However, recent studies of c- and v-erbA proteins function in cell differentiation, erythrocyte gene expression and *trans*-activation of a T3RE in the carbonic anhydrase gene (which is repressed by v-erbA and regulated by c-erbA/TRα in a hormone dependent fashion; see below) make it clear that c-erbA/TRα is dominant over v-erbA in erythroblasts if the two proteins are expressed at equivalent levels and that v-erbA is a repressor of c-erbA/TRα function only when expressed in excess stoichiometric levels (Disela *et al*, 1991).

## POSSIBLE TARGETS OF v- AND c-*erbA* ONCOGENE FUNCTION

The blocking of erythroid cell differentiation by v-erbA correlates with the constitutive suppression of three erythrocyte specific genes (carbonic anhydrase II [CAII]; the anion transporter Band 3 and the haem biosynthesis enzyme δ-aminolevulinate synthetase [ALA-S]) (Zenke *et al*, 1988, 1990). The expression of various other late erythrocyte differentiation genes (those encoding globins, Band 4.1) and of erythroid specific transcription factors (GATA 1, GATA 2) is not detectably affected (Zenke *et al*, 1988, and unpublished observations). Among these, only carbonic anhydrase type II has been identified so far as a direct target for erbA proteins (Pain *et al*, 1990; Disela *et al*, 1991), being hormonally regulated by c-erbA/TRα and repressed by v-erbA via a T3RE like element located about 700 bp upstream of the CAII gene promoter.

The relative contribution of the various domains of v-erbA to the regulation of the CAII, ALA-S, Band 3 or other target genes yet to be identified remains to be clarified. The identification of additional target genes will also help to resolve whether v-erbA acts as an antagonist of hormone regulated genes only or whether it also interferes with other regulatory circuits. Evidence indicates, for example, that both the liganded retinoic acid receptor α

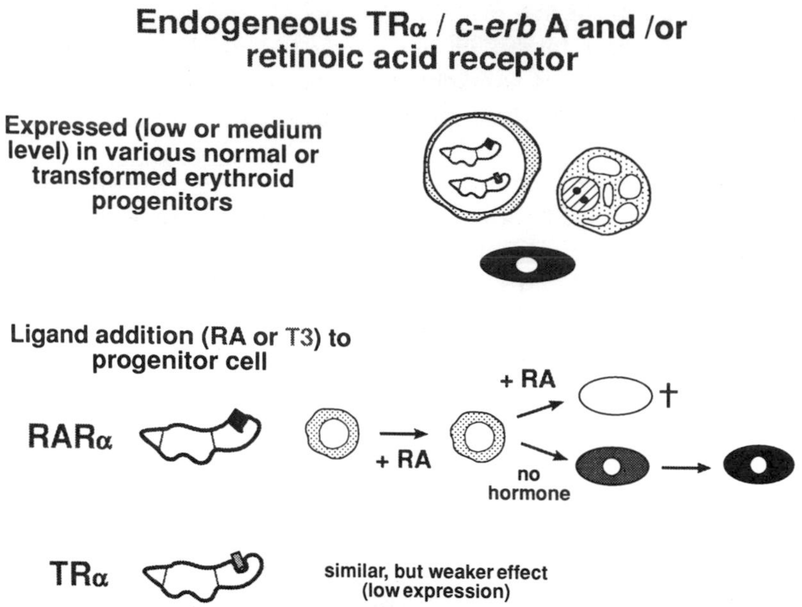

Endogeneous TRα / c-*erb* A and /or retinoic acid receptor

Expressed (low or medium level) in various normal or transformed erythroid progenitors

Ligand addition (RA or T3) to progenitor cell

+ RA

RARα

+ RA

no hormone

TRα

similar, but weaker effect (low expression)

**Fig. 1.** Endogenous c-erbA/TRα and/or retinoic acid receptor. Certain immature erythroid progenitors express endogenous nuclear thyroid hormone receptor α (TRα) and retinoic acid receptor (RARα) (top), and these receptors when activated by retinoic acid (RA) or thyroid hormone (T3) induce erythroid differentiation when added to early erythroid progenitors only (round cell with sparsely dotted cytoplasm) but cause cell death (empty cell, cross) if added later during erythroid differentiation (bottom)

(RARα) and c-erbA/TRα receptors interfere with activator protein 1 (AP1) mediated activation of transcription and that overexpression of v-erbA relieves this inhibition (Desbois *et al,* 1991a). This phenomenon is proposed to be the basis of the abrogation by an overexpressed v-erbA protein of the retinoic acid (RA) mediated inhibition of growth of avian fibroblasts (Desbois *et al,* 1991b). The importance of such a mechanism for the phenotype imposed by v-*erbA* on transformed erythroid cells remains to be investigated.

## DO c-erbA/TRα AND RELATED RECEPTORS HAVE A ROLE IN NORMAL ERYTHROID DIFFERENTIATION?

Although some features of the mechanisms whereby v-*erbA* inhibits erythroid differentiation and constitutively represses transcription of certain erythrocyte genes are emerging, little is known about the normal functions of c-*erbA*α proto-oncogene in erythroid cell proliferation and differentiation. Recent evidence suggests the existence of a normal pathway of c-erbA/TRα function that may well be a target for v-*erbA* oncogene action (Fig. 1). C-erbA/TRα and, even more strikingly, RARα modulated differentiation of normal TGF-α dependent erythroid progenitors, CFU-E colony forming cells and temperature sensitive protein tyrosine kinase oncogene transformed erythroblasts on activation by their cognate ligands. When RA was added in pulses to immature erythroid progenitors, differentiation was accelerated, whereas more mature cells underwent premature cell death. Thyroid hormone alone caused similar but weaker effects, most probably because of the very low expression of its receptors in the cells available in vitro. Notably, T3 strongly enhanced the action of RA, suggesting cooperative action of the two receptors in modulating erythroid differentiation (Schroeder *et al*, 1992).

Expression of the human RARα in receptor negative erythroblasts conferred RA induced regulation of differentiation to the otherwise unresponsive cells, demonstrating that RARα is essential for the RA effect. Likewise, enhanced expression of exogenous c-erbA/TRα in erythroblasts rendered them highly susceptible to modulation of differentiation by T3, suggesting a similar function of both receptors.

These results establish that endogenous c-erbA/TRα can have a role in regulating erythroid differentiation. Although the importance of this function in vivo through endogenous nuclear hormone receptors remains to be analysed, these results suggest that normal c-erbA dependent regulatory pathways may well be the target for v-*erbA* action.

## SUMMARY

The v-*erbA* oncogene of avian erythroblastosis virus alters the growth properties and arrests differentiation of chick erythroid progenitor cells. The v-erbA protein is a mutated, ligand independent version of the c-erbA/T3Rα chick receptor for T3, a ligand dependent transcriptional regulator. In reconstituted systems using idealized hormone responsive elements, overexpressed v-erbA acts as a dominant repressor of transcription mediated by liganded c-erbA/T3Rα. This property seems to account for at least part of the phenotype of AEV transformed erythroid cells and for the transcriptional repression of some erythrocyte specific genes. However, v-erbA is likely to interfere with regulatory circuits other than those directly regulated by T3 receptors. Aspects of this hypothesis are discussed in the context of available evidence for the role of T3 and other hormones in erythroid progenitor cells proliferation/differentiation.

## Acknowledgements

Grateful thanks are due to colleagues for communication of unpublished results, to Bernard Binetruy for critical reading of this manuscript and to Mrs Nathalie Frey for excellent secretarial assistance. Work from one of us (JG) is supported by funds from the Centre National de la Recherche Scientifique, Association pour la Recherche sur le Cancer (ARC) and Institut Curie.

## References

Beug H and Graf T (1989) Cooperation between viral oncogenes in avian erythroid and myeloid leukemia. *European Journal of Clinical Investigation* **19** 491–502

Beug H, Kahn P, Döderlein G, Hayman MJ and Graf T (1985) Characterization of hematopoietic cells transformed *in vitro* by AEV-H, an *erb*-containing avian erythroblastosis virus, In: Neth, Gallo, Greaves and Janka (eds). *Modern Trends in Human Leukaemia VI*, vol 29, pp 290–297, Springer Verlag, Berlin-Heidelberg

Bonde B, Sharif M and Privalski ML (1991) Ontogeny of the v-*erbA* oncoprotein from the thyroid hormone receptor: an alteration in the DNA-binding domain plays a role crucial for v-*erbA* function. *Journal of Virology* **65** 2037–2046

Burnside J, Darling DS and Chin WW (1990) A nuclear factor that enhances binding of thyroid hormone receptors to thyroid hormone response elements. *Journal of Biological Chemistry* **265** 2500–2504

Damm K, Thompson CC and Evans RM (1989) Protein encoded by v-*erbA* functions as a thyroid hormone receptor antagonist. *Nature* **339** 593–597

Danielsen M, Hinck L and Ringold GM (1989) Two amino-acids within the knuckle of the first finger specify DNA response element activation by the glucocorticoid receptor. *Cell* **57** 1131–1138

Desbois C, Aubert D, Legrand C, Pain B and Samarut J (1991a) A novel mechanism of action for the v-*erbA* oncogene: abrogation of the inactivation of AP1 transcription factor by retinoic acid receptor and thyroid hormone receptor. *Cell* **67** 731–740

Desbois C, Pain B, Guilhot C *et al* (1991b) V-*erbA* oncogene abrogates growth inhibition of chicken embryo fibroblasts induced by retinoic acid. *Oncogene* **6** 2129–2135

Disela C, Glineur C, Bugge T *et al* (1991) V-*erbA* overexpression is required to extinguish c-*erbA* function in erythroid cell differentiation and regulation of the *erbA* target gene CAII. *Genes and Development* **5** 2033–2047

Downward J, Yarden Y, Mayes E *et al* (1984) Close similarity of epidermal growth factor receptor and v-*erbB* oncogene protein sequence. *Nature* **307** 521–527

Evans RM (1988) The steroid and thyroid hormone receptor superfamily. *Science* **240** 889–896

Forman B, Yang CR, Au M, Casanova J, Ghysdael J and Samuels HH (1989) A domain containing leucine zipper like motifs mediate novel *in vivo* interactions between the thyroid hormone and retinoic acid receptors. *Molecular Endocrinology* **3** 1610–1626

Forman B, Casanova J, Raaka B, Ghysdael J and Samuels HH Half site spacing and orientation determines whether thyroid hormone and retinoic acid receptors and related factors bind to DNA response elements as monomers, homodimers, or heterodimers. *Molecular Endocrinology* (in press)

Forrest D, Sjöberg M and Vennström BV (1990) Contrasting developmental and tissue specific expression of alpha and beta thyroid hormone receptor genes. *EMBO Journal* **9** 1519–1528

Frykberg L, Palmieri S, Beug H, Graf T, Hayman MJ and Vennström B (1983) Transforming capacities of avian erythroblastosis virus mutants deleted in the *erbA* or *erbB* oncogenes. *Cell* **32** 227–238

Gandrillon O, Jurdic P, Benchaibi M, Xiao JH, Ghysdael J and Samarut J (1987) Expression of v-*erbA* oncogene in chicken embryo fibroblasts stimulates their proliferation *in vitro* and enhances tumor growth *in vivo*. *Cell* **49** 687–697

Gandrillon O, Jurdic P, Pain B *et al* (1989) Expression of the v-*erbA* product, an altered nuclear hormone receptor, is sufficient to transform erythrocytic cells *in vitro*. *Cell* **58** 115–121

Gilmore T, Declue JF and Martin GS (1985) Protein phosphorylation at tyrosine is induced by the v-*erbB* gene product *in vivo* and *in vitro*. *Cell* **40** 609–618

Glass CK, Holloway JM, Devary OV and Rosenfeld MG (1988) The thyroid hormone receptor binds with opposite transcriptional effects to a common sequence motif in thyroid hormone and estrogen response elements. *Cell* **54** 313–323

Glass CK, Lipkin SM, Devary OV and Rosenfeld MG (1989) Positive and negative regulation of gene transcription by a retinoic acid-thyroid hormone receptor heterodimer. *Cell* **59** 697–708

Glineur C, Bailly M and Ghysdael J (1989) The c-*erbA* $\alpha$-encoded thyroid hormone receptor is phosphorylated in its amino-terminal domain by casein kinase II. *Oncogene* **4** 1247–1254

Glineur C, Zenke M, Beug H and Ghysdael J (1991) Phosphorylation of the v-*erbA* protein is required for its function as an oncogene. *Genes and Development* **4** 1663–1686

Goldberg Y, Glineur C, Gesquiere JC *et al* (1988) Activation of protein kinase C or c-AMP-dependent protein kinase increases phosphorylation of the c-*erbA*-encoded thyroid hormone receptor and of the v-*erbA*-encoded protein. *EMBO Journal* **7** 2425–2433

Graf T and Beug H (1983) Role of the v-*erbA* and v-*erbB* oncogenes in erythroid cell transformation. *Cell* **34** 7–9

Hayman MJ, Ramsay GM, Savin K, Kitchener G, Graf T and Beug H (1983) Identification and characterization of the avian erythroblastosis virus *erbB* gene product as a membrane glycoprotein. *Cell* **32** 579–588

Holloway JH, Glass CK, Adler S, Nelson CA and Rosenfeld MG (1990) The C′-terminal interaction domain of the thyroid hormone receptor confers the ability of the DNA site to dictate positive or negative transcriptional activity. *Proceedings of the National Academy of Sciences of the USA* **87** 8160–8164

Izumo S and Mahdavi V (1988) Thyroid hormone receptor alpha isoforms generated by alternative splicing differentially activate myosin HC gene transcription. *Nature* **334** 539–542

Jansson M, Beug H, Gray C, Graf T and Vennström B (1987) Selective v-*erbB* genes can be complemented by v-*erbA* in erythroblasts and fibroblasts transformation. *Oncogene* **1** 167–173

Kahn P, Adkins B, Beug H and Graf T (1984) *Src* and *Fps*-containing avian sarcoma viruses transform chicken erythroid cells. *Proceedings of the National Academy of Sciences of the USA* **81** 7122–7126

Kahn P, Frykberg L, Brady C *et al* (1986) V-*erbA* cooperates with sarcoma oncogenes in leukemic cell transformation. *Cell* **45** 349–356

Khazaie K, Dull TJ, Graf T *et al* (1988) Truncation of the human EGF receptor leads to the differential transforming potentials in primary avian fibroblasts and erythroblasts. *EMBO Journal* **7** 3061–3071

Knight J, Zenke M, Disela Ch *et al* (1988) *Ts v-sea* transformed erythroblasts: a model system to study gene expression during erythroid differentiation. *Genes and Development* **2** 247–258

Koenig RJ, Warne RL, Brent GA, Harney JW, Larsen PR and Moore DD (1988) Isolation of a cDNA clone encoding a biologically active thyroid hormone receptor. *Proceedings of the National Academy of Sciences of the USA* **85** 5031–5035

Lax I, Johnson A, Howk R *et al* (1988) Chicken epidermal growth factor (EGF) receptor cDNA cloning, expression in mouse cells and differential binding of EGF and transforming growth factor α. *Molecular and Cellular Biology* **8** 1970–1978

Lazar MA, Hodin RA, Darling DS and Chin WW (1988) Identification of a rat c-*erbA* alpha related protein which binds deoxyribonucleic acid but does not bind thyroid hormone. *Molecular and Cellular Endocrinology* **2** 893–901

Lazar MA, Hodin RA, Darling DS and Chin WW (1989) A novel member of the thyroid/steroid hormone receptor family is encoded by the opposite strand of the rat c-*erbA* alpha transcriptional unit. *Molecular and Cellular Biology* **9** 1128–1136

Mader S, Kumar V, de Verneuil H and Chambon P (1989) Three amino-acids of the oestrogen receptor are essential to its ability to distinguish an oestrogen from a glucocorticoid-responsive element. *Nature* **338** 271–274

Mitsuhashi NT, Tennyson GE and Nikodem VM (1988) Alternative splicing generates messages encoding rat c-*erbA* proteins that do not bind thyroid hormone. *Proceedings of the National Academy Sciences of the USA* **85** 5804–5808

Miyajima N, Horluchi R, Shibuya Y *et al* (1989) Two *erbA* homologs encoding proteins with different T3 binding capacities are transcribed from opposite DNA stands of the same genetic locus. *Cell* **57** 31–39

Munoz A, Zenke M, Gehring U, Sap J, Beug H and Vennström B (1988) Characterization of the hormone binding domain of the chicken c-*erbA*/thyroid hormone receptor protein. *EMBO Journal* **7** 155–159

Näär AM, Boutin JM, Lipkin SM *et al* (1991) The orientation and spacing of core DNA-binding motifs dictate selective transcriptional responses to three nuclear receptors. *Cell* **65** 1267–1279

O'Donnell AL, Rosen ED, Darling DS and Koening RJ (1991) Thyroid hormone receptor mutations that interfered with transcriptional activation also interfere with receptor interaction with a nuclear protein. *Molecular Endocrinology* **5** 94–99

Pain B, Melet F, Jurdic P and Samarut J (1990) The carbonic anhydrase II gene, a gene regulated by thyroid hormone and erythropoietin, is repressed by the v-*erbA* oncogene in erythrocytic cells. *The New Biologist* **2** 284–294

Pain B, Woods CM, Saez J *et al* (1991) EGF-R as a hemopoietic growth factor receptor: the c-*erbB* product is present in chicken erythrocytic progenitors and controls their self renewal. *Cell* **65** 37–46

Privalsky ML, Boucher P, Koning A and Judelson C (1988) Genetic dissection of functional domains within the avian erythroblastosis virus *erbA* oncogene. *Molecular and Cellular Biology* **8** 4510–4517

Privalsky ML, Sharif M and Yamamoto KR (1990) The viral *erbA* oncogene protein, a constitutive repressor in animal cells, is a hormone-regulated activator in yeast. *Cell* **63** 1277–1286

Royer-Pokora B, Beug H, Claviez M, Winkhardt HJ, Friis RR and Graf T (1978) Transformation parameters in chicken fibroblasts transformed by AEV and MC29 avian leukemia viruses. *Cell* **13** 751–760

Sap J, Munoz A, Damm K *et al* (1986) The c-*erbA* protein is a high affinity receptor for thyroid hormone. *Nature* **324** 635–640

Sap J, Munoz A, Schmitt J, Stunnenberg H and Vennström B (1989) Repression of transcription mediated by a thyroid hormone response element by the v-*erbA* oncogene product. *Nature* **340** 242–244

Schroeder Ch, Raynoschek C, Fuhrmann U, Damm K, Vennstroem B and Beug H (1990) The v-*erb* A oncogene causes repression of erythroid-specific genes and an immature, aberrant differentiation phenotype in normal erythroid progenitors. *Oncogene* **5** 1445–1453

Schroeder C, Gibson L, Zenke M and Beug H (1992) Modulation of normal erythroid differentiation by the endogenous thyroid hormone- and retinoic acid receptors: a possible target for v-*erbA* oncogene action. *Oncogene* **7** 217–227

Sharif M and Privalski ML (1991) v-*erbA* oncogene function in neoplasia correlates with its ability to repress retinoic acid receptor action. *Cell* **66** 885–893

Umesono K and Evans RM (1989) Determinants of target specificity for steroid/thyroid hormone receptors. *Cell* **57** 1139–1146

Umesono K, Murakami KK, Thompson CC and Evans RM (1991) Direct repeats as selective response elements for the thyroid hormone, retinoic acid, and vitamin D3 receptors. *Cell* **65** 1255–1266

Vennstrom B, Fanshier L, Moscovici C and Bishop JM (1980) Molecular cloning of the avian erythroblastosis virus genome and recovery of oncogenic virus by transfection of chicken cells. *Journal of Virology* **36** 575–585

Weinberger C, Thompson CC, Ong ES, Lebo R, Gruol D and Evans RM (1986) The c-*erbA* gene encodes a thyroid hormone receptor. *Nature* **324** 641–646

Wood WM, Kao MY, Gordon DF and Ridgway EC (1989) Thyroid hormone regulates the mouse thyrotropin beta-subunit gene promoter in transfected primary thyrotropes. *Journal of Biological Chemistry* **264** 14840–14847

Zenke M, Kahn P, Disela C *et al* (1988) V-*erbA* specifically suppresses transcription of the avian erythrocyte anion transporter (Band 3) gene. *Cell* **52** 107–119

Zenke M, Munoz A, Sap J, Vennström B and Beug H (1990) V-*erbA* oncogene activation entails loss of hormone-dependent regulator activity of c-*erbA*. *Cell* **61** 1035–1049

The authors are responsible for the accuracy of the references.

# Retinoic Acid Receptors in Normal Growth and Development

**GILLIAN MORRISS-KAY**
*Department of Human Anatomy, South Parks Road, Oxford OX1 3QX*



## INTRODUCTION

All-trans-retinoic acid (RA) is a natural metabolite of vitamin A (all-trans-retinol). When given to vitamin A deficient animals, it can replace all of the natural functions of vitamin A except for vision, which specifically requires retinol. Therefore, with the exception of its role in vision, RA can be regarded as the active form of vitamin A at the cellular level.

The importance of vitamin A for normal growth and development was recognized in the 1930s, when Hale (1933, 1935, 1937) showed that vitamin A deficient pregnant sows produced malformed young. Through a series of back crosses and brother-sister matings, he was able to establish that the abnormalities were due purely to the nutritional deficiency, and were not hereditary. Mammalian embryos also develop abnormally under conditions of vitamin A excess (Cohlan, 1953; Giroud and Martinet, 1959). It is therefore clear that normal development depends on the availability of RA to the embryonic tissues and that the amount reaching the tissues must be within a specific range. Outside that range, the embryo is unable to develop normally, and severe excess or deficiency results in death.

Further clues to the complexity of the developmental function of RA are provided by comparisons of the patterns of malformations associated with vitamin A deficiency (Wilson *et al*, 1953) with those of RA excess (Shenefelt, 1972): the two patterns are quite distinct, indicating that different levels of RA are required by different embryonic tissues and developmental processes. The fact that embryonic RA requirements are tissue specific suggests that a mechanism for controlling RA levels must exist at the cellular level. There is good

*Cancer Surveys* Volume 14: *Growth Regulation by Nuclear Hormone Receptors*

evidence that this mechanism involves the cellular binding proteins for retinol and retinoic acid (CRBP and CRABP).

Recent progress in understanding the mechanism of action of RA in normal development has resulted mainly from the cloning of the nuclear retinoic acid receptors (RARs) and the elucidation of their spatiotemporal patterns of expression in embryos by in situ hybridization. These studies, together with equivalent data on the cellular binding proteins, have provided new and important insights into developmental mechanisms. This review will emphasize the progress that has been made in understanding the roles of RA in mammalian development; much has been published elsewhere on the role of retinoids, RARs and related genes in relation to pattern formation in the avian limb (reviewed in Tabin, 1991).

## RETINOIC ACID RECEPTOR FAMILIES, SUBTYPES AND ISOFORMS

Retinoic acid receptors are nuclear receptors that are members of the steroid/thyroid hormone receptor superfamily. Like the hormone receptors, they are ligand activated and have a direct effect on gene activity at the level of transcription (Green and Chambon, 1988; Evans, 1988; Ham and Parker, 1989; Diamond *et al,* 1990). Retinoic acid receptors are thought to activate gene expression through binding to a specific RA response element (RARE), although activation through binding to a thyroid hormone response element has also been observed (Umesono *et al,* 1988). In mammals, there are three homologous receptors, RAR-α, RAR-β and RAR-γ, all of which bind RA with high affinity (de Thé *et al,* 1987; Giguère *et al,* 1987; Petkovich *et al,* 1987; Benbrook *et al,* 1988; Brand *et al,* 1988; Krust *et al,* 1989; Zelent *et al,* 1989). The three genes encoding these receptor subtypes have been assigned to different chromosomes in the mouse (chromosomes 11, 14 and 15, respectively) and humans (chromosomes 17, 3 and 12, respectively) (Ishikawa *et al,* 1990; Mattei *et al,* 1991).

In common with the steroid and thyroid hormone nuclear receptors, RARs consist of six distinct functional domains, designated A to F. Differences between the three receptor subtypes are found mainly in the A and F regions and in the middle part of the D region, that is in the *trans*-activation domain (A) and in two regions whose functions are not understood (Leroy *et al,* 1992). The DNA binding C region, the RA binding E region and the adjacent parts of the D region are highly conserved between receptors. The very high degree of conservation between all regions of the same receptor from mouse and human tissues (Leroy *et al,* 1992) is highly significant from a developmental point of view, since it allows some confidence that studies on the receptor mediated roles of RA in the mouse embryo are also relevant to an understanding of human development.

Each of the receptor subtypes has a number of isoforms: seven for RAR-α (Leroy *et al,* 1991a), three for RAR-β (Zelent *et al,* 1991) and seven for RAR-β

(Kastner *et al,* 1990). The variation between isoforms is confined to the 5′ untranslated region and/or the A region of the receptor. In theory, the greater variety provided by the existence of so many isoforms should allow an even greater complexity of the developmental effects mediated by the receptors than would be possible with three unvariable subtypes. Northern blot analyses have shown differential distribution patterns for some isoforms in adult tissues (Kastner *et al,* 1990; Leroy *et al,* 1991a). There are some stage related differences in the level of isoform expression during early morphogenesis and organogenesis in whole mouse embryos (Leroy *et al,* 1992), but little information is available to indicate that there are relative differences in isoform expression between different embryonic tissues or regions at any specific stage. Retinoic acid receptor α2 and all of the RAR-β isoforms are RA inducible (de Thé *et al,* 1989; Zelent *et al,* 1989); an RARE has been identified in the RAR-α2 and RAR-β2 promoters (de Thé *et al,* 1990; Sucov *et al,* 1990; Leroy *et al,* 1991b).

In addition to the RARs, there is another family of nuclear receptors, the retinoid "X" receptors (RXRs), which show a lower affinity for all-trans-RA than the RARs (Mangelsdorf *et al,* 1990), and whose highest affinity ligand is 9-cis-RA (Mangelsdorf *et al,* 1992). Like the RARs, these receptors have α, β and γ subtypes, but they are less closely related to the RARs than is the thyroid hormone receptor (Manglesdorf and Evans, 1992). They have all been identified in mouse embryos, in which they show less specific patterns of expression than those of the RARs (Mangelsdorf *et al,* 1992). RXRs act co-operatively with RARs through the formation of heterodimers, selectively enhancing the binding of RARs to RA response elements on target genes (Leid *et al,* 1992).

## RETINOIC ACID RECEPTORS IN MAMMALIAN EMBRYOS

### Differential Quantitative Control of Ligand Availability

The ability of adjacent tissues to control free RA concentrations at different levels may be an important factor in pattern formation, since the influence of RARs on gene expression depends on the amount of RA available for their activation. There is good evidence to support the idea that cytoplasmic binding proteins play this part (Robertson, 1987; Maden *et al,* 1988, 1989; Dollé *et al,* 1990). Three closely related binding proteins are present in embryos: CRBP I, CRABP I and CRABP II. A second retinol binding protein, CRBP II, is not present in embryos; its role appears to be solely concerned with intestinal absorption of vitamin A (Crow and Ong, 1985). They are all similar in molecular structure, but CRBP I and II show high specificity for binding retinol, whereas CRABP I and II are highly specific for RA (see Chytil and Ong, 1987, for details and references).

In the mouse embryo, CRBP I is expressed in the absorptive cells (visceral endoderm) of the yolk sac, where it is well placed to bind retinol from the maternal blood (Ruberte *et al,* 1991). From here, retinol presumably reaches

protein that is not closely related to those discussed above has recently been discovered (Sani *et al,* 1991), but nothing is yet known about its developmental expression or function.

### Spatiotemporal Distribution Patterns

The distribution patterns of RARs, as revealed by in situ hybridization, have been described in mouse embryos at a series of stages (Dollé *et al,* 1989, 1990; Noji *et al,* 1989; Ruberte *et al,* 1990, 1991, 1992). The complementary DNA probes used in these studies distinguish between the different RAR subtypes but not their isoforms.

As in adult tissues, the distribution of RAR-α in the embryo is fairly ubiquitous. Northern blots indicate much higher levels of expression of RAR-α1 than of RAR-α2 in whole embryos (Leroy *et al,* 1992), suggesting that it is the pattern of expression of the α1 isoform that is seen with the in situ technique. It has been suggested that RAR-α1 may be a housekeeping gene, not only because of its widespread expression but also because the RAR-α1 P1 promoter exhibits characteristics of a housekeeping gene promoter (Brand *et al,* 1990; Leroy *et al,* 1991a). It is therefore particularly important to note that in early embryos, there are some tissues that do not express detectable levels of RAR-α. The clearest example of this is the neural tube at day 9, in which there is a distinct boundary between a domain of high level expression extending throughout the spinal cord and into the hindbrain as far as rhombomere 4, and a domain including the forebrain, midbrain and rhombomeres 1 to 3 of the hindbrain in which expression is undetectable (Fig. 1) (Ruberte *et al,* 1991). The level of this expression boundary is particularly interesting in relation to the patterns of expression of segmentation related genes and to the effects of RA excess on morphological and molecular aspects of hindbrain segmentation (see below). There is no correlation between RAR-α expression and the pattern of growth of the different regions of the brain, which shows the same cell cycle time in all regions at this and earlier stages (Tuckett and Morriss-Kay, 1985).

Unlike RAR-α, the receptors RAR-β and RAR-γ show restricted spatiotemporal patterns of expression during embryonic development, which are in general mutually exclusive (Fig. 1) (Dollé *et al,* 1990; Ruberte *et al,* 1991). This suggests that RA induced activation of each of these two receptors leads to distinctive effects on downstream genes and hence distinctive effects in terms of morphogenesis and differentiation. For instance, RAR-γ RNA is localized in precartilaginous condensations of mesenchyme and continues to be expressed during cartilage differentiation and in mature cartilage, being lost from these sites only when ossification begins (Ruberte *et al,* 1990). The mutual exclusiveness of RAR-β and RAR-γ shows particularly well in the developing hand or foot plate, in which RAR-γ is expressed in the developing skeletal elements of the digits, whereas RAR-β is expressed in the intervening mesenchyme (Dollé *et al,* 1990; Mendelsohn *et al,* 1991).

Another important difference between these two receptors is that RAR-β has an RARE within its promoter and is itself induced by RA (see above), whereas RAR-γ is not. This property of RAR-β has been used to interpret its ectopic expression in embryos, following local or general administration of exogenous RA, as an indicator of raised tissue RA levels (Mendelsohn *et al,* 1991; Noji *et al,* 1991; Rowe *et al,* 1991b). Ectopic expression of RAR-β following RA treatment of mouse embryos carrying an *RARElacZ* transgene has confirmed that this response involves the RARE (Rossant *et al,* 1991).

By analogy with the effects of excess RA on the pattern of RAR-β expression, the expression pattern of RAR-β in normal embryos may indicate tissues in which RA levels are higher than in adjacent non-expressing regions. For instance, the presence of RAR-β in the interdigital mesenchyme, and its exclusion from the cartilage forming regions, correlates well with the observation that lower RA levels are required for chondrogenesis than for the maintenance of a mesenchymal tissue morphology (Shapiro and Mott, 1981). In the interdigital mesenchyme, in the olfactory mesenchyme, and in many other sites, RAR-β is co-expressed with CRBP I (Ruberte *et al,* 1991, 1992; Dollé *et al,* 1990), adding further weight to the hypothesis that cells containing CRBP I are sites of RA synthesis and that RA levels are highest in these cells.

## Fine Local Differences in Receptor and Binding Protein Expression

The distribution patterns of RARs and binding proteins provide a mechanism for both quantitative control of RAR activation and qualitative differences in the effects of activated RARs on gene expression in different cell types. These differences of RAR and binding protein expression are clear at the whole tissue level, as described above, but can also be seen at a finer level, within a population of epithelial or mesenchyme cells. These differences may be important in defining subsets of cells that have begun to diverge towards different pathways of differentiation and morphogenesis. Similarly, differences between RAR and/or binding protein expression between an epithelium and its adjacent mesenchyme may play an important part in epithelial-mesenchymal interactions. For instance, the presumptive olfactory epithelium of the 14.5 day mouse fetus shows CRABP I expression in the epithelium only and CRBP I expression in the mesenchyme only, suggesting that free RA is controlled at different levels in the two tissues (Dollé *et al,* 1990). The middle layer of the olfactory epithelium consists of developing sensory receptor cells, with stem cell populations being located apically and basally; by day 15, the apical cells have differentiated to form supporting cells (Cuschieri and Bannister, 1975). CRABP I is present in the developing receptor cells but absent from the stem cells, whereas the apical (presumptive supporting) cells express RAR-β (Dollé *et al,* 1990). The pattern of gene expression is different in the adjacent respiratory portion of the nasal epithelium.

It has long been recognized that the pathway of epidermal differentiation is influenced by the level of RA available to the tissue during dermal-epidermal interaction stages of development (Hardy, 1983). Epidermis in which hair follicles would normally differentiate undergoes glandular morphogenesis when treated with retinoids (Hardy *et al,* 1990). This change is associated with the loss of RAR-γ expression from the dermal papilla and dermal sheath cells and its replacement by RAR-β, resulting in a change in the dermal-to-epidermal message that specifies hair or glandular development (Viallet *et al,* 1992).

## RETINOIC ACID RECEPTORS AND OTHER DEVELOPMENTAL CONTROL GENES

Homoeobox genes have fundamental roles in the organization of developmental pattern in insect and vertebrate embryos, as part of a complex regulatory network of genes (reviewed in Holland, 1990). They are associated particularly with the development of segmental pattern, for instance in the mammalian hindbrain and its derived neural crest (Hunt *et al,* 1991), but are also associated with limb development in both the anteroposterior and proximodistal axes (Dollé *et al,* 1989; Izpisúa-Belmonte *et al,* 1991; Yokouchi *et al,* 1991). The relationship between RARs and homoeobox genes is not well understood, and no homoeobox gene has yet been found to have an RARE. However, the observation that homoeobox genes in embryonal carcinoma (EC) cells are activated by RA (reviewed in Boncinelli, 1991) is very important in relation to the developmental functions of RARs. The activation of homoeobox genes by RA is related to their position on the chromosome, with the 3′ genes being activated first and to the greatest degree and the most 5′ genes being unresponsive. This linear response reflects the pattern of expression of the genes within the embryo, where the most 3′ genes are most rostrally expressed (Fig. 1) and the most 5′ genes are confined to more caudal levels of the embryonic axis. Furthermore, within the embryonic central nervous system, it is the rostral domains of homoeobox gene expression that are most susceptible to altered morphogenesis under conditions of RA excess.

The relationship between expression of *Hox-2.9* (the most 3′ gene of the *Hox-2* series) and hindbrain morphogenesis in RA treated embryos has been investigated, and the results confirm that the effects of RA on homoeogenes as observed in EC cells are relevant to developmental gene expression (Morriss-Kay *et al,* 1991). *Hox-2.9* is initially expressed in a relatively large domain of the early hindbrain neuroepithelium, but when rhombomere formation is complete, it is confined to rhombomere 4 (Fig. 1) (Murphy *et al,* 1989). Following exposure to excess RA, its domain of expression was extended, and the normal segmental pattern of the hindbrain was suppressed. Another segmentally expressed gene, *Krox-20,* failed to express in one of its normal domains and lost its segmental boundaries in the other. *Krox-20* is particularly interesting in its response to RA, because it is a transcription factor and binds to a sequence that is present in the upstream promoter region of human RAR-α (Brand *et al,* 1990).

The descriptive phase of study of RARs in relation to normal development is now approaching its end. There is enormous potential for investigations into the mechanisms through which RA activated RARs are involved in the control of normal developmental processes. In the near future, further studies on interactions between RA and homoeobox gene expression in relation to RA induced alterations in morphogenesis will yield useful information. Beyond that stage, progress will depend on the identification of more genes that have RAREs and are of clear developmental significance.

## SUMMARY

Retinoic acid receptors are ligand activated transcription factors that have a variety of important roles in normal development, through regulating the transcription of other developmental control genes. The great diversity of RA mediated developmental events is made possible through the existence of three receptor subtypes, each of which has a number of isoforms. The mechanism of gene regulation is thought to involve binding to an RA response element on the target gene, although in practice, few genes known to be activated by RA have so far been demonstrated to include such an element in the promoter region. Differential spatiotemporal expression patterns of the various subtypes have been described in mouse embryos, including very fine differences within different subpopulations of cells of the same tissue.

In addition to qualitative differences in the gene activation responses to RA through spatiotemporal differences in RAR expression patterns, local differences in RAR activation may be modulated by cytoplasmic binding proteins for retinol and RA, through quantitative control of ligand availability. The developmental control genes known to be activated by RA include the 3′ members of each of the homoeogene clusters, and it is likely that others will be identified in the near future.

## References

Bailey JS and Siu C-H (1988) Purification and partial characterization of a novel binding protein for retinoic acid from neonatal rat. *Journal of Biological Chemistry* **263** 9326–9332

Benbrook D, Lenhardt E and Pfahl M (1988) A new retinoic acid receptor identified from a hepatocellular carcinoma. *Nature* **333** 669–672

Boncinelli E, Simeone A, Acampora D and Mavilio F (1991) HOX gene activation by retinoic acid. *Trends in Genetics* **7** 329–334

Boylan JF and Gudas L (1991) Overexpression of the cellular retinoic acid binding protein I (CRABP-I) results in a reduction in differentiation-specific gene expression in F9 teratocarcinoma cells. *Journal of Cell Biology* **112** 965–979

Brand N, Petkovich M, Krust A *et al* (1988) Identification of a second human retinoic acid receptor. *Nature* **332** 850–853

Brand NJ, Petkovich M and Chambon P (1990) Characterization of a functional promoter for the human retinoic acid receptor-alpha (hRAR-α). *Nucleic Acid Research* **18** 6799–6806

Chytil F and Ong DE (1987) Intracellular vitamin A-binding proteins. *Annual Reviews of Nutrition* **7** 321–325

Cohlan SQ (1953) Excessive intake of vitamin A as a cause of congenital anomalies in the rat. *Science* **117** 535–536

Crow JA and Ong DE (1985) Cell-specific immunohistochemical localization of a cellular retinol-binding protein (type two) in the small intestine of the rat. *Proceedings of the National Academy of Sciences of the USA* **82** 4707–4747

Cuschieri A and Bannister LH (1975) The development of the olfactory mucosa in the mouse: light microscopy. *Journal of Anatomy* **119** 277–286

Dencker L, Annerwall E, Busch C and Eriksson U (1990) Localization of specific retinoid-binding sites and expression of cellular retinoic-acid-binding protein (CRABP) in the early mouse embryo. *Development* **110** 343–352

de Thé H, Marchio A, Tiollais P and Dejean A (1987) A novel steroid thyroid hormone receptor related gene inappropriately expressed in human hepatocellular carcinoma. *Nature* **330** 667–670

de Thé H, Marchio A, Tiollais P and Dejean A (1989) Differential expression and ligand regulation of the retinoic acid receptor α and β genes. *EMBO Journal* **8** 429–433

de Thé H, del Mar Vivanco-Ruiz M, Tiollais P, Stunnenberg H and Dejean A (1990) Identification of a retinoic acid response element in the retinoic acid receptor β gene. *Nature* **343** 177–180

Diamond MI, Miner JN, Yoshinaga SK and Yamamoto KR (1990) Transcription factor interactions: selectors of positive or negative regulation from a single DNA element. *Science* **249** 1266–1272

Dollé P, Ruberte E, Kastner P *et al* (1989) Differential expression of the genes encoding the retinoic acid receptors α, β, γ and CRABP in the developing limbs of the mouse. *Nature* **342** 702–705

Dollé P, Ruberte E, Leroy P, Morriss-Kay G and Chambon P (1990) Retinoic acid receptors and cellular binding proteins I A systematic study of their differential pattern of transcription during mouse organogenesis. *Development* **110** 1133–1151

Evans RM (1988) The steroid and thyroid hormone receptor superfamily. *Science* **240** 889–895

Giguère V, Ong ES, Segui P and Evans RM (1987) Identification of a receptor for the morphogen retinoic acid. *Nature* **330** 624–629

Giroud A and Martinet M (1959) Extension à plusieurs espèeces de mammifères des malformations embryonnaires par hypervitaminose A. *Comptes Rendus Société Biologique (Paris)* **153** 201–202

Green S and Chambon P (1988) Nuclear receptors enhance our understanding of transcription regulation. *Trends in Genetics* **4** 309–314

Hale F (1933) Pigs born without eyeballs. *Journal of Heredity* **24** 105–106

Hale F (1935) Relation of vitamin A to anophthalmos in pigs. *American Journal of Ophthalmology* **18** 1087–1093

Hale F (1937) Relation of maternal vitamin A deficiency to microphthalmia in pigs. *Texas State Journal of Medicine* **33** 228–232

Ham J and Parker MG (1989) Regulation of gene expression by nuclear hormone receptors. *Current Opinion in Cell Biology* **1** 503–511

Hardy M (1983) Vitamin A and the epithelial-mesenchymal interactions in skin differentiation, In: Sawyer RH and Fallon JH (eds). *Epithelial-Mesenchymal Interactions in Development,* pp 163–188, Praeger, New York

Hardy M, Dhouailly D, Törma H and Valquist A (1990) Either chick embryo dermis or retinoid treated mouse dermis can initiate glandular morphogenesis from mammalian epidermal tissue. *Journal of Experimental Zoology* **256** 279–289

Holland P (1990) Homeobox genes and segmentation: co-option, co-evolution and convergence. *Seminars in Developmental Biology* **1** 135–145

Hosler BA, LaRosa GJ, Grippo JF and Gudas LJ (1989) Expression of REX-1, a gene containing zinc finger motifs, is rapidly reduced by retinoic acid in F9 teratocarcinoma cells. *Molecular and Cellular Biology* **9** 5623–5629

Hunt P, Wilkinson D and Krumlauf R (1991) Patterning the vertebrate head: murine *Hox 2* genes mark distinct subpopulations of premigratory and migrating neural crest. *Development* **112** 43–50

Ishikawa T, Umesono K, Mangelsdorf DJ *et al* (1990) A functional retinoic acid receptor encoded by the gene on human chromosome 12. *Molecular Endocrinology* **4** 837–844

Izpisúa-Belmonte J-C, Falkenstein H, Dollé P, Renucci A and Duboule D (1991) Expression of the homeobox HOX-4 genes and the specification of positional information in chick wing development. *Nature* **350** 585–589

Kastner P, Krust A, Mendelsohn, C *et al* (1990) Murine isoforms of retinoic acid receptor γ with specific patterns of expression. *Proceedings of the National Academy of Sciences of the USA* **87** 2700–2704

Krust A, Kastner P, Petkovich M, Zelent A and Chambon P (1989) A third human retinoic acid receptor, hRAR-γ. *Proceedings of the National Academy of Sciences of the USA* **86** 5310–5314

Leid M, Kastner P, Lyons R *et al* (1992) Purification, cloning and RXR identity of the HeLa cell factor with which RAR or TR heterodimerizes to bind target sequences efficiently. *Cell* **68** 377–395

Leroy P, Krust A, Zelent A *et al* (1991a) Multiple isoforms of the mouse retinoic acid receptor α are generated by alternative splicing and differential induction by retinoic acid. *EMBO Journal* **10** 59–69

Leroy P, Nakshatri H and Chambon P (1991b) The mouse retinoic acid receptor alpha-2 (mRAR-α2) isoform is transcribed from a promoter that contains a retinoic acid response element. *Proceedings of the National Academy of Sciences of the USA* **88** 10138–10142

Leroy P, Krust A, Kastner P, Mendelsohn C, Zelent A and Chambon P (1992) Retinoic acid receptors, In: Morriss-Kay GM (ed). *Retinoids in Normal Development and Teratogenesis,* pp 7–25, Oxford University Press, Oxford

Maden M, Ong DE, Summerbell D and Chytil F (1988) Spatial distribution of cellular protein binding to retinoic acid in the chick limb bud. *Nature* **335** 733–735

Maden M, Ong DE, Summerbell D, Chytil F and Hirst DA (1989) Cellular retinoic acid-binding protein and the role of retinoic acid in the development of the chick embryo. *Developmental Biology* **135** 124–132

Maden M, Hunt P, Eriksson U, Kuroiwa A, Krumlauf R and Summerbell D (1991) Retinoic acid-binding protein, rhombomeres and the neural crest. *Development* **111** 35–44

Makover A, Soprano DR, Wyatt ML and Goodman DS (1989) An in situ-hybridization study of the localization of retinol-binding protein and transthyretin messenger RNAs during fetal development in the rat. *Differentiation* **40** 17–25

Mangelsdorf DJ and Evans RM (1992) Vitamin A receptors: new insights on retinoid control of transcription, In: Morriss-Kay GM (ed). *Retinoids in Normal Development and Teratogenesis,* pp 27–50, Oxford University Press, Oxford

Mangelsdorf DJ, Ong ES, Dyck JA and Evans RM (1990) Nuclear receptor that identifies a novel retinoic acid response pathway. *Nature* **345** 224–229

Mangelsdorf DJ, Borgmeyer U, Heyman RA *et al* (1992) Characterization of three RXR genes that mediate the action of 9-*cis*-retinoic acid. *Genes and Development* **6** 329–344

Mattei M-G, Rivière M, Krust S *et al* (1991) Chromosomal assignment of retinoic acid receptor (RAR) genes in the human, mouse and rat genomes. *Genomics* **10** 1061–1069

Mendelsohn C, Ruberte E, LeMeur M, Morriss-Kay G and Chambon P (1991) Developmental analysis of the retinoic acid-inducible RAR-β2 promoter in transgenic animals. *Development* **113** 723–734

Morriss GM (1972) Morphogenesis of the malformations induced in rat embryos by maternal hypervitaminosis A. *Journal of Anatomy* **113** 241–250

Morriss GM (1975) Abnormal cell migration as a possible factor in the genesis of vitamin A-induced craniofacial anomalies, In: Neubert D and Merker H-J (eds). *New Approaches to the Evaluation of Abnormal Embryonic Development,* pp 678–687, Georg Thieme Verlag,

Stuttgart

Morriss GM and Thorogood PV (1978) An approach to cranial neural crest cell migration and differentiation in mammalian embryos, In: MH Johnson (ed). *Development in Mammals* 3, pp 363-441, Elsevier-North Holland, Amsterdam

Morriss-Kay GM Retinoic acid and development. *Pathobiology* (in press)

Morriss-Kay GM, Murphy P, Hill RE and Davidson DR (1991) Effects of retinoic acid excess on expression of *Hox-2.9* and *Krox-20* and on morphological segmentation of the hindbrain of mouse embryos. *EMBO Journal* **10** 2985–2995

Murphy P, Davidson DR and Hill RE (1989) Segment-specific expression of a homeobox-containing gene in the mouse hindbrain. *Nature* **341** 156–159

Noji S, Yamaai T, Koyama E, Noho T and Taniguchi S (1989) Spatial and temporal expression pattern of retinoic acid receptor genes during mouse bone development. *FEBS Letters* **257** 93–96

Noji S, Nohno T, Koyama E *et al* (1991) Retinoic acid induces polarizing activity but is unlikely to be a morphogen in the chick limb bud. *Nature* **350** 83–86

Petkovich M, Brand N, Krust A and Chambon P (1987) A human retinoic acid receptor which belongs to the family of nuclear receptors. *Nature* **330** 444–450

Robertson M (1987) Towards a biochemistry of morphogenesis. *Nature* **330** 420–421

Rossant J, Zirngibl R, Cado D, Shago M and Giguère V (1991) Expression of a retinoic acid respons element-*hsplacZ* transgene defines specific domains of transcriptional activity during mouse embryogenesis. *Genes and Development* **5** 1333–1344

Rowe A, Richman JM and Brickell PM (1991) Retinoic acid treatment alters the distribution of retinoic acid receptor-β transcripts in the embryonic chick face. *Development* **111** 1007–1016

Ruberte E, Dollé P, Krust, A, Zelent A, Morriss-Kay G and Chambon P (1990) Specific spatial and temporal distribution of retinoic acid receptor gamma transcripts during mouse embryogenesis. *Development* **108** 213–222

Ruberte E, Dollé P, Chambon P and Morriss-Kay G (1991) Retinoic acid receptors and cellular binding proteins: II Their differential pattern of transcription during early morphogenesis in mouse embryos. *Development* **111** 45–60

Ruberte E, Nakshatri H, Kastner P and Chambon P (1992) Retinoic acid receptors and binding proteins in mouse limb development, In: Morriss-Kay GM (ed). *Retinoids in Normal Development and Teratogenesis,* pp 99–111, Oxford University Press, Oxford

Ruberte E, Friederich V, Morriss-Kay G and Chambon P Differential distribution patterns of CRABP I and CRABP II transcripts during mouse embryogenesis. *Development* (in press).

Sani BP, Reddy LG, Singh RK and Arello EA (1991) A new retinoid binding protein with a low molecular weight. *Biochemical and Biophysical Research Communications* **175** 1064–1069

Shapiro SS and Mott DJ (1981) Modulation of glycosaminoglycan biosynthesis by retinoids. *Annals of the New York Academy of Sciences* **359** 300–321

Shenefelt RE (1972) Morphogenesis of malformations in hamster caused by retinoic acid: relation to dose and stage at treatment. *Teratology* **5** 103–108

Strickland S, Smith KK and Marotti KR (1980) Hormonal induction of differentiation in teratocarcinoma stem cells: generation of parietal endoderm by retinoic acid and dibutyryl cAMP. *Cell* **21** 347–355

Sucov HM, Murakami KK and Evans RM(1990) Characterization of an autoregulated response element in the mouse retinoic acid receptor type β gene. *Proceedings of the National Academy of Sciences of the USA* **87** 5392–5396

Tabin CJ (1991) Retinoids, homeoboxes and growth factors: toward molecular models for limb development. *Cell* **66** 199–217

Tuckett F and Morriss-Kay GM (1985) The kinetic behaviour of the cranial neural epithelium during neurulation in the rat embryo. *Journal of Embryology and Experimental Morphology* **85** 111–119

Umesono K, Giguère V, Glass CK, Rosenfeld MG and Evans RM (1988) Retinoic acid and thyroid hormone induce gene expression through a common response element. *Nature* **336** 262–265

Vaessen M-J, Meijers JC, Bootsma D and van Kessel AG (1990) The cellular retinoic-acid- binding protein is expressed in tissues associated with retinoic acid-induced malformations. *Development* **110** 371–378

Viallet JP, Ruberte E, Krust A, Zelent A and Dhouailly D (1992) Expression of retinoic acid receptors and dermal-epidermal interactions during mouse skin morphogenesis, In: Morriss-Kay GM (ed). *Retinoids in Normal Development and Teratogenesis*, pp 199–210, Oxford University Press, Oxford

Webster WS Johnston MC, Lammer EJ and Sulik KK (1986) Isotretinoin embryopathy and the cranial neural crest: an in vivo and in vitro study. *Journal of Craniofacial Genetics and Developmental Biology* **6** 211–222

Wilson JG and Warkany J (1949) Aortic-arch and cardiac anomalies in the offspring of vitamin A deficient rats. *American Journal of Anatomy* **85** 113–155

Wilson JG, Roth CB and Warkany J (1953) An analysis of the syndrome of malformations induced by maternal vitamin A deficiency: effects of restoration of vitamin A at various times during gestation. *American Journal of Anatomy* **92** 189–217

Yokouchi Y, Sasaki H and Kuroiwa A (1991) Homeobox gene expression correlated with the bifurcation process of limb cartilage development. *Nature* **353** 443–445

Zelent A, Krust A, Petkovich M, Kastner P and Chambon P (1989) Cloning of murine α and β receptors and a novel receptor γ predominantly expressed in skin. *Nature* **339** 714–717

Zelent A, Mendelsohn C, Kastner P *et al* (1991) Differentially expressed isoforms of the mouse retinoic acid receptor β are generated by usage of two promoters and alternative splicing. *EMBO Journal* **10** 71–81

The author is responsible for the accuracy of the references.

# Retinoic Acid Receptor α in Acute Promyelocytic Leukaemia

**HUGUES DE THÉ • ANNE DEJEAN**

*Unité de Recombinaison et Expression Génétique (INSERM U163), Institut Pasteur, 28 rue du Dr Roux, 75724 Paris Cedex 15*



## THE RETINOIC ACID RECEPTORS

Retinoids and retinoic acid (RA) are a group of vitamin A derived substances that have striking effects on development, differentiation and proliferation in a wide variety of systems (reviewed in Brockes, 1990; Tabin, 1991). The identification of three retinoic acid receptors (RARα, β and γ) as members of the nuclear receptor superfamily led to important insights into the molecular mechanism of retinoid action (de Thé *et al,* 1987; Giguère *et al,* 1987; Petkovitch *et al,* 1987; Brand *et al,* 1988; Zelent *et al,* 1989). The RARs are very similar in their DNA and hormone binding domains, but the carboxy- and amino-terminal ends, which are highly conserved from mouse to humans, are distinct in each of the three RARs. The three receptors are expressed as isoforms, which are transcribed from different promoters and give rise to several proteins that differ by their aminoterminal end (Kastner *et al,* 1990; Leroy *et al,* 1991). This family of receptors activates or represses gene expression, in a ligand inducible fashion, through binding to hormone responsive elements located in the promoter regions of target genes (Evans, 1988; Beato, 1989). Analysis of retinoic acid response elements has shown that these, unlike steroid hormone response elements, appear to consist of direct repeats of a GGTTCA motif (de Thé *et al,* 1990a; Umesono *et al,* 1991). The receptors appear to have different affinities for RA (all-*trans*-RA), γ having the highest affinity and α the lowest. Conceivably, uncharacterized natural or synthetic retinoids may have different spectra of specificity.

The biological basis for RAR diversity is as yet unknown, although a different tissue distribution of the transcripts of the various RARs has been ex-

*Cancer Surveys* Volume 14: *Growth Regulation by Nuclear Hormone Receptors*
 0-87969-371-1/92. $3.00 + .00

tensively documented, suggesting that each receptor makes a distinct contributions to RA physiology (de Thé *et al,* 1989; Dollé *et al,* 1989). Moreover, it seems that RA regulation of molecules implicated in retinoid physiology is a common rule: the two major isoforms of RARβ and one of RARα, several proteins implicated in retinoid sequestration in the cytoplasm (cytoplasmic RA binding proteins) and enzymes implicated in retinoid synthesis are all directly regulated by RA. Such RA dependence of several components of the retinoid signal transduction pathway suggests that the molecular events triggered by RA administration may be considerably more complex than we think. This complexity is reinforced by the existence of another family of receptors, the RXRs, which respond to RA but do not bind all-*trans*-RA (Mangelsdorf *et al,* 1990). It is likely that they bind and respond to metabolites of RA.

## DIFFERENTIATION OF LEUKAEMIAS

One of the early and most extensively studied model for cell differentiation was the HL60 system, a human promyelocytic cell line that differentiates into granulocytes upon RA treatment (Breitman *et al,* 1980). Several studies have provided important insights into the molecular mechanism of HL60 differentiation, including the identification of the likely retinoid effector (RARα) and of some of the genes implicated (Collins *et al,* 1990). Human myeloid leukaemias represent a proliferation of malignant myeloid cells blocked at a specific stage of differentiation, and several studies have shown that they could continue their maturation in vitro, following treatment with selected inducers (reviewed in Sachs 1978; Koeffler, 1985). Retinoic acid, for example, consistently and specifically differentiates cultures of primary bone marrow cells from patients with acute promyelocytic leukaemia (APL) into mature granulocytes (Breitman *et al,* 1981; Chomienne *et al,* 1990a). These in vitro studies have been successfully transposed in vivo: treatment of APL patients with oral all-*trans*-RA induces morphological complete remissions by inducing differentiation of the malignant clone, thus providing one of the first examples of differentiation therapy (Huang *et al,* 1988; Castaigne *et al,* 1990; Warrell *et al,* 1991).

## MOLECULAR CLONING OF THE t(15;17) TRANSLOCATION AND CHARACTERIZATION OF THE FUSION GENE PRODUCTS

A highly specific t(15;17) balanced and reciprocal translocation is found in most patients with APL (Larson *et al,* 1984). The RARα gene maps to chromosome 17q21 (Mattei *et al,* 1988), close to the t(15;17) (q22-q21) translocation breakpoint. The spectacular RA sensitivity of APL cells and the gene location close to the t(15;17) breakpoint raised the question of a possible link between the two phenomena. The purpose of this review is to summarize the available data on RARα in APL and to discuss the current models for both transformation and RA induced differentiation in this disease.

The molecular cloning of the t(15;17) translocation in APL originated either by a direct study of the RARα gene structure, initiated on the basis of the (apparent) coincidence mentioned above (Chomienne *et al,* 1990b; de Thé *et al,* 1990b; Longo *et al,* 1990; Alcalay *et al,* 1991), or by chromosome walk analysis (Borrow *et al,* 1990). These were carried out either on patients' DNAs or on an APL cell line carrying the t(15;17) translocation and RA sensitive phenotype (NB4) (Lanotte *et al,* 1991). These early studies all showed (a) that RARα was the gene implicated on chromosome 17 and that the rearrangement always occurred in the second exon, (b) that the breakpoints on chromosome 15 appeared to be clustered in a single locus and (c) that this locus, initially named *myl* and now renamed *PML,* was a new gene expressed in APL cells, as a fusion PML/RARα mRNA. A detailed analysis of the genomic structure of the PML RARα rearrangement and fusion transcript in a large number of APL cases was published recently (Pandolfi *et al,* in press).

The PML gene located on chromosome 15 and fused to RARα was also characterized (de Thé *et al,* 1991; Goddart *et al,* 1991; Kakizuka *et al,* 1991; Pandolfi *et al,* 1991; Kastner *et al,* 1992). This previously unknown gene contains several interesting structural motifs including a proline rich region, a presumed DNA binding domain consisting of atypical zinc fingers, an α helical region containing a leucine zipper and an aminoterminal serine rich region containing presumed serine/threonine phosphorylation sites (Fig. 1). The cysteine rich region contains three clusters. The first cluster is similar to a recently described family of proteins (Freemont *et al,* 1991) in which the conserved metal binding residues are in the C3HC4 sequence. This family includes DNA binding proteins of varied functions: RAD18 (a yeast DNA repair enzyme), *RAG1* (a V(D)J recombination activating gene), viral *trans*-activating genes (*VZ61, IE110, CG30*), oncogenes (*bmi-1, T18*) and transcription factors (RFP, RPT1). A second cysteine rich cluster is similar to sequences present in RFP, RPT1 and T18, perhaps suggesting that PML also is a molecule involved in either transcriptional regulation or oncogenesis. PML is expressed as a family of transcripts that differ from each other by alternative splicing of the most 3′ coding exon (two additional coding exons may be omitted, but this does not alter the reading frame). Thus, the *PML* gene is expressed as a large number of proteins that contain the same DNA binding domain and leucine zipper but differ in the serine rich region.

Several abnormal proteins are present in APL cells as the consequence of the t(15;17) translocation. The first ones are the PML/RARα fusions transcribed from 15q+ derivatives, which have recently been characterized by several groups (Pandolfi *et al,* 1990; de Thé *et al,* 1991; Goddart *et al,* 1991; Kakizuka *et al,* 1991; Kastner *et al,* 1992). These proteins contain the aminoterminal part of PML, including the putative DNA binding domain and leucine zipper fused to the DNA binding domain and hormone binding domain of RARα (Fig. 1). Whereas the breakpoints on chromosome 17 always occur in intron 2, the breakpoints on chromosome 15 are clustered in two different introns of PML, leading to fusion proteins that contain a variable length

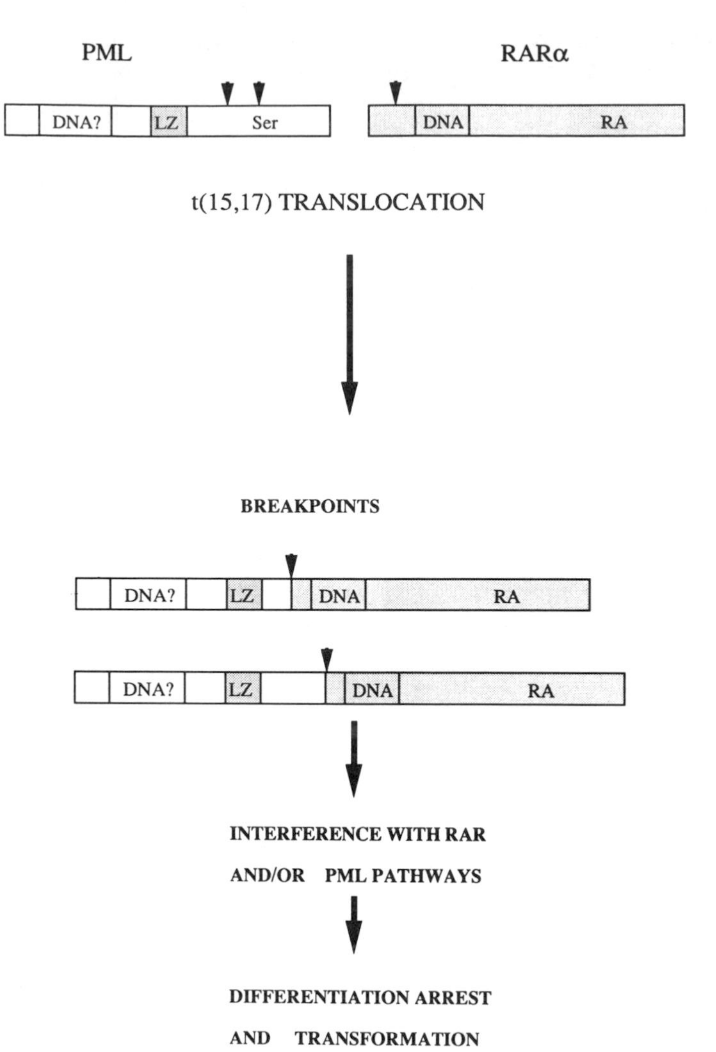


**Fig. 1.** Schematic representation of the structure of the PML and RARα proteins and of the PML/RARα fusion products generated by the t(15;17) translocation. LZ = leucine zipper; Ser = serine rich region

of PML sequences in the α-helical region. There is at present no consensus for the designation of the two major types of fusion products. These proteins have been detected in the blast cells of APL patients and, despite the fact that the PML/RARα fusion mRNA is clearly less abundant than the normal RARα transcript, the fusion proteins are in large excess to the normal receptors (Kastner *et al,* 1992; Pandolfi *et al,* in press). This observation suggests that RARα and PML/RARα are very different in their translational efficiencies or in their protein turnover rates and provides a molecular basis for a dominant effect of PML/RARα over RARα (see below). The t(15;17) also generates a carboxyterminus truncated PML gene product, which may have a role in leukaemogenesis. The 17p+ derived transcript was also identified and was shown to encode a small protein containing the aminoterminal region of RARα fused in frame to the carboxyterminal domain of PML. Since this fusion does not contain any significant functional domain (in particular, no DNA binding region), investigators have focused on the function of the product of the reciprocal translocation.

## FUNCTIONAL ANALYSIS OF PML/RARα

The fusion between a nuclear receptor and a presumptive transcription factor may alter either or both signal transduction pathways. Although several RA target genes have been identified to date, no PML activated genes are known. Thus, several groups have analysed the effect of the PML/RARα chimaera on RA responsive reporter systems. The results are somehow contradictory, very dependent on the reporter gene used and the cell type. Nevertheless, it appears that PML/RARα is not a functionally normal RAR. Moreover, deletion of the PML moiety of PML/RARα restores the receptors' *trans*-activating properties, pointing to the role of PML sequences in the impairment of the chimeric receptor's function (de Thé *et al,* 1991; Kakizuka *et al,* 1991; Kastner *et al,* 1992). Alteration of PML function has been elegantly documented by the demonstration that the immunofluorescence aspect of PML (speckled nuclear) was clearly distinct from the pattern of PML/RAR (cytoplasmic and nuclear) and that co-expression of PML/RARα could delocalize an epitope tagged PML (Kastner *et al,* 1992). In the presence of RA, PML/RARα becomes nuclear. Additional experiments showed that PML and PML/RARα could form dimers, presumably by a coiled/coiled interaction. Although these experiments do not elucidate PML function, they provide a molecular basis for a dominant effect of PML/RARα over PML.

## DISCUSSION

Alteration in transcriptional control has been proposed to be involved in oncogenesis (Lewin, 1991). For example, the oncogene v-*erbA*, whose cellular counterpart c-*erbA* encodes a nuclear receptor for thyroid hormone, acts as a constitutive repressor on thyroid hormone responsive genes. On this basis, it was proposed that the v-erbA product exerts its transforming effects by blocking thyroid hormone mediated erythrocytic differentiation (Damm *et al,* 1989; Zenke *et al,* 1990). One of the plausible models for APL leukaemogenesis would be that the PML-RARα protein, like v-erbA, functions as a dominant negative mutant. Assuming that, in promyelocytes, genes activated by RARs induce differentiation—which is strongly suggested by the effect of RA on APL or HL60 cells—the hybrid protein may contribute to leukaemogenesis by antagonizing differentiation. Support for this model comes from preliminary data from several groups suggesting that expression of the PML/RARα gene in HL60 cells impairs or blocks RA induced differentiation (Chomienne C, Farzaneh F and Pellici PG, personal communications).

One alternative or complementary model would assume that PML is a key regulatory gene implicated in promyelocytic differentiation or growth control. Fusion of a nuclear receptor hormone binding domain to a transcription factor or nuclear effector (Myc, E1A, MyoD) results in a hormone dependent function. Similarly, fusion of PML to the hormone binding domain of RARα might render PML function RA dependent. Indeed, the alteration in the intracellular localization of the PML/RARα fusion appears to be RA sensitive. However, in the absence of a known function for PML, it appears difficult to propose a model for PML mediated oncogenesis. Another suggestion for a role of PML in leukaemogenesis is the description of two typical cases of APL with translocations involving only chromosome 15 (Goddart *et al,* 1991; Chomienne C, personal communication).

The molecular cloning of the translocation and the identification of RARα as the target genes suggest that leukaemogenesis is somehow due an interference of the PML/RARα fusion with normal promyelocytic differentiation. The spectacular and highly specific effect of RA in this disease remains poorly understood. In particular, the observation that cells that carry a chimeric receptor (APL) or that do not (HL60) respond to RA in a comparable manner is most puzzling. Retinoic acid administration modifies the aberrant intracellular distribution of the PML/RARα chimaera, but it does not seem to restore the normal profile of PML. If PML/RARα represses RA sensitive genes, several suggestions can be made to account for the effect of RA. One of them would be that the chimaera becomes an efficient *trans*-activator of RA target genes at pharmacological RA concentrations. Alternatively, such high RA doses may activate other receptors (including RXRs) that would compete out the PML/RARα chimaera of the target DNA sequences. Recently, RARα has been shown to be RA inducible in promyelocytes (Chomienne *et al,* 1991).

Both *bcr* and *abl* gene products have been implicated in the pathogenesis of chronic myelogeneous leukaemia (Diekmann *et al,* 1991). Moreover, in the E26 induced leukaemias, fusion of two viral oncogenes (v-*myb* and v-*ets*) appears to be required: viruses carrying both v-*myb* and v-*ets* are only weakly leukaemogenic and become highly transformant only as variant viruses containing a v-Myb-Ets fusion protein are selected (Metz and Graf, 1991). It is conceivable that for the t(15;17) translocation, RARα and PML would be another illustration of this phenomenon.

## SUMMARY

Acute promyelocytic leukaemia has two highly specific particularities: a t(15;17) chromosomal translocation and the ability of a differentiation inducer all-*trans*-RA, to revert the malignant phenotype both in vitro and in vivo. Molecular characterization of the t(15;17) translocation has shown that it fuses a previously unknown zinc finger encoding gene, *PML,* to the RARα, suggesting a link between the molecular mechanism of transformation and of RA dependent differentiation. The PML/RARα fusion receptor—which is functionally altered—may block RA target genes, impair RA mediated differentiation and lead to transformation. Alternatively, or in addition, the PML transduction pathway may also be affected. Although it is clear that RA treatment must relieve APL cells from differentiation arrest, so far no model can satisfactorily account for this effect.

## References


Alcalay M, Zangrilli D, Pandolfi PP *et al* (1991) Translocation breakpoint of acute promyelocytic leukemia lies within the retinoic acid receptor α locus. *Proceedings of the National Academy of Sciences of the USA* **88** 1977–1981

Beato M (1989) Gene regulation by steroid hormones. *Cell* **56** 335–344

Borrow J, Goddard AD, Sheer D and Solomon E (1990) Molecular analysis of acute promyelocytic leukemia breakpoint cluster region on chromosome 17. *Science* **249** 1577–1580

Brand N, Petkovich M, Krust A *et al* (1988) Identification of a second human retinoic acid receptor. *Nature* **332** 850–853

Breitman T, Selonick SE and Collins SJ (1980) Induction of differentiation of the human promyelocytic cell-line HL60 by retinoic acid. *Proceedings of the National Academy of Sciences of the USA* **77** 2936–2940

Breitman TR, Collins SJ and Keene BR (1981) Terminal differentiation of promyelocytic leukaemic cells in response to retinoic acid. *Blood* **57** 1000–1004

Brockes J (1990) Reading the retinoid signals. *Nature* **3451** 766–768

Castaigne S, Chomienne C, Daniel MT *et al* (1990) All trans-retinoic acid as a differentiation therapy for acute promyelocytic leukaemia: I clinical results. *Blood* **76** 1704–1709

Chomienne C, Ballerini P, Balitrand N *et al* (1990a) All-trans retinoic acid in acute promyelocytic leukemias, II in vitro studies: structure-function relationship. *Blood* **76** 1710–1717

Chomienne C, Ballerini P, Balitrand N *et al* (1990b) The retinoic acid receptor a gene is rearranged in retinoic acid-sensitive promyelocytic leukemias. *Leukemia* **4** 802–807

Chomienne C, Ballerini P, Balitrand N *et al* (1991) All trans retinoic acid modulates RARα in promyelocytic cells. *Journal of Clinical Investigation* **88** 2150–2154

Collins SJ, Robertson KA and Mueller L (1990) Retinoic acid-induced gramlocytic differentiation of HL60 myeloid leukemia cells is mediated directly through the retinoic acid receptor α (RAR-α. *Molecular and Cellular Biology* **10** 2154–2163

Damm K, Thompson CC and Evans RM (1989) Protein encoded by v-*erbA* functions as a thyroid hormone receptor antagonist. *Nature* **339** 593–597

de Thé H, Marchio A, Tiollais P and Dejean A (1987) A novel steroid/thyroid hormone receptor-related gene inappropriately expressed in human hepatocellular carcinoma. *Nature* **330** 667–670

de Thé H, Marchio A, Tiollais P and Dejean A (1989) Differential expression and ligand regulation of the retinoic acid receptor a and b genes. *EMBO Journal* **8** 429–433

de Thé H, Chomienne C, Lanotte M, Degos L and Dejean A (1990a) The t(15;17) translocation of acute promyelocytic leukaemia fuses the retinoic acid receptor a gene to a novel transcribed locus. *Nature* **347** 558–561

de Thé H, Del Mar Vivanco-Ruiz M, Tiollais P, Stunnenberg H and Dejean A (1990b) Identification of a retinoic acid responsive element in the retinoic acid receptor β gene. *Nature* **343** 177–180

de Thé H, Lavau C, Marchio A, Chomienne C, Degos L and Dejean A (1991) The PML/RARα fusion mRNA generated by the t(15,17) translocation encodes a functionally altered retinoic acid receptor. *Cell* **66** 675–684

Diekmann D, Brill S, Garrett S *et al* (1991) *Bcr* encodes a GTPase-activating protein for $p21^{rac}$. *Nature* **351** 400–402

Dollé P, Ruberte E, Kastner P *et al* (1989) Differential expression of genes encoding α, β and γ retinoic acid receptors and CRABP in the developing limbs of the mouse. *Nature* **342** 702–705

Evans RM (1988) The steroid and thyroid hormone receptor superfamily. *Science* **240** 889–895

Freemont PS, Hanson IM and Trowsdale J (1991) A novel cysteine-rich sequence motif. *Cell* **64** 483–484

Giguère V, Ong ES, Segui P and Evans RM (1987) Identification of a receptor for the morphogen retinoic acid. *Nature* **330** 624–629

Goddart A, Borrow J, Freemont P *et al* (1991) Characterisation of a zinc finger gene disrupted by the t(15,17) translocation in acute promyelocytic leukemia. *Science* **254** 1371–1374

Huang ME, Yu-chen Y, Shu-rong C *et al* (1988) Use of all-trans retinoic acid in the treatment of acute promyelocytic leukemia. *Blood* **72** 567–572

Kakizuka A, Miller WH, Umesono K *et al* (1991) Chromosomal translocation t(15,17) in human acute promyelocytic leukemia fuses RARα with a novel putative transcription factor PML. *Cell* **66** 663–674

Kastner P, Krust A, Mendelson C *et al* (1990) Murine isoforms of the retinoic acid receptor γ with specific patterns of expression. *Proceedings of the National Academy of Sciences of the USA* **87** 2700–2704

Kastner P, Perez A, Lutz Y *et al* (1992) Structure, localization and transcriptional properties of two classes of retinoic acid receptor α fusion proteins in acute promyelocytic leukemia: stuctural similarities with a new family of oncoproteins. *EMBO Journal* **11** 629–642

Koeffler HP (1985) Study of differentiation and proliferation of leukemic cells using myeloid leukemia cell lines, In: Bloomfield CD (ed). *Chronic and Acute Leukemias in Adults*, pp 27–68, Martinus Nijhoff, Boston

Lanotte M, Martin V, Najman S, Ballerini P, Valensi S and Berger R (1991) NB4, a maturation inducible cell-line with t(15;17) marker isolated from a human promyelocytic leukaemia (M3). *Blood* **77** 1080–1086

Larson RA, Kondo K, Vardiman JW, Butler AE, Golomb HM and Rowley JD (1984) Evidence for a 15;17 translocation in every patient with acute promyelocytic leukemia. *American Journal of Medicine* **76** 827–841

Leroy P, Krust A, Zelent A *et al* (1991) Multiple isoforms of the mouse retinoic acid receptor a are generated by alternative splicing and differential induction by retinoic acid. *EMBO Journal* **10** 59–69

Lewin B (1991) Oncogenic conversion by regulatory changes in transcription factors. *Cell* **64** 303–312

Longo L, Pandolfi PP, Biondi A *et al* (1990) Rearrangement and aberrant expression of the retinoic acid receptor α in acute promyelocytic leukemia. *Journal of Experimental Medicine* **172** 1571–1575

Mangelsdorf DJ, Ong ES, Dyck JA and Evans RM (1990) Nuclear receptor that identifies a novel retinoic acid response pathway. *Nature* **345** 224–229

Mattei MG, Petkovich M, Mattei JF, Brand N and Chambon P (1988) Mapping of the human retinoic acid receptor to the q21 band of chromosome 17. *Human Genetics* **80** 186–188

Metz T and Graf T (1991) Fusion of the nuclear oncoproteins v-myb and v-ets is required for the leukemogenesis of E26 virus. *Cell* **66** 95–105

Pandolfi PP, Grignani F, Alcalay M *et al* (1991) Structure and origin of the acute promyelocytic leukemia myl/RARα cDNA and characterization of its retinoid binding and transactivation properties. *Oncogene* **6** 1285–1292

Pandolfi PP, Alcalay M, Fagioli M *et al* Genomic variability and alternative splicing generate multiple PML/RARα transcripts that encode aberrant PML proteins and PML/RARα isoforms in acute promyelocytic leukemias. *EMBO Journal* (in press)

Petkovich M, Brand NJ, Krust A and Chambon P (1987) A human retinoic acid receptor which

belongs to the family of nuclear receptors. *Nature* **330** 444–450
Sachs L (1978) Control of normal cell differentiation and the phenotypic reversion of malignancy in myeloid leukaemia. *Nature* **274** 535–539
Tabin H (1991) Retinoids, homeoboxes and growth factors towards molecular models for limb development. *Cell* **66** 199–217
Umesono K, Murakami KK, Tompson C, Evans R (1991) Direct repeats as selective response elements for the thyroid hormone, retinoic acid and vitamin D3 receptors. *Cell* **65** 1255–1266
Warrell RP, Frankel SR, Miller WH *et al* (1991) Differentiation therapy of acute promyelocytic leukemia by tretinoin (all trans retinoic acid). *New England Journal of Medicine* **324** 1385–1393
Zelent A, Krust A, Petkovich M, Kastner P and Chambon P (1989) Cloning of murine α and β retinoic acid receptors and a novel receptor γ predominantly expressed in skin. *Nature* **339** 714–717
Zenke M, Munoz A, Sap J, Vennström B and Beug H (1990) v-*erbA* oncogene activation entails the loss of hormone-dependent regulator activity of c-*erbA*. *Cell* **61** 1035–1049

The authors are responsible for the accuracy of the references.

# Mechanistic Interrelationships between Two Superfamilies: The Steroid/Retinoid Receptors and Transforming Growth Factor-β

**ANITA B ROBERTS • MICHAEL B SPORN**
*Laboratory of Chemoprevention, National Cancer Institute, Bethesda, Maryland 20892*



## INTRODUCTION

In this chapter, we summarize data demonstrating that various steroids and retinoids that act through a set of nuclear receptors can control the expression of transforming growth factor-β (TGF-β), a set of secreted peptides that act principally through cell membrane receptors. On the basis of these data, we propose that the effects of these steroids/retinoids on growth and differentiation of target cells or tissues are often mediated locally by one or more of the TGF-β isoforms.

*Cancer Surveys* Volume 14: *Growth Regulation by Nuclear Hormone Receptors*

## The Steroid/Retinoid Superfamily and Control of Cell Growth and Differentiation

Effects of steroid hormones as well as retinoids and vitamin $D_3$ on growth and differentiation of cells and tissues are well documented. As presented in detail elsewhere in this volume, the sex hormones, including oestrogen and testosterone and their analogues and antagonists, have profound effects on responsive tissues and cells derived from breast, ovary, prostate and testis; retinoids are important in controlling morphogenesis and in regulation of the growth and phenotype of a wide variety of cells including most epithelial cells (Roberts and Sporn, 1984), and vitamin $D_3$ metabolites are critical regulators of bone development and differentiation as well as of the growth and differentiation of diverse cell types (DeLuca *et al*, 1990). Yet only recently has it been appreciated that effects of these dissimilar lipophilic molecules are all mediated intracellularly by a common superfamily of nuclear ligand responsive receptors with structures related to the viral oncogene *erbA* (Evans, 1988; O'Malley, 1990); these receptors bind to specific motifs in enhancer regions of target genes, thereby regulating their expression. The diversity of the ligands belonging to this superfamily and the identification of so many "orphan" receptors for which ligands have as yet to be identified suggest that the set of molecules utilizing this common mechanism will be very large indeed (O'Malley, 1990).

Ultimately, retinoids, vitamin $D_3$ and steroid hormones are all thought to regulate the phenotype of cells by altering their pattern of gene transcription and expression of proteins. The important question then becomes one of identifying which genes are regulated by these molecules and their nuclear receptors and which specific mechanisms are utilized (Sporn and Roberts, 1991). Although the totality of their mechanisms of action in the control of cellular differentiation and proliferation remains to be elucidated, there is now substantial evidence that the target genes of these nuclear receptors include a set of secreted peptide growth factors that, acting through cell surface receptors, can mediate certain of the effects of the members of this superfamily on cell growth and differentiation. In breast cancer cells, for example, oestrogen has been shown to regulate expression of platelet derived growth factor, the insulin like growth factors, transforming growth factor-α and TGF-β (Cullen and Lippman, 1989). Of these, TGF-β is unique in that it often opposes or antagonizes the action of the other mitogenic growth factors and acts to inhibit cell growth (Roberts and Sporn, 1990), suggesting that it might have a unique role in mediating the growth inhibitory activities of the steroid/retinoid superfamily.

## TGF-β as Regulators of Cell Growth and Differentiation

For the past 10 years, our laboratory has been studying a family of growth factors called TGF-β; for recent reviews, see Roberts and Sporn (1990) and Massagué (1990). Three distinct isoforms of this peptide, TGF-βs 1, 2 and 3, are found, often differentially expressed, in mammalian species. The TGF-βs are produced by and act on most cell types, including mesenchymal cells, immune cells and epithelial cells, controlling not only their growth but also their phenotype. As an example, TGF-β strongly inhibits the growth of most epithelial cells in vitro (Moses *et al*, 1990) and exhibits a localization pattern in vivo, which strongly implicates it both in differentiation and maintenance of epithelial tissues and in interactions of epithelial and mesenchymal cells (Heine *et al*, 1990; Silberstein *et al*, in press). The data of Barnard *et al* (1989) suggest that endogenous TGF-β might function to suppress growth and regulate differentiation of epithelial cells in vivo; they have demonstrated an inverse relationship between mitotic labelling and expression of TGF-β1 messenger RNA (mRNA) in isolated intestinal enterocytes. Moreover, Silberstein and Daniels (1987) demonstrated inhibitory effects of exogenous TGF-β on growth and morphogenesis of mammary epithelium into end buds; the inhibition was reversible and localized to regions adjacent to implanted slow release pellets of the peptide.

The TGF-βs are multifunctional, the nature of their action on a target cell being critically dependent on many parameters including the cell type and its state of differentiation and the particular set of growth factors and hormones acting on the cell (Sporn and Roberts, 1988). Tissues exist in a dynamic state in which highly regulated mechanisms exist to control cellular differentiation and proliferation. That control depends on the immediate environment of the cells including extracellular effector peptides and cell-cell contact as well as cell-matrix interactions (Nathan and Sporn, 1991).

## TGF-βs AS MEDIATORS OF THE EFFECTS OF STEROIDS AND RETINOIDS ON CELL GROWTH

### Role of TGF-β in Effects of Oestrogens, Tamoxifen and Gestodene on Breast Cancer Cells

There is now substantial evidence that expression of growth factors by breast cancer cell lines is hormonally regulated and that the effects of the various hormones on cellular proliferation might be mediated in part by these peptides (Cullen and Lippman, 1989; Wakefield *et al*, 1990a). Thus, the ability of oestrogen to stimulate and tamoxifen, an anti-oestrogen, to inhibit the growth of several oestrogen receptor positive breast cancer cell lines can be correlated with their opposite effects on cellular secretion of the negative growth regulator, TGF-β. Treatment of T47-D cells with oestradiol results in decreased secretion of TGF-β (Arrick *et al*, 1990), whereas treatment of MCF-7 cells with tamoxifen results in increased secretion of TGF-β (Knabbe *et al*, 1987). Recently, a novel synthetic progestin, gestodene, which inhibits the growth of several breast cancer cell lines, has been shown to induce a large increase in the secretion of TGF-βs 1 and 2 from those cells; the ability of anti-TGF-β serum to reverse substantially the growth inhibitory effect of gestodene

**TABLE 1. Effects of members of the steroid/retinoid family of receptors on expression of TGF-β**

| | Inducer | | |
|---|---|---|---|
| | retinoic acid[a] | tamoxifen[b] | gestodene[c] |
| Cells | keratinocytes | fibroblasts | T47D |
| Isoform | TGF-β2 | TGF-β1 | TGF-β1 |
| Fold induction | >50 | >30 | >90 |
| % Active TGF-β | >30 | 70 | 100 |
| % Hormone activity due to TGF-β | >30 | nd | >60 |

nd = not determined
[a]Glick *et al*, 1990
[b]Colletta *et al*, 1990
[c]Colletta *et al*, 1991

on these cells shows that the secreted TGF-β acts in an autocrine fashion to mediate the inhibition of growth (Colletta *et al*, 1991). Taken together, these results demonstrate that one of the responses of breast cancer cells to treatment with steroid hormones that inhibit cell growth is to alter the secretion of a negative regulator of growth, TGF-β (Wakefield *et al*, in press).

Under certain circumstances, the secretion of TGF-β by stromal cells can also be regulated by steroid hormones. Using two different strains of human fetal fibroblasts, Colletta *et al* (1990) have shown that tamoxifen induces large increases in the secretion of type 1 TGF-β (see Table 1). No oestrogen receptor or oestrogen receptor mRNA was detected in these cells, suggesting that tamoxifen might be acting through a novel receptor.

One feature common to the effects of tamoxifen on both breast cancer cells and stromal fibroblasts is the secretion of TGF-β in the active form (see Table 1 and Fig. 1). Nearly all cells in culture, as well as platelets, induced to degranulate release TGF-β in a biologically inactive, latent form in which the mature, active TGF-β is complexed non-covalently with the remainder of its own precursor, called the latency associated protein or LAP (Wakefield *et al*, 1989). Activation of this latent complex is an important physiological control point regulating the activity of TGF-β in cells and tissues. The ability of these steroids to induce secretion of activated TGF-β is shared with another member of the steroid/retinoid family of nuclear receptors, namely retinoids, as will be discussed below (see Table 1).

## Role of TGF-β in Effects of Retinoids on Epithelial Cells In Vitro

Retinoic acid and TGF-β each inhibit DNA synthesis in cultured keratinocytes (Coffey *et al*, 1988a,b; Glick *et al*, 1989). Like the previous examples of the steroid hormones, primary murine keratinocytes treated with retinoic acid for several days secrete significantly increased quantities of TGF-β compared with control cultures (Glick *et al*, 1989). The concentration of retinoic acid half maximal for induction of TGF-β secretion was 0.3–0.4 μmol/l, similar to the

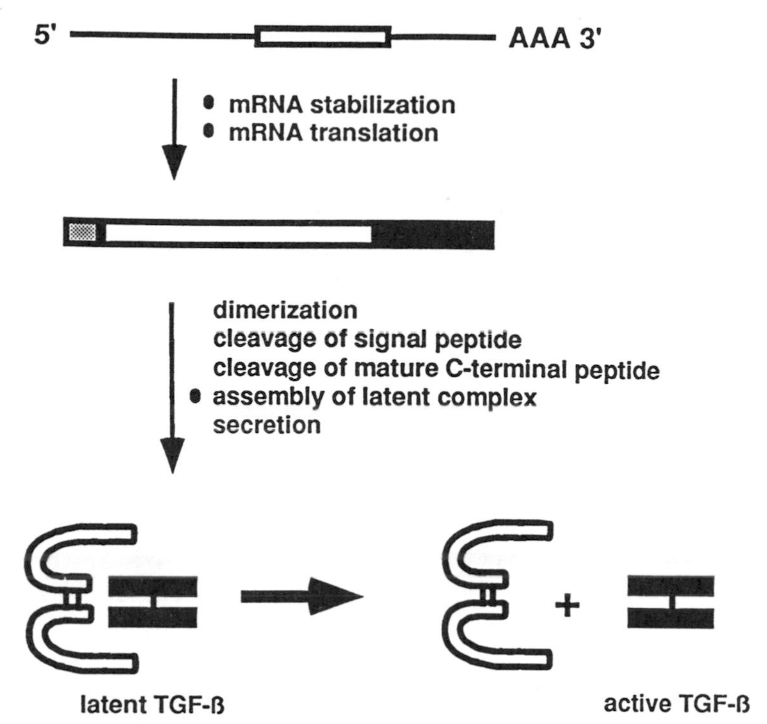


**Fig. 1.** Possible sites of post-transcriptional regulation of TGF-β secretion by members of the steroid/retinoid superfamily of receptors. Data suggest that stabilization of the TGF-β mRNAs, increased translational efficiency, altered assembly of the latent TGF-β complex and extracellular activation of the latent TGF-β complex are the most likely candidates for regulation by steroids/retinoids

median effective dose ($ED_{50}$) for inhibition of DNA synthesis by retinoic acid. Use of antisera specific for the different TGF-β isoforms demonstrated that the secreted TGF-β was nearly all the type 2 (Glick *et al*, 1989).

Highly significant biologically is the finding that the TGF-β2 secreted in response to treatment of keratinocytes with retinoic acid, like the TGF-β1 secreted in response to tamoxifen or gestodene, is in a biologically active form (Table 1) (Glick *et al*, 1989). Use of blocking antibodies to TGF-β2 showed that a minimum of 30% of the inhibition of keratinocyte DNA synthesis resulting from treatment with retinoic acid was mediated by the secreted TGF-β2 (Glick *et al*, 1989).

Danielpour *et al* (1991) have demonstrated that retinoic acid also induces secretion of TGF-β in established cell lines including human lung carcinoma A-549 cells and normal rat kidney NRK fibroblasts. As in the case of the keratinocytes, the isoform of TGF-β induced is TGF-β2. These cells are at least 100-fold more sensitive to induction by retinoic acid than are the keratinocytes, with $ED_{50}$ values of 1–10 nmol/l, but the magnitude of the induction of TGF-β2 was only 30–50% that of the keratinocytes. The biological importance of the increase in secretion in TGF-β2 in these cells is not known.

Jakowlew *et al* (1992) have studied the effects of retinoic acid on expression of the TGF-β isoforms in primary cultures of chondrocytes, myocytes and fibroblasts derived from chicken embryos. As in studies of mammalian cells, effects are targeted to expression of TGF-β2, with lesser effects on expression of TGF-β3. The direction of the effects was cell specific: in chondrocytes, where basal levels of TGF-β expression were low, retinoic acid significantly stimulated secretion of TGF-β2; in myocytes, where the basal levels of TGF-β expression were high, retinoic acid suppressed secretion of TGF-β2, and in fibroblasts retinoid treatment had no effect on TGF-β secretion.

## Role of TGF-β in Effects of Retinoids on Epithelia In Vivo

To ascertain that the observed induction of TGF-β2 secretion by retinoic acid was not restricted to in vitro culture, TGF-β expression was assessed in various target epithelia in vivo after either topical or systemic treatment of animals with retinoic acid. Immunohistochemical analysis of TGF-β staining in mouse skin demonstrated a specific and selective induction of TGF-β2 in the epidermis and in the follicular and interfollicular epithelium after topical application of retinoic acid (Glick *et al,* 1989). Since vitamin A deficiency is known to have profound effects on a variety of epithelial tissues, a more extensive study was undertaken to determine the immunohistochemical localization of TGF-βs 1, 2 and 3 in epithelial tissues of normal rats, vitamin A deficient rats and deficient rats supplemented with retinoic acid. These studies showed tissue specific alterations in the expression of the TGF-β isoforms dependent on the retinoid status, providing strong support for a prominent role of the TGF-βs in the mechanism of action of retinoids in control of epithelial cell differentiation (Glick *et al,* 1991). The specific findings in each tissue are summarized in Table 2. Overall, as had been observed in vitro, the levels of the type 2 isoform of TGF-β were most sensitive to changes in retinoid levels. However, strong induction of TGF-β3 was also seen in many epithelial tissues. By contrast, the levels of TGF-β1 were altered by the retinoid status in only the epidermis and vaginal epithelium (see Table 2). In every case, the effect of the retinoid status on the expression of the TGF-β isoforms was apparent within 24 to 48 hours after an oral dose of retinoic acid and was reversible, again suggesting an important interaction between the two families of effectors.

Retinoic acid can also differentially modulate the expression of the TGF-β isoforms in early mouse embryos exposed to excess retinoic acid (Mahmood R, Flanders KC and Morriss-Kay GM, unpublished). It is noteworthy that the described craniofacial and cardiac defects caused by teratogenic doses of retinoic acid are localized to tissues known to express high levels of TGF-β during morphogenesis (Heine *et al,* 1987; Lehnert and Akhurst, 1988; Pelton *et al,* 1989; Potts and Runyan, 1989; Akhurst *et al,* 1990; Schmid *et al,* 1991). In studies of early mouse embryos, administration of a single teratogenic dose of retinoic acid to pregnant rats on day 8 of pregnancy resulted in a transient decrease in expression of TGF-β1, had little effect on expression of TGF-β3,

**TABLE 2. Summary of effects of retinoids on expression of TGF-β in rats**

| Epithelium | TGF-β staining | |
|---|---|---|
| | vitamin A deficient | treated with retinoic acid |
| Epidermis | TGF-βs low | TGF-βs 1,2,3 increase |
| Tracheo-bronchial | TGF-βs low | TGF-βs 2,3 increase |
| Intestinal mucosa | TGF-βs low | TGF-βs 2,3 increase |
| Vaginal | TGF-βs high | TGF-βs 1,2,3 decrease |

Data from Glick *et al*, 1991

but completely ablated expression of TGF-β2 in many tissues including head mesenchyme and epithelial structures of the heart, again showing an interaction between these two families of effectors in vivo.

Although the above data solidly implicate endogenous TGF-β in epithelial cell differentiation resulting from changes in the tissue retinoid status, they do not by themselves implicate TGF-βs in the direct mechanism of action of retinoids. That is, in differentiating cells, the changes in expression of the TGF-βs could be a consequence of the sequence of events initiated by retinoid treatment, rather than a direct mediator of those effects. Thus, in embryonal carcinoma cells induced to differentiate with retinoic acid, expression of TGF-β2 is strikingly increased; however, TGF-β2 itself cannot bring about differentiation of the cells (Weima *et al,* 1989; Mummery *et al,* 1990). Moreover, in cultures of murine keratinocytes induced to differentiate by an increase in the $Ca^{2+}$ concentration of the medium, there is a selective increase in expression of TGF-β2, analogous to that seen after treatment of the cells with retinoic acid (Glick *et al,* 1990). In each of these systems, TGF-β2 is thought to mediate the inhibition of DNA synthesis associated with the differentiation but possibly not the differentiation itself (Glick *et al,* 1989, 1990).

### Relationships between Retinoids and TGF-β in Immune Cell Function

The interactions of TGF-β and retinoids on the proliferation and differentiation of immune cells are complex, often involving regulation of expression of the TGF-β receptors as well as of TGF-β itself. As an example, retinoic acid treatment of HL-60 human promyelocytic leukaemia cells induces their terminal differentiation into granulocytes and inhibits their growth, whereas TGF-β alone has no effect on the cells. However, treatment of cells with sub-optimal concentrations of retinoic acid induces the cells both to express TGF-β receptors and to secrete increased quantities of TGF-β, suggesting that the growth inhibitory effects of retinoic acid may be mediated, in part, by the autocrine action of TGF-β on the differentiating cells (Falk *et al,* 1991). In these cells, expression of TGF-β receptors, and the resulting response to inhibition of growth by exogenous TGF-β, can also be induced by other differentiating agents such as phorbol esters and dimethylsulphoxide, but unlike retinoic acid, these agents do not induce expression of TGF-β (Falk *et al,* 1991; Ruscetti *et al,* 1991). By contrast, in B lymphoma cells that lack detectable TGF-β receptors, treatment with phorbol esters results in differentiation and growth inhibition accompanied by increased expression of both TGF-β1 mRNA and its receptor; anti-TGF-β serum neutralizes the growth inhibition brought on by treatment with phorbol esters (Sing *et al,* 1990). In each of these examples, restricted expression of the TGF-β receptor on the differentiated cells limits the effects of TGF-β to inhibition of growth of the differentiated cell.

Recent studies of the regulation of human immunodeficiency virus expression in monocytic cells show that in certain instances, retinoic acid and TGF-β can have identical effects on cells in the absence of any identifiable interplay between them. Retinoic acid and TGF-β each suppress phorbol ester mediated, but not tumour necrosis factor-α mediated, virus production in a chronically infected promonocytic cell line, and both agents stimulate virus production when used to treat uninfected cells before infection (Poli *et al,* in press).

## Other Putative Interactions between Steroids and TGF-β

Other examples can be found, which suggest a mechanistic interplay between steroids and the TGF-βs, although the nature of the interactions has not been as thoroughly investigated. One common target of the steroid hormones and TGF-β is bone. Transforming growth factor-β is a mitogen for osteoblasts and a potent stimulator of bone growth (Robey *et al,* 1987; Joyce *et al,* 1990). Both androgens and oestrogens have been shown to be important in the maintenance of structural bone integrity and both have been shown to stimulate expression of TGF-β1 mRNA in human osteosarcoma cells (Komm *et al,* 1988; Benz *et al,* 1991). Moreover, vitamin D treatment of rat calvarial cells increases expression of TGF-β (Petkovich *et al,* 1987), and vitamin D deficiency, which has been shown to reduce the osteoinductive properties of bone matrix, results in selective reduction of TGF-β in rat bone (Finkelman *et al,* 1991). These data suggest that TGF-β might mediate the effects of these hormones on bone metabolism.

In the prostate, expression of both TGF-β1 and its receptor appear to be negatively regulated by androgens (Kyprianou and Isaacs, 1988; Kyprianou and Isaacs, 1989). That is, TGF-β1 mRNA levels increase as much as 40-fold after castration of rats and are strikingly decreased after androgen administration (Kyprianou and Isaacs, 1989); expression of TGF-β receptors is coordinately regulated (Kyprianou and Isaacs, 1988). These results suggest that TGF-β may have a role in the programmed cell death that occurs in the rat ventral prostate after castration induced androgen withdrawal.

## Antagonistic Effects of Steroids/Retinoids and TGF-β

The above data show that the TGF-βs mediate certain of the actions of steroids and retinoids on growth and differentiation of cells. However, examples can also be found in which the actions of these two families of molecules are antagonistic. One such system is the regulation of proteoglycan synthesis by cartilage explants. Retinoic acid inhibits cartilage proteoglycan synthesis and stimulates its degradation, whereas TGF-β has the opposite effect, enhancing proteoglycan synthesis and inhibiting its degradation (Morales and Roberts, 1992). Like previously described systems, treatment of explants with retinoic acid selectively induces synthesis of TGF-β2; however, the amount of TGF-β2 synthesized either is too little or is sequestered in the matrix and unavailable to the chondrocytes (Morales and Roberts, 1992). In a more complex biological system, it has been shown that retinoic acid and TGF-β have distinctly different effects on subsequent limb formation when implanted on beads in early stage chick limb buds: retinoic acid induces duplication of the digits (Thaller and Eichele, 1987), whereas TGF-β treatment results in foreshortening or deletion of specific skeletal elements of the developing limb (Hayamizu *et al,* 1991).

## Implications for Treatment of Disease

Epithelial tissues that depend on retinoids or steroids for appropriate cell differentiation and growth account for more than three quarters of primary cancers in both men and women, including organ sites and tissues such as the bronchi and trachea, stomach, intestine, uterus, bladder, testis, breast, prostate, pancreas and skin (Sporn and Roberts, 1991). Extensive data have been obtained on the use of retinoids for prevention or treatment of epithelial carcinogenesis in experimental animals (reviewed in Moon and Itri, 1984) and significant clinical results have now been obtained in prevention of head and neck cancer, as well as skin cancer, with 13-*cis*-retinoic acid (Kraemer *et al,* 1988; Hong *et al,* 1990). The ability of retinoids to delay the preneoplastic progression of carcinogen treated cells in vitro and to modify the differentiation and proliferation of both preneoplastic and neoplastic cells was an important conceptual advance in our understanding of the mechanism of action of retinoids in the suppression of carcinogenesis (Lasnitzki 1955; Merriman and Bertram 1979; Breitman 1982).

Transforming growth factor-βs are also potent inhibitors of epithelial cell proliferation (Moses *et al,* 1990), and there is now an extensive literature showing that loss of responsiveness of cells to TGF-β is often associated with neoplastic progression (Wakefield and Sporn, 1990). However, many tumour cells remain responsive to inhibition by TGF-β, especially in the early stages of malignant progression. The demonstration that retinoids can induce expression of TGF-β as well as TGF-β receptors (Glick *et al,* 1989, 1990; Ruscetti *et al,* 1991) suggests that local enhancement of responsiveness to TGF-β may have a role in the mechanism of chemoprevention by retinoids.

In a broader sense, the data on retinoids, together with the observed local induction of TGF-β activity by two other members of the steroid/retinoid superfamily of nuclear receptors, tamoxifen and gestodene, open new horizons in therapy and prevention of epithelial cancers, suggesting that these agents could induce a localized response to TGF-β in specific target tissues (Wakefield and Sporn, 1990; Wakefield *et al*, in press). The induction of TGF-β expression by breast cancer cells treated with gestodene (Colletta *et al*, 1991) and by putative tumour stromal cells treated with tamoxifen (Colletta *et al*, 1990) suggests that combinations of these agents might prove effective. In a rat tumour model for human breast cancer, studies now show that combined treatment with a retinoid and tamoxifen, after removal of the primary tumour, was consistently more effective than either agent alone in suppressing the postsurgical appearance of new mammary tumours (Ratko *et al*, 1989); our data would suggest that multiple isoforms of TGF-β might be induced by this regimen to act on the tumour cells in both autocrine and paracrine fashion.

## MOLECULAR MECHANISMS OF INTERACTION BETWEEN STEROIDS/RETINOIDS AND TGF-β

### TGF-β Regulates the Synthesis of Steroids

Thus far, we have discussed control of TGF-β expression by the steroids/retinoids, but reciprocal regulation by TGF-β of steroid synthesis in the adrenal cortex, the ovary and the testis has also been studied extensively. Nothing is known about possible effects of TGF-β on retinoid metabolism. The effects of TGF-β on steroidogenesis are independent of proliferative effects and are cell specific: TGF-β inhibits steroidogenesis in adrenocortical cells (Feige *et al*, 1987; Hotta and Baird, 1987; Rainey *et al*, 1988), testicular Leydig cells (Avallet *et al*, 1987; Morera *et al*, 1988; Benahmed *et al*, 1989) and ovarian thecal cells (Magoffin *et al*, 1989; Hernandez *et al*, 1990) but enhances the activity of follicle stimulating hormone (FSH) on aromatase activity in granulosa cells (Hutchinson *et al*, 1987; Adashi *et al*, 1989). Although most studies have focused on effects of exogenous TGF-β on these cells, it is clear that the peptide also acts endogenously as an autocrine or paracrine mediator of cell function, modulating the steroidogenic activity of circulating hormones on their target cells.

In adrenocortical cells, TGF-β inhibits both basal and adrenocorticotropin stimulated cortisol production and decreases angiotensin II stimulated cortisol synthesis, decreasing the expression of angiotensin II receptors on the cell by half (Feige *et al*, 1987). Transforming growth factor-β also has direct effects on steroidogenesis, decreasing the availability of exogenous lipoprotein (Hotta and Baird, 1987) and reducing the expression of 17α-hydroxylase, a key enzyme in the synthesis of 17α-hydroxylated products such as cortisol (Feige *et al*, 1987; Rainey *et al*, 1988). Immunohistochemical analysis localizes TGF-β to the inner zones of the cortex that are involved in cortisol synthesis (Thompson *et al*, 1989).

Transforming growth factor-β also has both direct and indirect effects on steroid synthesis by gonadal cells. Transforming growth factor-β secreted by ovarian thecal cells is thought to act in a paracrine manner on target granulosa cells to amplify the effects of FSH on aromatase activity and progesterone synthesis (Hutchinson *et al*, 1987; Adashi *et al*, 1989); effects are also seen on ovarian androgen synthesis, where TGF-β acts in an autocrine fashion to inhibit thecal cell androsterone synthesis again at the level of 17α-hydroxylase (Magoffin *et al*, 1989; Hernandez *et al*, 1990). In the testis, TGF-β secreted by Sertoli cells in response to stimulation with oestradiol, dexamethasone, or thyroxine (Benahmed *et al*, 1988) is proposed to act in a paracrine fashion to control Leydig cell steroidogenesis (Morera *et al*, 1988; Benahmed *et al*, 1989). Transforming growth factor-β acts both indirectly by downregulating expression of (human) chorionic gonadotropin receptors by Leydig cells and directly by inhibiting synthesis at a site prior to pregnenolone synthesis.

### Steroids/Retinoids Regulate the Activity of TGF-β

As discussed above, one feature common to the actions of steroids and retinoids on induction of TGF-β expression is the secretion of TGF-β in its biologically active form (see Fig. 1). This could be important in the use of these agents to induce local expression of TGF-β, such as in the proposed chemoprevention of carcinogenesis, for it circumvents a major regulatory step in control of TGF-β action. Moreover, the pharmacokinetics of the latent and active forms of TGF-β in vivo differ strikingly, with the latent form having a substantially prolonged half life and larger volume of distribution than active TGF-β (Wakefield *et al*, 1990b). This could have significant implications for the local action of the TGF-β induced by retinoids or steroid hormones. It is noteworthy that TGF-β1 secreted by activated alveolar macrophages in a bleomycin induced model of pulmonary fibrosis (Khalil N, personal communication) and by T lymphocytes involved in suppression of experimental autoimmune encephalomyelitis (Miller *et al*, 1992) is also in the active form. Whether the mechanisms of activation of TGF-β in these two disease models are similar to or distinct from those involved in the activation of TGF-β secreted in response to steroid/retinoid treatment remains to be found. In neither case has it been determined whether activation occurs intracellularly or extracellularly. Two candidate sites of action would be either alteration of the ability of the LAP protein to dimerize (Brunner *et al*, 1989) or stimulation of extracellular and possibly cell membrane associated protease activity (Lyons *et al*, 1990; Dennis and Rifkin, 1991; Rifkin *et al*, 1991).

### Post-transcriptional Regulation of TGF-β by Steroids/Retinoids

The specific mechanisms underlying the induction of TGF-β expression by steroids and retinoids are still unknown. Induction of TGF-β1 synthesis by tamoxifen (Colletta *et al*, 1990) and gestodene (Colletta *et al*, 1991) and of TGF-β2 synthesis by retinoids in both keratinocytes (Glick *et al*, 1989) and chicken embryo chondrocytes (Jakowlew *et al*, 1992) has been shown to occur principally by post-transcriptional mechanisms. These limited observations, together with the notable absence of defined steroid/retinoid response elements in the promoters of the TGF-β genes, suggest that the TGF-β genes them-

Glick AB, Flanders KC, Danielpour D, Yuspa SH and Sporn MB (1989) Retinoic acid induces transforming growth factor-β2 in cultured keratinocytes and mouse epidermis. *Cell Regulation* **1** 87–97

Glick AB, Danielpour D, Morgan D, Sporn MB and Yuspa SH (1990) Induction and autocrine receptor binding of transforming growth factor-beta 2 during terminal differentiation of primary mouse keratinocytes. *Molecular Endocrinology* **4** 46–52

Glick AB, Abdulkarem N, Flanders KC, Lumadue JA, Smith JM and Sporn MB (1991) Complex regulation of TGFβ expression by retinoic acid in the vitamin A-deficient rat. *Development* **111** 1081–1086

Hayamizu TF, Sessions SK, Wanek N and Bryant SV (1991) Effects of localized application of transforming growth factor beta 1 on developing chick limbs. *Developmental Biology* **145** 164–173

Heine U, Munoz EF, Flanders KC *et al* (1987) Role of transforming growth factor-beta in the development of the mouse embryo. *Journal of Cell Biology* **105** 2861–2876

Heine UI, Munoz EF, Flanders KC, Roberts AB and Sporn MB (1990) Colocalization of TGF-beta 1 and collagen I and III, fibronectin and glycosaminoglycans during lung branching morphogenesis. *Development* **109** 29–36

Hernandez ER, Hurwitz A, Payne DW, Dharmarajan AM, Purchio AF and Adashi EY (1990) Transforming growth factor-beta 1 inhibits ovarian androgen production: gene expression, cellular localization, mechanisms(s), and site(s) of action. *Endocrinology* **127** 2804–2811

Hong WK, Lippman SM, Itri LM, *et al* (1990) Prevention of second primary tumors in squamous cell carcinoma of the head and neck with 13-*cis*-retinoic acid. *New England Journal of Medicine* **323** 795–801

Hotta M and Baird A (1987) The inhibition of low density lipoprotein metabolism by transforming growth factor-beta mediates its effects on steroidogenesis in bovine adrenocortical cells in vitro. *Endocrinology* **121** 150–159

Hutchinson LA, Findlay JK, de Vos FL and Robertson DM (1987) Effects of bovine inhibin, transforming growth factor-beta and bovine activin-A on granulosa cell differentiation. *Biochemical and Biophysical Research Communications* **146** 1405–1412

Jakowlew SB, Cubert J, Danielpour D, Sporn MB and Roberts AB (1992) Differential regulation of the expression of transforming growth factor-β mRNAs by growth factors and retinoic acid in chicken embryo chondrocytes, myocytes and fibroblasts. *Journal of Cellular Physiology* **150** 377–385

Joyce ME, Roberts AB, Sporn MB and Bolander ME (1990) Transforming growth factor-beta and the initiation of chondrogenesis and osteogenesis in the rat femur. *Journal of Cell Biology* **110** 2195–2207

Knabbe C, Lippman ME, Wakefield LM *et al* (1987) Evidence that transforming growth factor-beta is a hormonally regulated negative growth factor in human breast cancer cells. *Cell* **48** 417–428

Komm BS, Terpening CM, Benz DJ *et al* (1988) Estrogen binding, receptor mRNA, and biologic response in osteoblast-like osteosarcoma cells. *Science* **241** 81–84

Kraemer KH, DiGiovanna JJ, Moshell AN, Tarone RE and Peck GL (1988) Prevention of skin cancer in xeroderma pigmentosa with the use of oral isotretinoin. *New England Journal of Medicine* **318** 1633–1637

Kyprianou N and Isaacs JT (1988) Identification of a cellular receptor for transforming growth factor-beta in rat ventral prostate and its negative regulation by androgens. *Endocrinology* **123** 2124–2131

Kyprianou N and Isaacs JT (1989) Expression of transforming growth factor-beta in the rat ventral prostate during castration-induced programmed cell death. *Molecular Endocrinology* **3** 1515–1522

Lasnitzki I (1955) The influence of A-hypervitaminosis on the effect of 20-methyl cholanthrene on mouse prostate glands grown in vitro. *British Journal of Cancer* **9** 434–441

Lehnert SA and Akhurst RJ (1988) Embryonic expression pattern of TGF beta type-1 RNA suggests both paracrine and autocrine mechanisms of action. *Development* **104** 263–273

Lyons RM, Gentry LE, Purchio AF and Moses HL (1990) Mechanism of activation of latent recombinant transforming growth factor beta 1 by plasmin. *Journal of Cell Biology* **110** 1361–1367

Magoffin DA, Gancedo B and Erickson GF (1989) Transforming growth factor-beta promotes differentiation of ovarian thecal-interstitial cells but inhibits androgen production. *Endocrinology* **125** 1951–1958

Massagué J (1990) The transforming growth factor-beta family. *Annual Reviews of Cell Biology* **6** 597–641

Merriman RL and Bertram JS (1979) Reversible inhibition by retinoids of 3-methyl cholanthrene-induced neoplastic transformation in C3H/10T1/2 CL8 cells. *Cancer Research* **39** 1661–1666

Miller A, Lider O, Roberts AB, Sporn MB and Weiner HL (1992) Suppressor T cells generated by oral tolerization to myelin basic protein suppress both in vitro and in vivo immune responses by the release of TGF-β following antigen specific triggering. *Proceedings of the National Academy of Sciences of the USA* **89** 421–425

Moon RC and Itri LM (1984) Retinoids and cancer, In: Sporn MB, Roberts AB and Goodman DS (eds). *The Retinoids* vol 2, pp 327–371, Academic Press, New York

Morales TI and Roberts AB (1992) The interaction between retinoic acid and the transforming growth factors-β in calf articular cartilage organ cultures. *Archives of Biochemistry and Biophysics* **293** 79–84

Morera AM, Cochet C, Keramidas M, Chauvin MA, de Peretti E and Benahmed M (1988) Direct regulating effects of transforming growth factor beta on the Leydig cell steroidogenesis in primary culture. *Journal of Steroid Biochemistry* **30** 443–447

Moses HL, Yang EY and Pietenpol JA (1990) TGF-beta stimulation and inhibition of cell proliferation: new mechanistic insights. *Cell* **63** 245–247

Mummery CL, Slager H, Kruijer W *et al* (1990) Expression of transforming growth factor beta 2 during the differentiation of murine embryonal carcinoma and embryonic stem cells. *Developmental Biology* **137** 161–170

Nathan C and Sporn M (1991) Cytokines in context. *Journal of Cell Biology* **113** 981–986

O'Malley B (1990) The steroid receptor superfamily: more excitement predicted for the future. *Molecular Endocrinology* **4** 363–369

Pelton RW, Nomura S, Moses HL and Hogan BL (1989) Expression of transforming growth factor β2 RNA during murine embryogenesis. *Development* **106** 759–767

Petkovich PM, Wrana JL, Grigoriadis AE, Heersche JN and Sodek J (1987) 1,25-Dihydroxyvitamin D3 increases epidermal growth factor receptors and transforming growth factor beta-like activity in a bone-derived cell line. *Journal of Biological Chemistry* **262** 13424–13428

Poli G, Kinter AL, Justement JS, Bressler P, Kehrl JH and Fauci AS Retinoic acid mimics transforming growth factor-β in the regulation of human immunodeficiency virus expression in monocytic cells. *Proceedings of the National Academy of Sciences of the USA* (in press)

Potts JD and Runyan RB (1989) Epithelial-mesenchymal cell transformation in the embryonic heart can be mediated, in part, by transforming growth factor beta. *Developmental Biology* **134** 392–401

Rainey WE, Viard I, Mason JI, Cochet C, Chambaz EM and Saez JM (1988) Effects of transforming growth factor beta on ovine adrenocortical cells. *Molecular and Cellular Endocrinology* **60** 189–198

Ratko TA, Detrisac CJ, Dinger NM, Thomas CF, Kelloff GJ and Moon RC (1989) Chemopreventive efficacy of combined retinoid and tamoxifen treatment following surgical excision of a primary mammary cancer in female rats. *Cancer Research* **49** 4472–4476

Rifkin DB, Moscatelli D, Flaumenhaft R, Sato Y, Saksela O and Tsuboi R (1991) Mechanisms controlling the extracellular activity of basic fibroblast growth factor and transforming growth factor. *Annals of the New York Academy of Sciences* **614** 250–258

Roberts AB and Sporn MB (1984) Cellular biology and biochemistry of the retinoids, In: Sporn MB, Roberts AB and Goodman DS (eds). *The Retinoids*, vol 2, pp 209–286, Academic Press, New York

Roberts AB and Sporn MB (1990) The transforming growth factors-β, In: Sporn MB and Roberts AB (eds). *Handbook of Experimental Pharmacology: Peptide Growth Factors and Their Receptors I* vol 95, pp 419–472, Springer-Verlag, Berlin

Robey PG, Young MF, Flanders KC *et al* (1987) Osteoblasts synthesize and respond to transforming growth factor-type β (TGF-β) in vitro. *Journal of Cell Biology* **105** 457–463

Ruscetti FW, Dubois C, Falk LA *et al* (1991) In vivo and in vitro effects of TGF-beta 1 on normal and neoplastic haemopoiesis. *Ciba Foundation Symposium* **157** 212–227

Schmid P, Cox D, Bilbe G, Maier R and McMaster GK (1991) Differential expression of TGF beta 1, beta 2 and beta 3 genes during mouse embryogenesis. *Development* **111** 117–130

Silberstein GB and Daniel CW (1987) Reversible inhibition of mammary gland growth by transforming growth factor-beta. *Science* **237** 291–293

Silberstein GB, Flanders KC, Roberts AB and Daniel CW Regulation of mammary morphogenesis: evidence for extracellular matrix-mediated inhibition of ductal budding by transforming growth factor-β1. *Developmental biology* (in press)

Sing GK, Ruscetti FW, Beckwith M *et al* (1990) Growth Inhibition of a human lymphoma cell line: induction of a transforming growth factor-β-mediated autocrine negative loop by phorbol myristate acetate. *Cell Growth and Differentiation* **1** 549–557

Sporn MB and Roberts AB (1988) Peptide growth factors are multifunctional. *Nature* **332** 217–219

Sporn MB and Roberts AB (1991) Interactions of retinoids and transforming growth factor-beta in regulation of cell differentiation and proliferation. *Molecular Endocrinology* **5** 3–7

Thaller C and Eichele G (1987) Identification and spacial distribution of retinoids in the developing chick limb bud. *Nature* **327** 625–628

Thompson NL, Flanders KC, Smith JM, Ellingsworth LR, Roberts AB and Sporn MB (1989) Expression of transforming growth factor-beta 1 in specific cells and tissues of adult and neonatal mice. *Journal of Cell Biology* **108** 661–669

Wakefield LM and Sporn MB (1990) Suppression of carcinogenesis: a role for TGF-beta and related molecules in prevention of cancer, In: Klein G (ed). *Tumor Suppressor Genes*, pp 217–243, Marcel Dekker Inc, New York

Wakefield LM, Smith DM, Broz S, Jackson M, Levinson AD and Sporn MB (1989) Recombinant TGF-beta 1 is synthesized as a two-component latent complex that shares some structural features with the native platelet latent TGF-beta 1 complex. *Growth Factors* **1** 203–218

Wakefield L, Kim SJ, Glick A, Winokur T, Colletta A and Sporn M (1990a) Regulation of transforming growth factor-beta subtypes by members of the steroid hormone superfamily. *Journal of Cell Science* **13 (Supplement)** 139–148

Wakefield LM, Winokur TS, Hollands RS, Christopherson K, Levinson AD and Sporn MB (1990b) Recombinant latent transforming growth factor beta 1 has a longer plasma half-life in rats than active transforming growth factor beta 1 and a different tissue distribution. *Journal of Clinical Investigation* **86** 1976–1984

Wakefield LM, Colletta AA, McCune BK and Sporn MB Roles for transforming growth factors-β in the genesis, prevention and treatment of breast cancer, In: Dickson RB and Lippman ME (eds). *Breast Cancer: Cellular and Molecular Biology*, Kluwer Publishers (in press)

Weima SM, van Rooijen MA, Feijen A *et al* (1989) Transforming growth factor-beta and its receptor are differentially regulated in human embryonal carcinoma cells. *Differentiation* **41** 245–253

The authors are responsible for the accuracy of the references.

# Peroxisome Proliferators: A Model for Receptor Mediated Carcinogenesis

**STEPHEN GREEN**

*Imperial Chemical Industries, Central Toxicology Laboratory, Cell and Molecular Biology Section, Alderley Park, Macclesfield, Cheshire, SK10 4TJ*



## PEROXISOME PROLIFERATORS

Peroxisomes are single membrane cytoplasmic organelles first discovered in the kidney where they were referred to as microbodies (Rhodin, 1954). These microbodies were termed peroxisomes when they were found to contain oxidases that produce hydrogen peroxide (de Duve *et al,* 1966). A wide variety of chemicals when administered to rats and mice produce a striking increase in both the size and number of peroxisomes. The first such peroxisome proliferator to be identified was the hypolipidaemic drug clofibrate (Paget, 1963; Hess *et al,* 1965). However, the list (Table 1) also includes several other hypolipidaemic drugs as well as herbicides, leukotriene antagonists and plasticizers (reviewed in Reddy and Lalwani, 1983; Lock *et al,* 1989; Moody *et al,* 1991). The most potent peroxisome proliferators are hypolipidaemic drugs (eg Wy-14, 643), whereas plasticizers such as di-(2-ethyhexyl) phthalate (DEHP) are much weaker. In rats and mice, these compounds produce hepatomegaly as a result of both liver hyperplasia (Hess *et al,* 1965; Reddy *et al,* 1979; Styles *et al,* 1988) and increase in peroxisomal volume (Hess *et al,* 1965; Lazarow and de Duve, 1976). Although the extent of peroxisome proliferation is greatest in the liver, it is also found in the proximal convoluted tubular epithelium of the kidney.

The levels of several peroxisomal enzymes (Lazarow and de Duve, 1976; Osumi and Hashimoto, 1978), as well as members of the microsomal cyto-


 0-87969-371-1/92. $3.00 + .00

**TABLE 1. Peroxisome proliferators**[a]

| | |
|---|---|
| Fibrate hypolipidaemic drugs | Phthalates |
| Clofibrate | Di-2-ethylhexylphthalate |
| Nafenopin | Di-2-ethylhexyladipate |
| Methylclofenapate | 2-Ethylhexanoic acid |
| Gemfibrozil | |
| Bezafibrate | Other environmental |
| Ciprofibrate | 2,4-Dichlorophenoxy acetic acid |
| Fenofibrate | 2,4,5-Trichlorophenoxy acetic acid |
| Clobuzarit | Tricholoroacetic acid |
| | Lactofen |
| Non-fibrate hypolipidaemic drugs | |
| Wy-14 643 | Physiological |
| Tibric Acid | Cold acclimatization |
| BR-931 | High fat diet |
| Tiadenol | Thyroxine |
| | Triiodothyronime |
| | Dehydroepiandrosterone |
| Other drugs | |
| Acetylsalicyclic acid | |
| Ly-171883 | |
| Valproic acid | |

[a]The list includes clinically important hypolipidaemic drugs, plasticizers (eg di-2(ethylhexyl)phthalate), herbicides (eg lactofen) and leukotriene antagonists (eg Ly-171883)

chrome P450 IV gene family (Orton and Parker, 1982), are increased 10–30-fold in response to administration of peroxisome proliferator. These increases are parallelled by induction of their respective messenger RNAs as early as 2 hours after peroxisome proliferator administration and reflect an increase in transcription (Reddy *et al,* 1986; Hardwick *et al,* 1987). The peroxisomal enzymes studied in most detail include acyl CoA oxidase, bifunctional enzyme and thiolase. Together, these enzymes cause the peroxisomal β-oxidation of long chain fatty acids. Acyl CoA oxidase is the key enzyme in this pathway since it is the rate limiting step and in addition produces the hydrogen peroxide that some have implicated in the hepatocarcinogenic process (see below).

## RODENT HEPATOCARCINOGENESIS

Clofibrate was one of the first peroxisome proliferators to be identified as a rat liver carcinogen (Reddy and Qureshi, 1979). Further studies examining other hypolipidaemic drugs suggested that peroxisome proliferators represent a novel class of rodent hepatocarcinogens (Reddy *et al,* 1980). The basic mechanism(s) by which peroxisome proliferators induce tumours in rats and mice is unknown. The carcinogenic potency of the various peroxisome proliferators differs considerably. For example, 0.1% Wy-14, 643 fed to rats will produce a 100% liver tumour incidence after 60 weeks compared with only about 10% in male rats fed 1.2% DEHP for 2 years (Table 2) (Marsman *et al,* 1988).

## TUMOUR INITIATION

Peroxisome proliferators are termed non-genotoxic carcinogens since they fail to cause DNA damage directly when tested with a number of genotoxic assays, for example the Ames *Salmonella* mutagenicity assay (Warren *et al,* 1980). Peroxisome proliferators therefore differ from classic chemical mutagens and carcinogens such as diethylnitrosamine and N-methylnitrosourea. The oxidative stress hypothesis was proposed by Reddy to explain how peroxisome proliferators may cause cancer and is based on a correlation between the ability of a compound to stimulate peroxisome proliferation and induce tumours (Reddy and Lalwani, 1983). It is proposed that hydrogen peroxide, produced by the increase in peroxisomal fatty acid β-oxidation, results in oxidative stress leading to DNA damage and possibly tumour initiation. This model is supported by the demonstration of an increase in hydroxyl radicals in rats given peroxisome proliferators (Elliott *et al,* 1986). In addition, DNA damage can be detected, as evidenced by both an approximate doubling in the number of 8-OH-deoxyguanosine lesions in the liver of rats chronically fed a diet containing the potent peroxisome proliferator ciprofibrate (Kasai *et al,* 1989) and DNA alterations detected with the $^{32}$P post-labelling assay (Randerath *et al,* 1991). Furthermore, co-administration of the antioxidant ethoxyquin with the peroxisome proliferator ciprofibrate inhibited hepatic tumorigenesis possibly as a result of minimizing an increase in hydroxyl radicals (Rao *et al,* 1984). There is also a correlation between the potency of weak (eg DEHP) and potent (eg Wy-14, 643) carcinogens and the level of accumulated lipofuscin that is used as a putative indicator of oxidative damage (Reddy *et al,* 1982; Conway *et al,* 1989). Against this evidence for oxygen radical production and DNA damage are data from an animal study demonstrating that the peroxisome proliferator Wy-14 643 fails to initiate growth selectable foci (Cattley *et al,* 1989).

## TUMOUR PROMOTION

Although some DNA damage is apparent in peroxisome proliferator treated animals, perhaps the best evidence that peroxisome proliferation alone cannot account for tumorigenesis is an animal study comparing the effects of DEHP and Wy-14, 643 over 12 months. Tumours arose in all the rats fed Wy-14, 643 but in none of the rats fed DEHP, despite the fact that DEHP produced only 25% less peroxisome proliferation than Wy-14, 643 (Marsman *et al,* 1988). Since peroxisome proliferators are also liver mitogens (Hess *et al,* 1965; Styles *et al,* 1988), the effect of Wy-14, 643 and DEHP on DNA replication was determined. Because the carcinogenicity of DEHP and Wy-14, 643 correlated better with mitogenesis than with peroxisome proliferation, it was suggested that peroxisome proliferators induce tumours by influencing the growth of initiated lesions and therefore act as tumour promoters (Marsman *et al,* 1988; Cattley and Popp, 1989). The origin of these initiated lesions is not clear. One possibility is that peroxisome proliferators have mixed tumour initiation (oxygen radical production) and tumour promotion (liver mitogenesis) activity, and this would explain their ability to act as complete carcinogens. Another possibility is that initiated lesions arise spontaneously, and this is supported by the observation of greater numbers of tumours in older rats than in younger rats given the same dose of peroxisome proliferators for the same time (Cattley *et al,* 1991, Kraupp-Grasl *et al,* 1991). Other studies also suggest that peroxisome proliferators act as tumour promoters rather than initiators. For example, significantly more and larger tumours are present in animals given an initiating dose of the genotoxin aflatoxin B1 followed by treatment with the peroxisome proliferator nafenopin than in animals receiving either compound alone (Kraupp-Grasl *et al,* 1990). It is also interesting to compare the promoting effect of peroxisome proliferators with that of the classical liver tumour promoter phenobarbitone following initiation with diethylnitrosamine. Peroxisome proliferators increase the mean volume of foci, whereas phenobarbitone increases their number (Cattley and Popp, 1989). Moreover, foci induced by peroxisome proliferators are unusual in that they are γ-glutamyltranspeptidase negative and weakly basophilic after staining with haematoxylin and eosin (Cattley and Popp, 1989; Kraupp-Grasl *et al,* 1990). This may indicate that the cellular and molecular mechanism of tumour promotion by peroxisome proliferators differs from that of phenobarbitone.

## PEROXISOME PROLIFERATOR ACTIVATED RECEPTORS

We and others have speculated on the existence of specific receptors that could mediate the action of peroxisome proliferators (Reddy and Rao, 1986; Issemann and Green, 1990). Reddy's group originally proposed the existence of a peroxisome proliferator binding protein (PPBP) in rat liver (Lalwani *et al,* 1983, 1987). This 70 kDa binding protein was purified from rat liver on a nafenopin affinity column, and the antibodies raised against it were used to identify complementary DNA (cDNA) clones (Alvares *et al,* 1990). The sequence of these clones indicates that PPBP is related to the heat shock protein hsp72. It should be noted, however, that the potent peroxisome proliferator Wy-14, 643 does not bind to PPBP (Lalwani *et al,* 1987) and that PPBP is ubiquitously expressed at high levels, making it unlikely to be a mediator of peroxisome proliferator action.

The possibility that the true biological mediator of peroxisome proliferator action could be a member of the steroid hormone receptor superfamily prompted us to screen a mouse liver cDNA library using a probe derived from the combined sequences of several such receptors. This led to the identification of four new members of the hormone receptor family (Issemann and Green, 1990, 1991). One of these receptors can be activated by various peroxisome proliferators including hypolipidaemic drugs and a metabolite, mono 2-ethylhexyl phthalate (MEHP), of the plasticizer DEHP (Table 1) (Issemann and Green, 1990). We have termed this receptor the peroxisome proliferator activated receptor (PPAR). PPAR is postulated to be a ligand activated transcription factor because it shares a common primary organization with the other nuclear hormone receptors and contains regions homologous to both the DNA and ligand binding domains (reviewed in Green and Chambon, 1988).

Three *Xenopus* receptors have been described that can also be activated by peroxisome proliferators (Wahli W and Dreyer C, personal communication). One of these (xPPARα) is highly homologous to the mouse PPAR, and the other two are more distantly related. It will be of interest to determine whether additional PPARs exist in rodents and humans and to define their possible role in mediating peroxisome proliferator action.

## PPAR MEDIATES PEROXISOME PROLIFERATOR ACTION

A chimaeric receptor (ER-PPAR) constructed from the DNA binding domain of the oestrogen receptor (ER) and the putative ligand binding domain of PPAR can activate an oestrogen responsive gene in the presence of peroxisome proliferators (Issemann and Green, 1990). When ER-PPAR was tested with several different peroxisome proliferators, a good correlation was found between their ability to activate ER-PPAR and their potency either as peroxisome proliferators or as rat liver carcinogens (Table 2). For example, Wy-14, 643 was more potent in the chimeric receptor assay than MEHP, the primary DEHP metabolite. These data suggest that PPAR could mediate the biological effects of peroxisome proliferators. Further support comes from a study of the tissue specific expression of PPAR that compares well with the tissue specific induction of acyl CoA oxidase by peroxisome proliferators (Nemali *et al*, 1988).

A comparison of the primary aminoacid sequence reveals complete identity within the proximal box of the DNA binding domain of PPAR and other nuclear hormone receptors that bind to a DNA sequence related to TGACCT or TGAACT (Schwabe *et al*, 1990; Umesono and Evans, 1989; de The *et al*, 1990). We therefore predicted that PPAR recognizes a similar motif (Issemann and Green, 1990). Examination of the promoter region of the rat acyl CoA oxidase gene (Osumi *et al*, 1987) reveals several such motifs, and we now have evidence that a sequence consisting of a direct repeat of the motifs TGACCT and TGTCCT positioned 570 nucleotides upstream of the transcription initiation start site binds PPAR and is important in mediating the response to peroxisome proliferators (Tugwood *et al*, 1992). Other studies support the conclusion that this region of acyl CoA oxidase is important in conferring responsiveness to peroxisome proliferators (Osumi *et al*, 1991). These data therefore indicate that PPAR mediates the induction of the key marker of the peroxisome proliferator response.

**TABLE 2. Comparison of the ability of peroxisome proliferators to activate the chimaeric receptor ER-PPAR, induce acyl CoA oxidase (ACO) or produce liver tumours[a]**

| Compound | ER-PPAR ($EC_{50}$ mmol/l) | ACO ($EC_{50}$ mmol/l) | Reference | Dose | Period (months) | Tumour incidence | Reference |
|---|---|---|---|---|---|---|---|
| Wy 14 643 | 2 | <50 | 1 | 0.1% | 12 | 100% | 4 |
| Nafenopin | 10 | 50–100 | 1,2 | 0.1% | 18–25 | 80% | 5 |
| Methylclofenapate | 55 | | | 0.1% | 18 | 100% | 6 |
| DEHP/MEHP | 50 | 100–250 | 1,3 | 1.2% | 24 | 10% | 7 |
| Clofibrate | 310 | 300 | 1,3 | 0.5% | 24–28 | 91% | 8 |
| TCA | 8000 | >5000 | 3 | 5g/l | 14 | 32% | 9 |

[a]Activation of ER-PPAR is measured with a transient expression system established in monkey kidney cells (Issemann and Green, 1990). Recent experiments indicate the full length PPAR to be about 8-fold more sensitive to Wy-14 643 ($EC_{50}$ 250 nmol/l). Induction of acyl CoA oxidase is measured with primary rat hepatocytes. Rodent bioassays were done with male Fischer (F344) rats except for TCA, for which male B6C3F1 mice were used. The animals were chronically administered the peroxisome proliferator either in the feed or drinking water (TCA). The percentage of animals bearing liver tumours (tumour incidence) is indicated

References: (1) Gray *et al*, 1983; (2) Bieri *et al*, 1984; (3) Mitchell *et al*, 1984; (4) Marsman *et al*, 1988; (5) Reddy and Rao, 1977; (6) Reddy *et al*, 1982; (7) National Toxicology Program, 1983; (8) Reddy and Qureshi, 1979; (9) Herren-Freund *et al*, 1987

## NATURAL LIGANDS

Given the diversity of peroxisome proliferators that activate PPAR, it is interesting to speculate on whether they bind to PPAR directly or modulate its activity through some indirect mechanism. One possibility for an indirect mechanism of action would be if peroxisome proliferators cause a perturbation in lipid metabolism that induces the natural PPAR ligand. Alternatively, peroxisome proliferators could cause changes in the phosphorylation of signal transduction proteins, perhaps including PPAR, as has been suggested to explain how dopamine activates the chicken ovalbumin upstream promoter nuclear hormone receptor (Power *et al,* 1991).

Since natural factors such as a high fat diet (Nilsson *et al,* 1986; Flatmark *et al,* 1988), fatty acids (Intrasuksri and Feller, 1991) and the steroid dehydroepiandrosterone (Frenkel *et al,* 1990) can also induce peroxisome proliferation, it is probable that the phenomenon represents a physiological response to some natural biological stimulus and that a natural ligand for PPAR exists. Peroxisomes are important in the metabolism of long chain fatty acids and the production of cholic acid from cholesterol (Kase *et al,* 1986). The natural inducer of peroxisome proliferation could be a steroid, fatty acid or cholesterol metabolite. The identification of the natural PPAR ligand should yield valuable information concerning the role of PPAR and its link with cancer.

## SPECIES COMPARISONS

It is intriguing that peroxisome proliferation is seen in species such as mouse, rat and hamster but not in guinea pig and monkey (Elcombe and Mitchell, 1986; Eacho *et al,* 1986; Lake *et al,* 1989). Importantly, it is not seen in humans (Blumcke, *et al,* 1983; Hanefeld *et al,* 1983) or in human hepatocytes in culture (Elcombe and Mitchell, 1986). The basis for the species differences in response to peroxisome proliferators is unlikely to be differences in metabolism or pharmacokinetics (Elcombe and Mitchell, 1986) and appears to reflect differences in the way that hepatocytes respond to the peroxisome proliferator stimulus. If peroxisome proliferation is mediated by PPAR, then such species differences could reflect either variation in PPAR or in the gene networks regulated by PPAR. We are therefore examining the level of PPAR expression as well as the activity of PPAR in response to peroxisome proliferators in different species, since this has important consequences in assessing the hazard peroxisome proliferators pose to humans.

## FUTURE MODELS

In view of the similarity between PPAR and steroid hormone receptors, will peroxisome proliferator induced liver tumours behave similarly to other endocrine cancers such as mammary and prostate cancer? In this respect, it is notable that the maintenance of clofibrate induced liver tumours may depend on the presence of clofibrate (Svoboda and Azarnoff, 1979), a property similar to the oestrogen dependence of N-methylnitrosourea initiated mammary tumours in rats (Sukumar, 1990). It will also be interesting to determine whether cell lines can be established from peroxisome proliferator induced tumours. These could be valuable research models since it is possible that their growth in athymic nude mice will depend on the presence of peroxisome proliferators. Such cell lines may therefore be similar to oestrogen receptor positive breast cancer cell lines such as MCF-7, whose growth is oestrogen dependent in nude mouse models (Jordan *et al,* 1989).

Because PPAR can be activated by a wide range of peroxisome proliferators and regulates the transcription of the acyl CoA oxidase gene, it is probable that PPAR and possibly additional PPAR like receptors are the mediators of peroxisome proliferator action. By analogy with the mechanism of action of other nuclear hormone receptors, PPAR would be activated by the binding of a peroxisome proliferator, recognize specific DNA sequence motifs located upstream of peroxisome proliferator target genes and activate specific gene transcription. Such target genes would include those of the β-oxidation enzymes and P450 IVA1 but could also include genes important in hyperplasia and carcinogenesis. If the oxidative stress hypothesis were correct, then regulation of acyl CoA oxidase expression by PPAR could explain the carcinogenic effects of peroxisome proliferators (see above). Alternatively, tumour formation could result from important changes in cellular growth and differentiation (Gerbacht *et al,* 1990). Therefore, PPAR may alter the expression of key genes relevant to growth and differentiation such as oncogenes (Bentley *et al,* 1988; Cherkaoui-Malki *et al,* 1990), growth factors or their receptors (Gupta *et al,* 1988). The identification of the carcinogenic mechanism by which peroxisome proliferators and other non-genotoxic carcinogens such as 2,3,7,8 tetrachlorodibenzo-p-dioxin influence tumorigenesis is an important area of research that should yield valuable insights into the mechanisms of non-genotoxic carcinogenesis.

Are all the effects of peroxisome proliferators receptor mediated? If they are then learning more about the role and function of PPAR and possibly additional closely related receptors provides an exciting and unique opportunity to understand more about the role of peroxisome proliferators in peroxisome proliferation and cancer.

## SUMMARY

Peroxisome proliferators are a diverse group of rodent hepatocarcinogens, which include hypolipidaemic drugs, plasticizers and herbicides. A member of the nuclear hormone receptor superfamily has been identified that can be activated by peroxisome proliferators. The receptor is therefore termed the peroxisome proliferator activated receptor (PPAR). Three *Xenopus* nuclear hormone receptors have been described that are closely related to the mouse PPAR and can all be activated by peroxisome proliferators. Therefore, a family of PPARs that is possibly important in mediating the action of some hypolipidaemic drugs may exist in many species including mice and humans. The most widely used marker of peroxisome proliferator action is the peroxisomal β-oxidation enzyme acyl CoA oxidase. PPAR recognizes a specific peroxisome proliferator response element located in the acyl CoA oxidase gene promoter. It is therefore argued that the hypolipidaemic and carcinogenic effects of peroxisome proliferators may be mediated by PPAR in a manner similar to that of steroid hormone action.

## References


Alvares K, Carrillo A, Yuan PM, Kawano H, Morimoto RI and Reddy JK (1990) Identification of cytosolic peroxisome proliferator binding protein as a member of the heat shock protein HSP70 family. *Proceedings of the National Academy of Sciences of the USA* **87** 5293–5297

Bentley P, Bieri F, Muakkassah-Kelly S, Staubli W and Waechter F (1988) Mechanisms of tumour induction by peroxisome proliferators. *Archives in Toxicology* **12 (Supplement)** 240–247

Bieri F, Bentley P, Waechter F and Staubli W (1984) Use of primary cultures of adult rat hepatocytes to investigate mechanisms of action of nafenopin, a heptocarcinogenic peroxisome proliferator. *Carcinogenesis* **5** 1033–1039

Blumcke S, Schwartzkopff W, Lobeck H, Edmondson NA, Prentice DE and Blane GF (1983) Influence of fenofibrate on cellular and subcellular liver structure in hyperlipidemic patients. *Atherosclerosis* **46** 105–116

Cattley RC and Popp JA (1989) Differences between the promoting activities of the peroxisome proliferator Wy-14,643 and phenobarbital in rat liver. *Cancer Research* **49** 3246–3251

Cattley RC, Marsman DS and Popp JA (1989) Failure of the peroxisome proliferator Wy-14,643 to initiate growth-selectable foci in rat liver. *Toxicology* **56** 1–7

Cattley RC, Marsman DS and Popp JA (1991) Age-related susceptibility to the carcinogenic effect of the peroxisome proliferator Wy-14,643 in rat liver. *Carcinogenesis* **12** 469–473

Cherkaoui-Malki M, Lone YC, Corral-Debrinski M and Latruffe N (1990) Differential proto-oncogene mRNA induction from rats treated with peroxisome proliferators. *Biochemical and Biophysical Research Communications* **173** 855–861

Conway JG, Tomaszewski KE, Olson MJ, Cattley RC, Marsman DS and Popp JA (1989) Relationship of oxidative damage to the hepatocarcinogenicity of the peroxisome proliferators di(2-ethylhexyl)phthalate and Wy-14,643. *Carcinogenesis* **10** 513–519

de Duve C (1966) Peroxisomes (microbodies and related particles). *Physiology Reviews* **46** 319–375

de The H, Vivanco-Ruiz MM, Tiollais P, Stunnenberg H and Dejean A (1990) Identification of a retinoic acid responsive element in the retinoic acid receptor beta gene. *Nature* **343** 177–180

Eacho PI, Foxworthy PS, Johnson WD, Hoover DM and White SL (1986) Hepatic peroxisomal changes induced by a tetrazole-substituted alkoxyacetophenone in rats and comparison with other species. *Toxicology and Applied Pharmacology* **83** 430–437

Elcombe CR and Mitchell AM (1986) Peroxisome proliferation due to di(2-ethylhexyl) phthalate (DEHP): species differences and possible mechanisms. *Environmental Health Perspectives* **70** 211–219

Elliott BM, Dodd NJ and Elcombe CR (1986) Increased hydroxyl radical production in liver peroxisomal fractions from rats treated with peroxisome proliferators. *Carcinogenesis* **7** 795–799

Flatmark T, Nilsson A, Kvannes J *et al* (1988) On the mechanism of induction of the enzyme systems for peroxisomal beta-oxidation of fatty acids in rat liver by diets rich in partially hydrogenated fish oil. *Biochimica et Biophysica Acta* **962** 122–130

Frenkel RA, Slaughter CA, Orth K *et al* (1990) Peroxisome proliferation and induction of peroxisomal enzymes in mouse and rat liver by dehydroepiandrosterone feeding. *Journal of Steroid Biochemistry* **35** 333–342

Gerbacht U, Bursch W, Kraus P *et al* (1990) Effects of hypolipidemic drugs nafenopin and clofibrate on phenotypic expression and cell death (apoptosis) in altered foci of rat liver. *Carcinogenesis* **11** 617–624

Gray TJB, Lake BG, Beamand JA, Foster JR and Gangolli SD (1983) Peroxisome proliferation in primary cultures of rat hepatocytes. *Toxicology and Applied Pharmacology* **67** 15–25

Green S and Chambon P (1988) Nuclear receptors enhance our understanding of transcription regulation. *Trends in Genetics* **4** 309–314

Gupta C, Hattori A and Shinozuka H (1988) Suppression of EGF binding in rat liver by the hypolipidemic peroxisome proliferators, 4-chloro-6-(2,3-xylidino)-2-pyrimidinylthio- (N-beta-hydroxyethyl) acetamide and di(2-ethylhexyl) phthalate. *Carcinogenesis* **9** 167–169

Hanefeld M, Kemmer C and Kadner E (1983) Relationship between morphological changes and lipid-lowering action of p-chlorphenoxyisobutyric acid (CPIB) on hepatic mitochondria and peroxisomes in man. *Atherosclerosis* **46** 239–246

Hardwick JP, Song BJ, Huberman E and Gonzalez FJ (1987) Isolation, complementary DNA sequence, and regulation of rat hepatic lauric acid omega-hydroxylase (cytochrome P-450LA omega): identification of a new cytochrome P-450 gene family. *Journal of Biological Chemistry* **262** 801–810

Herren-Freund SL, Pereira MA, Khoury MD and Olson G (1987) The carcinogenicity of trichloroethylene and its metabolites, trichloroacetic acid and dichloroacetic acid, in mouse liver. *Toxicology and Applied Pharmacology* **90** 183–189

Hess R, Staubli W and Reiss W (1965) Nature of the hepatomegalic effect produced by ethyl-chlorophenoxy-isobutyrate in the rat. *Nature* **208** 856–859

Intrasuksri U and Feller DR (1991) Comparison of the effects of selected monocarboxylic, dicarboxylic and perfluorinated fatty acids on peroxisome proliferation in primary cultured rat hepatocytes. *Biochemical Pharmacology* **42** 184–188

Issemann I and Green S (1990) Activation of a member of the steroid hormone receptor superfamily by peroxisome proliferators. *Nature* **347** 645–650

Issemann I and Green S (1991) Cloning of novel members of the steroid hormone receptor superfamily. *Journal of Steroid Biochemistry and Molecular Biology* **40** 263–269

Jordan VC, Gottardis MM, Robinson SP and Friedl A (1989) Immune-deficient animals to study "hormone-dependent" breast and endometrial cancer. *Journal of Steroid Biochemistry* **34** 169–176

Kasai H, Okadi Y, Nishimura S, Rao MS and Reddy JK (1989) Formation of 8-hydroxydeoxyguanosine in liver DNA of rats following long-term exposure to a peroxisome proliferator. *Cancer Research* **49** 2603–2605

Kase BF, Prydz K, Björkhem I and Pedersen JI (1986) In vitro formation of bile acids from di- and trihydroxy-5 beta-cholestanoic acid in human liver peroxisomes. *Biochimica et Biophysica Acta* **877** 37–42

Kraupp-Grasl B, Huber W, Putz B, Gerbracht U and Schulte-Hermann R (1990) Tumour promotion by the peroxisome proliferator nafenopin involving a specific subtype of altered foci in rat liver. *Cancer Research* **50** 3701–3708

Kraupp-Grasl B, Huber W, Taper H and Schulte-Hermann R (1991) Increased susceptibitlity of aged rats to hepatocarcinogenesis by the peroxisome proliferator nafenopin and the possible involvement of altered liver foci occurring spontaneously. *Cancer Research* **51** 666–671

Lalwani ND, Fahl WE and Reddy JK (1983) Detection of a nafenopin-binding protein in rat liver cytosol associated with the induction of peroxisome proliferation by hypolipidemic compounds. *Biochemical and Biophysical Research Communications* **116** 388–393

Lalwani ND, Alvares K, Reddy MK, Reddy MN, Parikj I and Reddy JK (1987) Peroxisome proliferator-binding protein: identification and partial characterization of nafenopin-, clofibric acid-, and ciprofibrate-binding proteins from rat liver. *Proceedings of the National Academy of Science USA* **84** 5242–5246

Lake BG, Evans JG, Gray TJB, Korosi SA and North CJ (1989) Comparative studies on nafenopin-induced hepatic peroxisome proliferation in the rat, Syrian hamster, guinea pig, and marmoset. *Toxicology and Applied Pharmacology* **99** 148–160

Lazarow PB and de Duve C (1976) A fatty acyl-CoA oxidizing system in rat liver peroxisomes; enhancement by clofibrate, a hypolipidemic drug. *Proceedings of the National Academy of Sciences of the USA* **73** 2043–2046

Lock EA, Mitchell AM and Elcombe CR (1989) Biochemical mechanisms of induction of hepatic peroxisome proliferation. *Annual Reviews of Pharmacology and Toxicology* **29** 145–163

Marsman DS, Cattley RC, Conway JG and Popp JA (1988) Relationship of hepatic peroxisome proliferation and replicative DNA synthesis to the hepatocarcinogenicity of the peroxisome proliferators di(2-ethylhexyl)phthalate and (4-chloro-6-(2,3-xylidino)-2-pyrimidinylthio) acetic acid (Wy-14,643) in rats. *Cancer Research* **48** 6739–6744

Mitchell AM, Bridges JW and Elcombe CR (1984) Factors influencing peroxisome proliferation in cultured rat hepatocytes. *Archives in Toxicology* **55** 239–246

Moody DE, Reddy JK, Lake BG, Popp JA and Reese DH (1991) Peroxisome proliferation and nongenotoxic carcinogenesis: commentary on a symposium. *Fundamental and Applied Toxicology* **16** 233–248

National Toxicology Program (1983) Carcinogenesis bioassay of di(2-ethylhexyl)phthalate (CAS No 117-81-7) in F344 rats and B6C3F1 mice. *NTP Technical Report Series* No 217, National Institutes of Health, Bethesda

Nemali MR, Usuda N, Reddy MK *et al* (1988) Comparison of constitutive and inducible levels of expression of peroxisomal beta-oxidation and catalase genes in liver and extrahepatic tissues of rat. *Cancer Research* **48** 5316–5324

Nilsson A, Arey H, Pedersen JI and Christiansen EN (1986) The effect of high-fat diets on microsomal lauric acid hydroxylation in rat liver. *Biochimica et Biophysica Acta* **879** 209–214

Orton TC and Parker GL (1982) The effect of hypolipidemic agents on the hepatic microsomal drug-metabolizing enzyme system of the rat: induction of cytochrome(s) P-450 with specificity toward terminal hydroxylation of lauric acid. *Drug Metabolism and Disposition* **10** 110–115

Osumi T and Hashimoto T (1978) Enhancement of fatty acyl-CoA oxidizing activity in rat liver peroxisomes by di-(2-ethylhexyl)phthalate. *Journal of Biochemistry* **83** 1361–1365

Osumi T, Ishii N, Miyazawa S and Hashimoto T (1987) Isolation and structural characterization of the rat acyl-CoA oxidase gene. *Journal of Biological Chemistry* **262** 8138–8143

Osumi T, Wen JK and Hashimoto T (1991) Two cis-acting regulatory sequences in the peroxisome proliferator-responsive enhancer region of the rat acyl CoA oxidase gene. *Biochemical and Biophysical Research Communications* **175** 866–871

Paget GE (1963) Experimental studies on the toxicity of Atromid with particular reference to fine structural changes in the liver of rodents. *Journal of Atherosclerosis Research* **3** 729–736

Power RF, Lydon JP, Conneely OM and O'Malley BW (1991) Dopamine activation of an orphan of the steroid receptor superfamily. *Science* **252** 1546–1548

Randerath E, Randerath K, Reddy R, Danna TF, Rao MS and Reddy JK (1991) Induction of rat liver DNA alterations by chronic administration of peroxisome proliferators as detected by 32P-postlabeling. *Mutation Research* **247** 65–76

Rao MS, Lalwani ND, Watanabe TK and Reddy JK (1984) Inhibitory effect of antioxidants ethoxyquin and 2(3)-tert-butyl-4-hydroxyanisole on hepatic tumorigenesis in rats fed ciprofibrate, a peroxisome proliferator. *Cancer Research* **44** 1072–1076

Reddy JK and Rao MS (1977) Malignant tumours in rats fed nafenopin, a hepatic peroxisome proliferator. *Journal of the National Cancer Institute* **59** 1645–1650

Reddy JK and Qureshi SA (1979) Tumorigenicity of the hypolipidemic peroxisome proliferator ethyl-α-p-chlorophenoxyisobutyrate (clofibrate) in rats. *British Journal of Cancer* **40** 476–482

Reddy JK, Rao MS, Azarnoff DL and Sell S (1979) Mitogenic and carcinogenic effects of a hypolipidemic peroxisome proliferator, (4-chloro-6-(2,3-xylidino)-2-pyrimidinylthio)acetic acid (Wy-14,643), in rat and mouse liver. *Cancer Research* **39** 152–161

Reddy JK, Azarnoff DL and Hignite CE (1980) Hypolipidemic hepatic peroxisome proliferators form a novel class of chemical carcinogens. *Nature* **283** 397–398

Reddy JK, Lalwani ND, Reddy, MK and Qureshi SA (1982) Excessive accumulation of autofluorescent lipofuscin in the liver during hepatocarcinogenesis by methyl clofenapate and other hypolipidemic peroxisome proliferators. *Cancer Research* **42** 259–266

Reddy JK and Lalwani ND (1983) Carcinogenesis by hepatic peroxisome proliferators: evaluation of the risk of hypolipidemic drugs and industrial plasticizers to humans. *Critical Reviews in Toxicology* **12** 1–58

Reddy JK and Rao MS (1986) Peroxisome proliferators and cancer: mechanisms and implications. *Trends in Pharmacological Science* **7** 438–443

Reddy JK, Goel SK, Nemali MR *et al* (1986) Transcription regulation of peroxisomal fatty acyl-CoA oxidase and enoyl-CoA hydratase/3-hydroxyacyl-CoA dehydrogenase in rat liver by peroxisome proliferators. *Proceedings of the National Academy of Sciences USA* **83** 1747–1751

Rhodin J (1954) Correlation of ultrastructural organization and functions in normal and experimentally changed proximal convoluted tubule cells of the mouse kidney. PhD Thesis, Karolinska Institute, Stockholm, Sweden

Schwabe JWR, Neuhaus D and Rhodes D (1990) Solution structure of the DNA-binding domain of the oestrogen receptor. Nature **348** 458–461

Styles JA, Kelly M, Pritchard NR and Elcombe CR (1988) A species comparison of acute hyperplasia induced by the peroxisome proliferator methylclofenapate: involvement of the binucleated hepatocyte. *Carcinogenesis* **9** 1647–1655

Sukumar S (1990) An experimental analysis of cancer: role of ras oncogenes in multistep carcinogenesis. *Cancer Cells* **2** 199–204

Svoboda DJ and Azarnoff DL (1979) Tumours in male rats fed ethyl chlorophenoxyisobutyrate, a hypolipidaemic drug. *Cancer Research* **39** 34191–13428

Tugwood J, Issemann I, Anderson R, Bundell K, McPheat W and Green S (1992) The mouse peroxisome proliferator activated receptor recognizes a response element in the 5′ flanking sequence of the rat acyl CoA oxidase gene. *EMBO J* **11** 433–439

Warren JR, Simmon VF and Reddy JK (1980) Properties of hypolipidemic peroxisome proliferators in the lymphocyte (3H) thymidine and Salmonella mutagenesis assay. *Cancer Research* **40** 36–41

Umesono K and Evans RM (1989) Determinants of target gene specificity for steroid/thyroid hormone receptors. *Cell* **57** 1139–1146

The author is responsible for the accuracy of the references.

# Biographical Notes

**Hartmut Beug** was trained as a biologist at the University of Freiburg, Germany. He did his PhD on cellular slime moulds in Tübingen and joined the group of Dr Thomas Graf, with whom he collaborated on RNA tumour viruses in Tübingen and Heidelberg. As a group leader at the EMBL Heidelberg, he concentrated on leukaemogenesis by the avian erythroblastosis virus, and he is now continuing this work as a senior scientist at the Research Institute of Molecular Pathology in Vienna, Austria.

**Albert Brinkmann,** PhD, is associate professor in the department of endocrinology and reproduction of the Erasmus University, Rotterdam. He graduated in chemistry from the University of Utrecht and received his PhD on the cellular uptake of steroids from Erasmus University in 1972. From 1973 to 1978 he was assistant professor in the department of cell biology at the University of Leiden. His research then was focused on biochemical aspects of male sexual differentiation. In 1978 he joined the department of biochemistry at Erasmus University. Since 1981 he has worked on the biochemistry of the action of androgens in normal and pathological cell systems.

**Gary C Chamness,** PhD, studied at the California Institute of Technology and the University of California, Berkeley. Since 1970 he has been a member of the department of medicine/oncology at the University of Texas Health Science Center, San Antonio, working on steroid receptor assays and mechanisms, the synthesis of fluorescent and affinity ligands for steroid receptors and prognostic factors in breast cancer. He is also editor of *Breast Cancer Research and Treatment.*

**Krishna Chatterjee** qualified in medicine in 1982 from the University of Oxford and then trained in clinical endocrinology. After a period of research at Massachusetts General Hospital, Boston, supported by an MRC travelling fellowship and then an NIH Fogarty fellowship, he went to the department of medicine at Cambridge University as a Wellcome senior clinical research fellow. His research interests are the molecular basis of thyroid hormone action and clinical thyroid hormone resistance states.

**Anne Dejean** was born in Cholet, France, in 1957 and obtained her PhD in 1988 under Pierre Tiollais at the Pasteur Institute in Paris. She is now director of research at INSERM.

**Hugues de Thé** was born in Marseilles in 1959 and obtained his MD in 1989. He trained in molecular biology in Pierre Tiollais' laboratory, gaining a PhD in 1990. He is a research associate at INSERM.

**Suzanne A W Fuqua** received a master's degree in microbiology from the University of Houston and a PhD in tumour virology from the University of Texas Graduate School of Biomedical Sciences at Houston. She is currently an assistant professor of medicine at the University of Texas Health Science Center in San Antonio, Texas.

**Jacques Ghysdael,** PhD, graduated in biological chemistry at the Free University of Brussels, Belgium, and was a postdoctoral student at the department of microbiology, University of Southern California, Los Angeles. In 1983 he joined the Centre National de la Recherche Scientifique (CNRS), France, to pursue studies on the biochemical function of oncogenic proteins and their proto-oncogenic version. He worked at the Pasteur Institute in Lille and is currently a group leader at the Curie Institute in Orsay.

**Stephen Green** was an undergraduate and graduate student in the biochemistry department of Liverpool University. In 1984 he moved to Pierre Chambon's group in Strasbourg, where he cloned and characterized the human oestrogen receptor. In 1988 he became a senior scientist at

ICI and has established a molecular biology group interested in elucidating the mechanism of action of non-genotoxic carcinogens.

**Anne Guiochon-Mantel** graduated in medicine at the University of Paris in 1983. She trained in endocrinology as resident of Paris Hospitals (1979–1984) and became assistant of biochemistry at Paris University (1985–1991). Presently, she is maître de conferences of biochemistry at Paris University. She obtained her PhD in molecular endocrinology in 1990. She joined the INSERM unit on hormones and reproduction (U135) in 1982. She worked on the uteroglobin gene and progesterone receptor gene expression. Her work is now focused on the molecular mechanisms of the nuclear localization of the progesterone receptor.

**Kathryn B Horwitz** is a graduate of Barnard College, Columbia University, in New York City. She received her PhD from Southwestern Medical School in Dallas, Texas, in 1975, for studies of progesterone receptors in breast cancer performed in the laboratory of WL McGuire at the University of Texas School of Medicine, San Antonio. In 1979 she joined the faculty of the University of Colorado School of Medicine, where she is now professor, with appointments in medicine, pathology, and molecular biology. Her research focuses on the molecular actions of oestrogen receptors and progesterone receptors in breast cancer.

**Shun-Yuan Jiang** was an undergraduate student in the department of biology, National Taiwan Normal University, Taiwan, Republic of China. She obtained an MSc in microbiology and immunology at the National Defense Medical Center in Taiwan before joining the Human Cancer Biology graduate programme in the department of human oncology, University of Wisconsin. She is supported by scholarships from the National Science Council and the National Defense Medical Center, Taiwan, and will take up an appointment at the National Defense Medical Center after completing her PhD in 1992.

**V Craig Jordan,** a graduate of Leeds University Medical School, was awarded a PhD by Leeds University in 1972 for anti-oestrogen research. After 2 years at the Worcester Foundation for Experimental Biology, Massachusetts, where he conducted the first laboratory studies with tamoxifen as an anti-tumour agent, he returned to Leeds as lecturer in pharmacology. In 1979 he headed the endocrine unit of the Ludwig Institute for Cancer Research in Berne, Switzerland, and since 1980 has been at the University of Wisconsin (Madison), where his research programme has focused on the laboratory and clinical applications of anti-oestrogens. In 1985 he was awarded a DSc from the University of Leeds and in 1989 the BF Cain award from the American Association for Cancer Research. He is professor of human oncology and pharmacology and director of the breast cancer programme for the University of Wisconsin Comprehensive Cancer Center.

**Roger J B King** graduated in biochemistry from London University and after completing his PhD at Edinburgh University spent a year at Harvard before joining the Imperial Cancer Research Fund. His work has centred on the mechanism of action of steroid hormones at both basic science and clinical levels.

**William L McGuire,** MD, studied at Western Reserve University in Cleveland, Ohio, and the National Cancer Institute in Bethesda, Maryland. Since 1975 he has been professor and chief of medical oncology at the University of Texas Health Science Center, San Antonio. His research interest is biological factors relating to prognosis in breast cancer.

**Edwin Milgrom** graduated in medicine and biochemistry at Paris University and took up an internship and residency in endocrinology at the Paris University Hospitals. He is now professor of biochemistry in the faculty of medicine, University Paris-Sud and head of the INSERM unit in hormones and reproduction (U135).

**Gillian Morriss-Kay** (née Morriss) is a university lecturer in human anatomy in the University of Oxford. She read zoology at Newcastle University and worked for her PhD at Cambridge. She has held a junior research fellowship and teaching posts at Cambridge and a senior research fellowship at Oxford. Her research interests lie in the field of mammalian morphogenesis and are particularly concerned with the interactions between retinoids, extracellular matrix and cell behaviour in intact embryos.

**Gerald C Mueller,** MD, PhD, is professor of oncology at the McArdle Laboratory, University of Wisconsin, Madison, Wisconsin. He carried out PhD studies on the metabolism of carcinogenic azo-dyes under Dr James A Miller and in 1950 was appointed assistant professor at the McArdle Laboratory for Cancer Research, where he engaged in a study of growth regulation by steroid hormones and tumour promoting phorbol esters. His research interests have included oestrogen receptor action, phorbol ester action, nuclear processes relating to chromatin and DNA replication and the role of lipid metabolism in cell replication. His recent work has concentrated on decision events leading to cell replication or terminal differentiation.

**Malcolm Parker** graduated in biochemistry at the University of Manchester Institute of Science and Technology and obtained his PhD at the University of Leicester. His interest in the role of hormones in gene expression began when he joined Dr Bert O'Malley's laboratory as a postdoctoral fellow. He then worked at the Imperial Cancer Research Fund and the Agricultural Research Council's Institute of Animal Physiology before taking up his present position as head of the molecular endocrinology laboratory at ICRF.

**Martine Perrot-Applanat,** DSc, obtained degrees from the University of Paris and "Ecole Normale Supérieure" in physiology and biochemistry. She was lecturer of biochemistry at the University of Paris-Sud. Her early work, at the INSERM steroid hormones laboratory (U33) under Professor EE Baulieu and the INSERM unit on hormones and reproduction (U135) under Professor E Milgrom, was on steroid binding plasma proteins. As research associate (1977) then director of research (1988) at the Centre National de la Recherche Scientifique, France, she has carried out research on cellular biology of steroid receptors.

**Anita B Roberts** is deputy chief of the laboratory of chemoprevention of the National Cancer Institute, Bethesda, where she has worked since 1976. She received her PhD in biochemistry from the University of Wisconsin for studies on the metabolism of retinoic acid. Her current research interests focus on aspects of the action of transforming growth factor β.

**Michael B Sporn** is chief of the laboratory of chemoprevention of the National Cancer Institute, Bethesda, a post he has held since 1973. He received his MD degree from the University of Rochester and his research training at the National Institutes of Health. The research interests of his laboratory are currently focused on the use of retinoids as chemoprevention agents and aspects of the activity and mechanisms of action of transforming growth factor β.

**Jamshed R Tata,** after undergraduate studies in India, obtained his doctorate in biochemistry from the Collège de France, Paris. Following postdoctoral fellowships in the United States, England and Sweden, he joined the Medical Research Council in 1962 as a junior scientist at the National Institute for Medical Research, London, and has remained there as head of the laboratory of developmental biochemistry, which he founded in 1973. His current research interests mainly concern the hormonal regulation of development.

**Jan Trapman,** PhD, is a staff investigator in the department of pathology, Erasmus University, Rotterdam. He graduated in chemistry at the Technical University, Delft, and received his PhD at the Free University, Amsterdam, in 1975.

**Alan E Wakeling** graduated from St Andrews University and received his PhD from the University of Nottingham. Following postdoctoral work at Cornell University and the Upjohn Company in the United States and in the department of biochemistry, Glasgow University, he joined ICI Pharmaceuticals, where his main interest has been in the discovery of new hormone antagonists and their application to cancer therapy.

**George Wilding,** MD, is assistant professor of human oncology at the University of Wisconsin Clinical Cancer Center and chief of the oncology section of the Middleton Veterans Administration Hospital. He received an MS in pharmacology at Pennsylvania State University in 1976 and an MD from the University of Massachusetts in 1980. He trained in medical oncology at the National Cancer Institute and remained in the breast cancer section under Marc Lippman until 1988, when he moved to the University of Wisconson. His research focuses on the growth control of human prostate cancer and prostate carcinogenesis.

# Index

# LIST OF PREVIOUS ISSUES

## VOLUME 1 1982

**No. 1: Inheritance of Susceptibility to Cancer in Man**
Guest Editor: W F Bodmer

**No. 2: Maturation and Differentiation in Leukaemias**
Guest Editor: M F Greaves

**No. 3: Experimental Approaches to Drug Targeting**
Guest Editors: A J S Davies and M J Crumpton

**No. 4: Cancers Induced by Therapy**
Guest Editor: I Penn

## VOLUME 2 1983

**No. 1: Embryonic & Germ Cell Tumours in Man and Animals**
Guest Editor: R L Gardner

**No. 2: Retinoids and Cancer**
Guest Editor: M B Sporn

**No. 3: Precancer**
Guest Editor: J J DeCosse

**No. 4: Tumour Promotion and Human Cancer**
Guest Editors: T J Slaga and R Montesano

## VOLUME 3 1984

**No. 1: Viruses in Human and Animal Cancers**
Guest Editors: J Wyke and R Weiss

**No. 2: Gene Regulation in the Expression of Malignancy**
Guest Editor: L Sachs

**No. 3: Consistent Chromosomal Aberrations and Oncogenes in Human Tumours**
Guest Editor: J D Rowley

**No. 4: Clinical Management of Solid Tumours in Childhood**
Guest Editor: T J McElwain

**VOLUME 4 1985**

**No. 1: Tumour Antigens in Experimental and Human Systems**
Guest Editor: L W Law

**No. 2: Recent Advances in the Treatment and Research in Lymphoma and Hodgkin's Disease**
Guest Editor: R Hoppe

**No. 3: Carcinogenesis and DNA Repair**
Guest Editor: T Lindahl

**No. 4: Growth Factors and Malignancy**
Guest Editors: A B Roberts and M B Sporn

**VOLUME 5 1986**

**No. 1: Drug Resistance**
Guest Editors: G Stark and H Calvert

**No. 2: Biochemical Mechanisms of Oncogene Activity: Proteins Encoded by Oncogenes**
Guest Editors: H E Varmus and J M Bishop

**No. 3: Hormones and Cancer: 90 Years after Beatson**
Guest Editor: R D Bulbrook

**No. 4: Experimental, Epidemiological and Clinical Aspects of Liver Carcinogenesis**
Guest Editor: E Farber

**VOLUME 6 1987**

**No. 1: Naturally Occurring Tumours in Animals as a Model for Human Disease**
Guest Editors: D Onions and W Jarrett

**No. 2: New Approaches to Tumour Localization**
Guest Editor: K Britton

**No. 3: Psychological Aspects of Cancer**
Guest Editor: S Greer

**No. 4: Diet and Cancer**
Guest Editors: C Campbell and L Kinlen

**VOLUME 7 1988**

**No. 1: Pain and Cancer**
Guest Editor: G W Hanks

**No. 2: Somatic Cell Genetics and Cancer**
Guest Editor: L M Franks

**No. 3: Prospects for Primary and Secondary Prevention of Cervix Cancer**
Guest Editors: G Knox and C Woodman

**No. 4: Tumour Progression and Metastasis**
Guest Editor: I Hart

## VOLUME 8 1989

**No. 1: Colorectal Cancer**
Guest Editor: J Northover

**No. 2: Nitrate, Nitrite and Nitroso Compounds in Human Cancer**
Guest Editors: D Forman and D E G Shuker

**No. 3: A Critical Assessment of Cancer Chemotherapy**
Guest Editor: A H Calvert

**No. 4: The Use of Cytokines in Cancer Therapy**
Guest Editors: F R Balkwill and W Fiers

## VOLUME 9 1990

**No. 1: Haemopoietic Growth Factors: Their Role in the Treatment of Cancer**
Guest Editor: M Dexter

**No. 2: Germ Cell Tumours of the Testis: A Clinico-Pathological Perspective**
Guest Editors: P Andrews and T Oliver

**No. 3: Genetics and Cancer— Part I**
Guest Editors: W Cavenee, B Ponder and E Solomon

**No. 4: Genetics and Cancer— Part II**
Guest Editors: W Cavenee, B Ponder and E Solomon

## VOLUME 10 1991

**Cancer, HIV and AIDS**
Guest Editors: V Beral, H W Jaffe and R A Weiss

## VOLUME 11 1991

**Prostate Cancer: Cell and Molecular Mechanisms in Diagnosis and Treatment**
Guest Editor: J T Isaacs

**VOLUME 12 1992**

**Tumour Suppressor Genes, the Cell Cycle and Cancer**
Guest Editor: A J Levine

**VOLUME 13 1992**

**A New Look at Tumour Immunology**
Guest Editors: A J McMichael and W F Bodmer